AF865653

Advances in Geophysical Exploration

Advances in Geophysical Exploration

Guest Editors

Paulo T. L. Menezes
Valeria C F Barbosa

MDPI

Basel • Beijing • Wuhan • Barcelona • Belgrade • Novi Sad • Cluj • Manchester

Guest Editors
Paulo T. L. Menezes
Departamento de Geologia
Aplicada da Faculdade de Geologia
Universidade do Estado do Rio de
Janeiro
Rio de Janeiro
Brazil

Valeria C F Barbosa
Department of Geophysics
Observatório Nacional
Rio de Janeiro
Brazil

Editorial Office
MDPI AG
Grosspeteranlage 5
4052 Basel, Switzerland

This is a reprint of the Special Issue, published open access by the journal *Applied Sciences* (ISSN 2076-3417), freely accessible at: https://www.mdpi.com/journal/applsci/special_issues/361SFC5784.

For citation purposes, cite each article independently as indicated on the article page online and as indicated below:

Lastname, A.A.; Lastname, B.B. Article Title. *Journal Name* **Year**, *Volume Number*, Page Range.

ISBN 978-3-7258-8133-8 (Hbk)
ISBN 978-3-7258-8134-5 (PDF)
https://doi.org/10.3390/books978-3-7258-8134-5

© 2026 by the authors. Articles in this reprint are Open Access and distributed under the Creative Commons Attribution (CC BY) license. The reprint as a whole is distributed by MDPI under the terms and conditions of the Creative Commons Attribution-NonCommercial-NoDerivs (CC BY-NC-ND) license (https://creativecommons.org/licenses/by-nc-nd/4.0/).

Contents

About the Editors

Paulo T. L. Menezes

Paulo T. L. Menezes holds a degree in Geology from UERJ (1986), a master's degree in Geophysics from UFPA (1990), and a doctorate in Geophysics from MCT-ON (1996). He is currently an associate professor at UERJ and a Master Geophysicist at PETROBRAS. He has worked in Applied Geophysics since 1982, on both academic and commercial projects in mineral prospecting, hydrogeophysics, geothermal exploration, environmental studies, and hydrocarbon exploration. He worked on implementing a calibration methodology for airborne gamma-spectrometric data in Brazil using calibration pads. Since 2006, he has been working with integrated interpretation of multiphysics and seismic methods in hydrocarbon exploration.

Valeria C F Barbosa

Valeria C F Barbosa holds a degree in Geology from Universidade Federal Rural do Rio de Janeiro (1986), a master's degree in Geophysics from UFPA (1990), and a doctorate in Geophysics from Universidade Federal do Pará (1998). Barbosa has experience in Geosciences, focusing on Applied Geophysics, with expertise in gravity inversion, Euler deconvolution, magnetic inversion, and sedimentary basins.

Preface

Applied geophysics has advanced significantly in recent years, driven by innovations in acquisition, processing, and interpretation that improve subsurface imaging and characterization. These developments are vital for resource exploration, environmental monitoring, and sustainable energy. This Special Issue "Advances in Geophysical Exploration", published in *Applied Sciences*, brings together fourteen original papers showcasing state-of-the-art research and practical applications.

This reprint addresses three main themes:

1. Algorithmic and Computational Advances: Innovations in forward and inverse modeling for reliable subsurface imaging.
2. Exploratory Case Studies: Applications in mineral, hydrocarbon, hydrogen, and geothermal exploration, integrating geophysical, geological, and engineering insights.
3. Hydrogeological and Environmental Applications: Geophysical methods for groundwater management and basin characterization.

A key highlight in this collection is the integration of Artificial Intelligence (AI) and Machine Learning (ML) into geophysical workflows. Studies demonstrate how these technologies enhance predictive modeling and data processing, complementing physics-based approaches. Other contributions present breakthroughs in electromagnetic modeling, gravity gradient processing, magnetotelluric imaging, seismic inversion, and magnetic property estimation from borehole logs, addressing challenges such as noise suppression and large-scale data inversion.

This reprint reflects a collaborative effort to advance applied geophysics. We sincerely thank all the authors for their high-quality contributions and the reviewers for their valuable feedback, which ensured the success of this collection. We hope it serves as a valuable resource for exploration and the energy transition.

Paulo T. L. Menezes and Valeria C F Barbosa
Guest Editors

applied sciences

MDPI

Editorial

Editorial for the Special Issue "Advances in Geophysical Exploration"

Paulo T. L. Menezes [1,2,*] and Valéria C. F. Barbosa [3]

1 Departamento de Geologia Aplicada, Faculdade de Geologia, Universidade do Estado do Rio de Janeiro, Rua São Francisco Xavier, 524, Rio de Janeiro 20950-000, Brazil
2 Petroleo Brasileiro S.A. Avenida Henrique Valadares, 28, Rio de Janeiro 20231-030, Brazil
3 Observatório Nacional, Rua General José Cristino, 77, Rio de Janeiro 20921-400, Brazil; valcris@on.br
* Correspondence: paulo.menezes@uerj.br

Recent advances in applied geophysics have greatly enhanced our ability to accurately image and understand the Earth's physical characteristics and geological processes. These innovations include cutting-edge technologies and methodologies that offer deeper insights into subsurface structures and materials. A key priority in modern geophysical exploration is the commitment to minimizing the environmental impacts of exploration activities. This growing awareness is shaping practices across diverse exploration domains, including mineral resources, hydrocarbon extraction, groundwater management, environmental assessments, hydrogen production, and geothermal energy exploitation.

To promote knowledge sharing and collaboration, this Special Issue provides a dedicated platform to discuss the latest advancements in geophysical exploration. Authors are invited to contribute their research and present their findings to a diverse and engaged geoscience community. The issue highlights key developments across geophysical methods, ranging from established techniques to cutting-edge innovations. We particularly welcome interpretative case studies illustrating practical applications of these methods. Contributions may also focus on developing new analytical or computational methodologies, as well as the integration of interdisciplinary strategies that offer novel insights. We propose dividing the themes into three main areas of concentration:

1. The recent advancements in forward and inverse modeling algorithms. These developments are essential for accurately interpreting geophysical data and generating reliable subsurface images.
2. Developments and case studies in mineral, hydrocarbon, hydrogen, and geothermal exploration, emphasizing successful projects, innovative techniques, and the integration of geophysical methods with other scientific disciplines.
3. Developments and case studies in hydrogeological and environmental studies, showcasing applied geophysics projects that have achieved significant results. This area also highlights innovative acquisition, processing, and interpretation techniques, as well as the integration of geophysical methods with other scientific disciplines.

We received diligent responses from the geoscience community, with fourteen papers addressing various key aspects of these three areas of concentration.

Xue et al. [1] propose an integrated approach combining multi-scale static data, including geological information, well logging, and seismic surveys, with dynamic production data to improve connectivity evaluations in fracture–cavity reservoirs. Their methodology uses structural gradient tensors and stratigraphic continuity to delineate caves and dissolution holes, RGB fusion analysis to characterize fault development, fracture simulation to map fracture zones, 3D visualization to assess spatial relationships, and dynamic data to verify interwell connectivity. Applied to the S91 unit of the Tarim Oilfield, China, this approach revealed that fracture–cavity units exhibit discontinuous patterns controlled by large-scale faults and dense fracture networks. Three-dimensional visualization identified potential connectivity channels, and production data confirmed effective interwell connectivity, highlighting residual oil potential and supporting optimized well placement and injection–production strategies.

Wang and Song [2] develop a parallel adaptive high-order finite-element framework for large-scale 3D marine controlled-source electromagnetic (MCSEM) modeling. Their approach combines coordinate transformations based on the form invariance of Maxwell's equations with a goal-oriented adaptive mesh refinement strategy, allowing complex geometries, thin sedimentary layers, and multi-scale heterogeneities to be efficiently handled. By mapping the physical domain to a computationally favorable domain with equivalent anisotropic materials, the method preserves the mathematical structure of Maxwell's equations while reducing computational complexity. The framework is fully parallelized and validated through synthetic models and the Marlim R3D benchmark dataset, demonstrating high accuracy and efficiency across different water depths and frequency ranges. It shows particular advantages in shallow-water settings and models with depth-dependent conductivity variations, as well as in handling realistic seafloor topographies. This work highlights the potential for large-scale, high-resolution CSEM surveys, supporting accurate imaging and inversion of offshore resources, with possible extensions to other electromagnetic modeling applications such as plane-wave EM modeling.

Wang et al. [3] propose a Kalman–FIR fusion (K-F) filtering method to overcome the challenges of noise suppression and spectral fidelity in high-dynamic airborne gravimetry. Conventional finite impulse response (FIR) filters often attenuate genuine gravity signals when suppressing platform-induced noise, while Kalman filtering lacks explicit spectral control, creating difficulties for downstream processing. To address these issues, the authors develop a dual-stage framework: first, a state-space Kalman filter that incorporates triaxial acceleration error modeling to compensate for horizontal motion coupling interference; second, a window-optimized FIR filter that ensures explicit spectral control and compatibility with terrain correction. Engineering tests on the GIPS-1A airborne gravimeter show that the K-F method achieves a repeat-line internal consistency of 0.558 mGal at 0.01 Hz—an accuracy improvement of more than 65% compared to standalone FIR filtering—and enhances spatial resolution to 2.5 km (half-wavelength). Beyond improving noise suppression under turbulent conditions, the method enables the recovery of gravity anomaly data segments that were previously lost to airflow disturbances. The method has been implemented in the GIPS-1A gravimetry system and deployed across various platforms, including UAVs, helicopters, and fixed-wing aircraft, for industrial applications.

Wang et al. [4] investigated the genesis of thermal anomalies in the southwestern A'nan Sag, Erlian Basin, using 3D magnetotelluric (MT) inversion to construct a high-resolution resistivity model. They identified an elliptical low-resistivity anomaly (0–5 Ω m) at depths of 10–15 km, interpreted as hypersaline fluids serving as the primary heat source. Faults F1 and F3 act as medium–low resistivity conduits (10–40 Ω m) connecting Anomaly C to mid-depth reservoirs D1/D2 (5 km, 5–10 Ω m) and shallow low-resistivity layers, while an overlying high-resistivity caprock (40–100 Ω m) seals the system, forming a convective "source–conduit–reservoir–cap" architecture. Integrated seismic S-wave analysis indicates that thermal energy from partial melt beneath the Abaga volcanic zone supplements Anomaly C via conduction, combining with mantle-derived heat to form a composite source. This study elucidates the coupled conduction–convection mechanisms sustaining regional thermal anomalies and provides a quantitative framework for geothermal exploration and resource assessment in the A'nan Sag.

Piauilino et al. [5] demonstrate the use of a convolutional equivalent layer for jointly processing large gravity-gradient datasets efficiently. By assuming a single fictitious physical property distribution on a planar layer, all components of the gravity-gradient tensor remain physically consistent. However, this approach becomes computationally prohibitive for large datasets due to high demands. To overcome this limitation, they exploit the block-Toeplitz Toeplitz-block structure of the sensitivity matrix for data arranged on regular grids, significantly reducing computational costs. Tests on synthetic and real datasets show that the method can process over 290,000 AGG (Falcon airborne gravity gradiometer) data points in seconds and handle up to 1,250,000 AGG points in a few minutes. The methodology was also successfully applied to real FTG (Full Tensor Gradiometry) datasets.

Wang and Zuo [6] introduce MagTCN, an Aeromagnetic compensation method based on temporal convolutional networks, developed to address the limited generalization and compensation accuracy of existing techniques, particularly as Aeromagnetic sensors reach higher sensitivity levels. Unlike earlier neural network approaches that relied on standard convolutional layers or hybrid methods such as wavelet decomposition combined with long short-term memory (LSTM) networks, MagTCN leverages depthwise separable convolutions with residual connections, which improve feature extraction efficiency, reduce computational cost, and stabilize gradient propagation. In addition, the authors design a gradient correction loss function grounded in the Tolles–Lawson (T–L) model, enabling the network to better capture the temporal dependencies between Aeromagnetic interference and auxiliary flight parameters while preventing overfitting and enhancing convergence stability. Validation on both synthetic datasets and the DAF-MIT AIA Open Flight Data shows that MagTCN achieves a significantly lower standard deviation of residual magnetic interference than traditional convolutional and wavelet–LSTM methods, thereby demonstrating superior compensation accuracy, robustness, and generalization capability for modern Aeromagnetic survey applications.

Lyrio et al. [7] propose a novel methodology to estimate magnetic susceptibility and remanent magnetization directly from the magnetic field information acquired by orientation tools in borehole imaging logs, data typically used only for spatial orientation of images. This approach addresses a critical gap in hydrocarbon exploration, where direct measurements of susceptibility and remanence are rare due to the cost and complexity of acquiring oriented core samples, despite their importance for forward magnetic modeling and risk reduction in Pre-Salt exploration, such as in the Campos Basin, Brazil. The method involves correcting the acquired magnetic field data for tool rotation and diurnal variation, followed by the application of simple equations under reasonable assumptions to derive rock magnetic parameters. Application to a Pre-Salt well demonstrated that estimated susceptibility values for carbonates and basalts closely matched laboratory measurements, while remanent magnetization of basalts was also obtained with consistent results. The methodology shows promise for providing reliable magnetic property estimates without additional acquisition costs, as imaging logs are routinely recorded in exploration wells. Moreover, it offers potential applications beyond hydrocarbon settings, notably in mining exploration, where susceptibility aids lithological identification, and possibly in magnetostratigraphy, provided improvements in resolution and corrections are achieved.

Zhang et al. [8] propose a novel inversion methodology for tilted transverse isotropic (TTI) media affected by inclined fractures, addressing key challenges in seismic parameter estimation for fractured reservoirs. The study derives a new reflection coefficient equation for TTI media with rotationally invariant inclined fractures, based on seismic scattering theory and the steady-phase method, providing improved accuracy and theoretical consistency compared to previous models. To enhance inversion stability, the authors introduce a scale normalization technique that transforms parameters into a uniform scale, mitigating errors caused by inconsistent parameter magnitudes during inversion. The methodology is validated with synthetic models and field data, showing strong agreement with well logs and geological interpretations. The results demonstrate that the proposed approach significantly improves inversion accuracy and reliability in fractured TTI media, offering practical value for reservoir characterization, well placement optimization, and fractured reservoir development.

Monroe et al. [9] present a methodology for enhanced porosity prediction in subsurface reservoirs by integrating Bayesian Linearized Inversion (BLI) of seismic and well-log data with both an empirical model (Han's equation) and a Support Vector Regression (SVR) machine learning model. The BLI framework is employed to optimally estimate elastic parameters, including compressional and shear wave velocities (V_P,V_S) and density, mitigating uncertainties inherent in the initial seismic data and improving the reliability of subsequent porosity predictions. By incorporating sparse spike wavelets into the inversion process, the methodology refines seismic resolution and enhances the extraction of petrophysical properties. The study applies this workflow to data from seventeen wells in the Norne field, North Sea, using two primary wells for inversion and training fifteen secondary wells for predictive modeling. Evaluation metrics demonstrate that the SVR-based porosity predictions achieve superior accuracy, with a Pearson correlation coefficient of 0.90 and improved R^2 and RMSE values compared to Han's empirical equation, highlighting the capacity of machine learning to capture complex nonlinear relationships in geophysical data. The authors emphasize that the workflow, including wavelet selection, hyperparameter tuning, and integration of BLI outputs, is adaptable to different geological contexts, providing a robust framework for high-precision reservoir characterization and predictive modeling in hydrocarbon exploration.

Wang et al. [10] address the limitations of conventional Transient Electromagnetic Method (TEM) modeling and inversion, which often assume non-magnetic subsurface conditions, by developing a three-dimensional (3-D) TEM framework that simultaneously accounts for variations in resistivity and magnetic permeability. Recognizing that neglecting magnetic susceptibility can lead to misinterpretation in regions with igneous rock accumulations or ferromagnetic minerals, the study employs an edge-based finite element method on unstructured grids combined with a second-order implicit backward Euler scheme for accurate forward modeling. The forward algorithm is rigorously validated against analytical solutions for homogeneous half-space models and 1D numerical solutions, demonstrating its ability to capture the influence of non-uniform magnetic permeability on TEM responses. Building on this, a 3D inversion methodology using the Limited-memory Broyden–Fletcher–Goldfarb–Shanno (L-BFGS) optimization algorithm is implemented, enabling the simultaneous recovery of resistivity and magnetic permeability anomalies. Synthetic case studies, including separate and joint inversion scenarios, show that the proposed approach accurately recovers the position and amplitude of targets, whereas inversion that neglects magnetic permeability produces spurious anomalies, underscoring the necessity of including magnetic parameters in TEM analysis. Although the current framework is limited to large-loop magnetic sources, isotropic resistivity models, and does not account for induced polarization effects, it significantly enhances the applicability and flexibility of 3D TEM inversion, offering a robust tool for quantitative interpretation of TEM data in geologically complex and magnetically active regions.

Chen and Dong [11] propose an integrated approach for improved velocity prediction in carbonate rocks, addressing the challenges posed by complex pore structures and strong matrix heterogeneity. The methodology combines a modified squirt flow model, which accounts for local fluid flow between microcracks with different aspect ratios, with a mixed random medium model representing matrix heterogeneity. Unlike David and

Zimmerman's conventional inversion method, the microcrack inversion assigns each group of microcracks a distinct equivalent medium, improving inversion accuracy by reflecting the sequential closure of microcracks. The modified squirt flow model is further integrated with the Gassmann model to capture attenuation effects caused by local flow between microcracks. The random medium model simulates inhomogeneous rock matrices by adjusting parameters such as autocorrelation lengths, rounding coefficients, and orientation angles. Validation against ultrasonic experimental data from four limestone samples and one dolomite core sample demonstrates that the new model achieves P-wave prediction errors below 5%, outperforming the Biot model, particularly in samples with uniform micropore aspect ratio distributions, smaller pore radii, and more homogeneous structures. The study highlights the effectiveness of combining microcrack-specific squirt flow modeling with heterogeneous matrix representations to enhance the accuracy and adaptability of petrophysical predictions in carbonate cores.

Correa and Regis [12] present a method for constructing 3D resistivity models from marine Controlled-Source Electromagnetic (MCSEM) data by spatially constraining 1D inversions in the Common Midpoint (CMP) domain. The methodology formulates a least-squares optimization problem where each 1D inversion column contributes to a 3D resistivity cube, and inter-column correlations are enforced via Tikhonov-style smoothing regularization applied in all three spatial directions. The iterative algorithm updates the model to minimize the misfit between the measured multi-component electric field data and synthetic responses computed from the layered model columns, while also allowing for the incorporation of hard constraints, such as well log resistivities. Synthetic tests, including scenarios with resistive targets at varying depths and resistive bodies embedded within conductive layers, demonstrate that the approach accurately reconstructs subsurface resistivity distributions. Application to real MCSEM data from a gas hydrate accumulation area off the southeast coast of Brazil shows that the method effectively generates a preliminary 3D resistivity model, which serves as both a rapid interpretive tool and a computationally efficient initial guess for subsequent full 3D inversions. The spatial constraints enhance the stability and resolution of the inversion, though the reliability of the model depends on the geological complexity; layered structures yield robust results, whereas localized features such as faults introduce lateral variability that may reduce model fidelity.

Valente et al. [13] propose a computationally efficient method for the forward modeling and inversion modeling of 3D marine Controlled-Source Electromagnetic (CSEM) data. The approach employs a time-domain finite-difference solution of Maxwell's equations in a fictitious dielectric medium, with the correspondence principle mapping results to the conductive medium. This enables multi-frequency simulations in a single run and allows gradient computation via the adjoint-state method. Model reparameterization smoothing is used to stabilize inversion without requiring trial-and-error regularization. Two optimization strategies are tested: steepest descent with line search and a modified momentum method inspired by the Adam algorithm, using a single momentum term, and both are benchmarked against the conventional L-BFGS-B (Limited-memory Broyden–Fletcher–Goldfarb–Shanno with Bound constraints) algorithm. Validation on three synthetic datasets—including two resistive targets, a MARE-inspired model, and a turbidite reservoir analog—demonstrates that the momentum method, despite requiring more iterations, produces models comparable to L-BFGS-B, fitting observed data slightly better while consuming less memory. The study presents an effective framework for 3D CSEM inversion, yielding high-quality resistivity models with reduced computational cost, which is particularly beneficial for large-scale geophysical applications.

Rey et al. [14] integrate geological mapping and borehole data with geophysical techniques, including seismic reflection, time-domain electromagnetic (TDEM) surveys, and magnetic measurements, to characterize the northern boundary of the Bailén basin, southeastern Spain. Focusing on the Baños de la Encina fault, seismic reflection revealed lithological variations and two faults causing local subsidence. TDEM surveys provided estimates of the Palaeozoic basement depth and fault locations, exploiting resistivity contrasts between sedimentary infill and basement rocks, while magnetic data identified fault-related anomaly alignments (N30E and N60E), including previously unmapped fractures. Calibration with borehole data confirmed consistency among the datasets, allowing detailed mapping of vertical and lateral bedrock variations. The study demonstrates the complementarity of seismic, electromagnetic, and magnetic methods for resolving complex faulting, basin structure, and aquifer characteristics.

Author Contributions: Conceptualization, P.T.L.M. and V.C.F.B.; writing—original draft preparation, P.T.L.M.; writing—review and editing, V.C.F.B. All authors have read and agreed to the published version of the manuscript.

Acknowledgments: We would like to express our gratitude to all the authors who submitted their papers to this Special Issue. We also appreciate the contributions of the reviewers, whose feedback significantly enhanced the quality of the manuscripts and contributed substantially to the success of this Special Issue. Lastly, we want to extend our thanks to the editors of *Applied Sciences* for their support throughout the publication process.

Conflicts of Interest: P.T.L.M was employed by the company Petrobras. The remaining authors declare that the research was conducted in the absence of any commercial or financial relationships that could be construed as a potential conflict of interest.

References

1. Xue, Y.; Gao, Y.; Peng, X. Connectivity Evaluation of Fracture-Cavity Reservoirs in S91 Unit. *Appl. Sci.* **2025**, *15*, 9738. [CrossRef]
2. Wang, F.; Cheng, S. Efficient Three-Dimensional Marine Controlled-Source Electromagnetic Modeling Using Coordinate Transformations and Adaptive High-Order Finite Elements. *Appl. Sci.* **2025**, *15*, 9626. [CrossRef]
3. Wang, G.; Xiong, S.; Yan, F.; Luo, F.; Wang, L.; Zhou, X. Kalman–FIR Fusion Filtering for High-Dynamic Airborne Gravimetry: Implementation and Noise Suppression on the GIPS-1A System. *Appl. Sci.* **2025**, *15*. 9363. [CrossRef]
4. Wang, S.; Xu, W.; Guo, T.; Duan, W.; Wang, Z. Analysis of the Genetic Mechanism of Thermal Anomaly in the A'nan Sag, Erlian Basin Based on 3D Magnetotelluric Imaging. *Appl. Sci.* **2025**, *15*, 85. [CrossRef]
5. Piauilino, L.S.; Oliveira Junior, V.C.; Barbosa, V.C.F. A Scalable and Consistent Method for Multi-Component Gravity-Gradient Data Processing. *Appl. Sci.* **2025**, *15*, 8396. [CrossRef]
6. Wang, H.; Zuo, B. An Aeromagnetic Compensation Algorithm Based on a Temporal Convolutional Network. *Appl. Sci.* **2025**, *15*, 3105. [CrossRef]
7. Lyrio, J.C.S.O.; Laier, A.P.C.; Junior, J.C.; Rodrigues, A.N.G.; Martins, L.d.S. Susceptibility and Remanent Magnetization Estimates from Orientation Tools in Borehole Imaging Logs. *Appl. Sci.* **2025**, *15*, 2873. [CrossRef]
8. Zhang, G.; Dai, S.; Li, H.; Hao, H.; Chen, T. Inversion of Elastic and Fracture Parameters in Tilted Transverse Isotropic Media with Parameter Standardization. *Appl. Sci.* **2025**, *15*, 2792. [CrossRef]
9. Teruya Monroe, J.A.; de Figueiredo, J.J.S.; Amanajas, C.E.S. Seismic Prediction of Porosity in the Norne Field: Utilizing Support Vector Regression and Empirical Models Driven by Bayesian Linearized Inversion. *Appl. Sci.* **2025**, *15*, 616. [CrossRef]
10. Wang, C.; Dong, Y.; Gao, J.; Tan, H.; Wang, Y.; Dong, W. Three-Dimensional Forward Modeling and Inversion Study of Transient Electromagnetic Method Considering Inhomogeneous Magnetic Permeability. *Appl. Sci.* **2024**, *14*, 11660. [CrossRef]
11. Chen, Y.; Dong, P. Modeling of Characteristics of Complex Microstructure and Heterogeneity at the Core Scale. *Appl. Sci.* **2024**, *14*, 1385. [CrossRef]
12. Correa, J.L.; Régis, C. Spatially Constrained 1D Inversions of Common-Midpoint Marine Controlled-Source Electromagnetic Data to Create a 3D Electrical Model. *Appl. Sci.* **2024**, *14*, 11281. [CrossRef]

13. Valente, A.; Nascimento, D.; Costa, J. CSEM Optimization Using the Correspondence Principle. *Appl. Sci.* **2024**, *14*, 8846. [CrossRef]
14. Rey, J.; Mendoza, R.; Hidalgo, M.C.; Marinho, B. Testing the Efficacy of Indirect Methods on Characterization of Sedimentary Basins by Correlation of Direct Data and Geophysical Techniques. *Appl. Sci.* **2024**, *14*, 7308. [CrossRef]

Disclaimer/Publisher's Note: The statements, opinions and data contained in all publications are solely those of the individual author(s) and contributor(s) and not of MDPI and/or the editor(s). MDPI and/or the editor(s) disclaim responsibility for any injury to people or property resulting from any ideas, methods, instructions or products referred to in the content.

applied sciences

MDPI

Article

Connectivity Evaluation of Fracture-Cavity Reservoirs in S91 Unit

Yunlong Xue [1], Yinghan Gao [2] and Xiaobo Peng [1,*]

[1] College of Geophysics and Petroleum Resources, Yangtze University, Wuhan 430100, China; 201900786@yangtzeu.edu.cn
[2] Sinopec Group Northwest Petroleum Branch Oil Production Plant No. 3, Kuqa 842209, China; gaoyh.xbsj@sinopec.com
* Correspondence: pengxiaobo@yangtzeu.edu.cn

Abstract: Carbonate fracture–cavity reservoirs are significant oil and gas reservoirs globally, and their efficient development is influenced by the connectivity between fracture–cavity units within the reservoir. These reservoirs primarily consist of large caves, dissolution holes, and natural fractures, which serve as the primary storage and flow spaces. The S91 unit of the Tarim Oilfield is a karstic fracture–cavity reservoir with shallow coverage. It exhibits significant heterogeneity in the fracture–cavity reservoirs and presents complex connectivity between the fracture–cavity bodies. The integration of static and dynamic data, including geology, well logging, seismic, and production dynamics, resulted in the development of a set of static and dynamic connectivity evaluation processes designed for highly heterogeneous fracture–cavity reservoirs. Methods include using structural gradient tensors and stratigraphic continuity attributes to delineate the boundaries of caves and holes; performing RGB fusion analysis of coherence, curvature, and variance attributes to characterize large-scale fault development features; applying ant-tracking algorithms and fracture simulation techniques to identify the distribution and density characteristics of fracture zones; utilizing 3D visualization technology to describe the spatial relationship between fracture–cavity units and large-scale faults and fracture development zones; and combining dynamic data to verify interwell connectivity. This process will provide a key geological basis for optimizing well network deployment, improving water and gas injection efficiency, predicting residual oil distribution, and formulating adjustment measures, thereby improving the development efficiency of such complex reservoirs.

Keywords: Unit S91; fracture cave type reservoir; RGB attribute fusion; crack simulation; connectivity evaluation

1. Introduction

Carbonate fracture–cavity reservoirs are significant global oil and gas resources. Their development is primarily controlled by the connectivity of fracture–cavity units. These reservoirs, consisting of cavities, dissolution pores, and fractures that collectively function as storage and flow channels, exhibit strong heterogeneity and spatial dispersion [1–6]. The spatial architecture and flow capacity of fracture–cavity units directly control fluid migration, reservoir utilization, and recovery efficiency. These units are therefore critical geological factors that influence development outcomes. The S91 unit of the Tarim Oilfield was formed through shallow karstification. It is characterized by pronounced heterogeneity and complex connectivity. Previous studies have delineated its fracture–cavity system and clarified the development patterns of composite karst reservoirs. Although integrated management has enhanced recovery, substantial residual oil remains. This highlights the need for additional recovery efforts. The complex multi-scale structure of fracture–cavity systems, ranging from microscopic throats to large caves, along with their highly uncertain spatial distribution, presents significant challenges for accurate characterization and quantitative evaluation of connectivity. Conventional static geological data (e.g., core samples and well logs) provide only limited information near the wellbore, making it difficult to characterize the fracture–cavity network structure between wells [7–10]; geophysical methods, such as seismic techniques, can identify large fracture–cavity targets; however, they lack the resolution to detect small fractures and passages that govern fluid flow and provide only limited direct evidence of connectivity [11–13]. Dynamic production data analysis methods—such as well testing, interference testing, tracer monitoring, and production response analysis—provide quantitative insights into reservoir connectivity and dynamic parameters. However, their application during the early stages of development is limited by implementation complexity, delayed data acquisition, and high operational costs [14–16].

To address these limitations, this study proposes an integrated approach that combines multi-scale static data with dynamic response analysis to enhance the reliability and practicality of connectivity evaluations in fracture–cavity reservoirs. Structural gradient tensors and stratigraphic continuity attributes enhance the accuracy of karst boundary delineation. They also help reduce strong reflection interference. RGB fusion combines coherence, curvature, and variance attributes, thereby highlighting the spatial distribution and development of large-scale faults. Ant colony tracking and fracture simulation applied to post-stack data enable efficient and accurate identification of fracture networks and their density distributions. Three-dimensional visualization elucidates the spatial relationships between fracture–cavity units and fracture networks. Dynamic data further validate interwell connectivity. This approach synthesizes information from geological studies, well logs, seismic data, and production dynamics, establishing a comprehensive connectivity evaluation framework that integrates both static and dynamic data. The framework supports well-network design, injection–production optimization, and residual oil prediction, ultimately enhancing the development efficiency of complex fracture–cavity reservoirs.

2. Connectivity Evaluation Process

Structural gradient tensors and stratigraphic continuity properties are employed to refine the delineation of cave and pore boundaries. This approach effectively mitigates the interference of strong reflective interfaces in seismic data. Compared with traditional amplitude attributes, the present method provides greater clarity and accuracy in identifying areas containing karst caves and pores. It thereby more precisely depicts their spatial distribution. RGB attribute fusion analysis combines coherence, curvature, and variance attributes to facilitate a comprehensive assessment of large-scale fault development. These three attributes are integrated using the RGB color model. This integration complements and cross-validates structural information. It also emphasizes the comprehensive features of large-scale faults, including their spatial distribution, structural details, and development intensity. As a result, interpretation accuracy is enhanced. Ant-tracking algorithms and fracture simulation techniques are employed to identify fracture zone distributions and characterize fracture density. This method relies solely on post-stack data, avoiding the extensive effort required for pre-stack processing. It accurately depicts the spatial distribution of fracture networks and provides quantitative characterization of fracture density. The approach enhances the accuracy and reliability of fracture prediction, while dynamic data are used to evaluate connectivity between wells.

This study integrates static geological information with dynamic production data to establish a dynamic–static connectivity evaluation process tailored for heterogeneous fracture–cavity reservoirs (Figure 1). The resulting framework supports well-network optimization, improves water and gas injection efficiency, predicts residual oil distribution, and guides adjustment strategies. Collectively, these measures enhance the development efficiency of complex reservoirs.

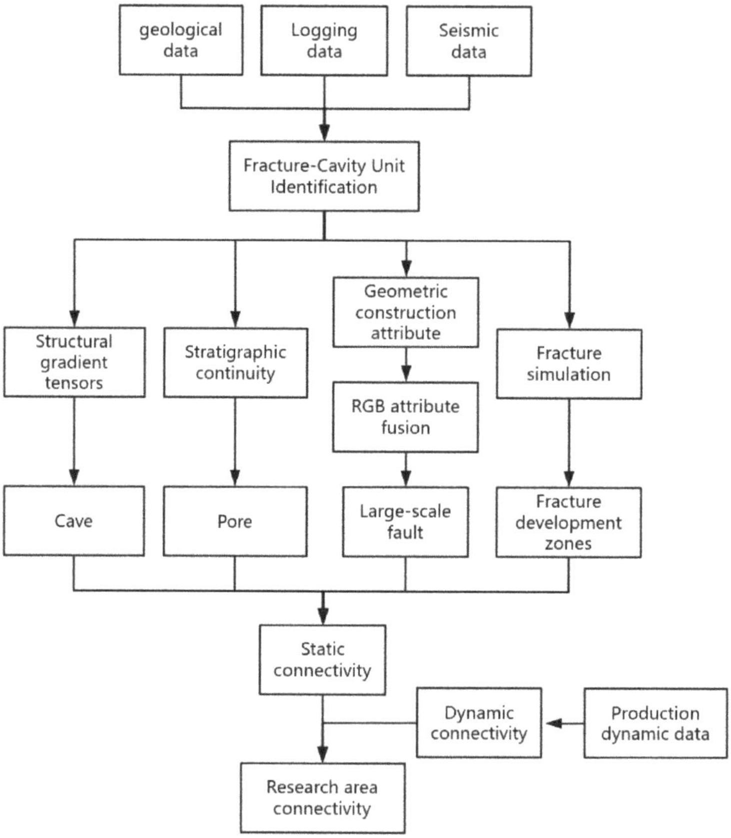

Figure 1. Connectivity evaluation process.

3. Geological Characteristics of the Study Area

The S91 unit is located in the central–western part of the eighth district of the Tarim River Oilfield (Figure 2). This district hosts a karstic carbonate reservoir within the Tarim Basin. The S91 unit forms a highly heterogeneous system with an extensive fracture–cavity network. This network developed through multiple tectonic episodes, mainly during the Caledonian and Early Hercynian paleokarst periods. The Tarim Basin is a multi-cycle composite basin. It evolved under alternating extensional and compressional stresses, resulting in a "three ridges and four troughs" tectonic pattern. The Tarim Oilfield is situated on the Akekule Uplift, with the eighth district located on its south-west-dipping segment. Vertically, karst caves exhibit distinct layered distributions. These layers are primarily aligned with the ancient water table. The S91 unit lies within the karstic steep-slope zone of the Akekule Uplift. A near east–west-trending, nose-shaped structure defines the central part of the local upper unit. Overall, the structural high displays a gentle slope.

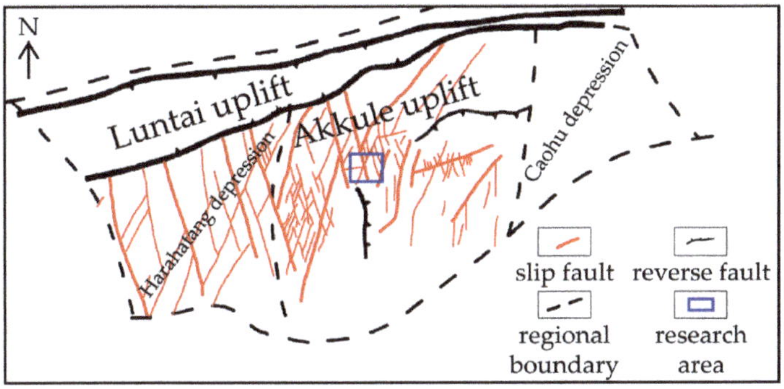

Figure 2. Location and geological features of the study area.

Unit S91 spans the Ordovician strata from bottom to top, including the Yingshan, Yijianfang, Qiaerbake, Lianglitage, and Santamu groups. The primary oil reservoir in the study area is hosted within the carbonate rocks of the Yijianfang and Yingshan groups. Reservoir spaces include five types: dissolution pores, intercrystalline pores, karst caves, weathering fractures, and tectonic fractures, with fracture–cavity systems representing the primary storage and permeability domains. The specific lithological combinations of the Ordovician strata are summarized in Table 1.

Table 1. Summary table of Ordovician strata in Unit S91.

Stratigraphic System					Lithological Description
Erathem	**System**	**Series**	**Group**	**Code**	
Paleozoic	Ordovician	Middle	Santamu	O_{3S}	The upper part consists of yellowish-gray mudstone, mudstone, and gray mudstone, gray mudstone, and reddish-brown mudstone interbedded.
			Lianglitage	O_{3L}	The upper part is light gray mudstone, and the lower part is light gray powdery mudstone interbedded with sandy mudstone.
			Qiaerbake	O_{3q}	The upper section is mainly reddish-brown mudstone, while the lower section is gray mudstone.
		Lower	Yijianfang	O_{2yj}	The upper part consists of gray mudstone and yellowish-gray sandy mudstone, while the lower part consists of yellowish-gray microcrystalline limestone.
			Yingshan	$O_{1\text{–}2y}$	Yellowish-gray mudstone, yellowish-gray microcrystalline limestone.

4. Fracture–Cavity Unit Identification

Carbonate bedrock typically exhibits the "three-high" characteristics: high density, high seismic velocity, and high wave impedance. Its considerable burial depth, substantial thickness, and uniform composition result in minimal velocity variations. Consequently, seismic profiles often display weak reflection amplitudes. In contrast, fractures, dissolution cavities, and large karst cave systems reduce rock density, seismic velocity, and impedance. This reduction increases the contrast between the fracture–cavity body and the surrounding dense rock. It produces strong reflections in seismic offset data, which often appear as bead-like patterns. These patterns reflect the combined seismic response of large karst cavities and dense pore systems. Their identification and characterization typically rely on seismic attributes such as root-mean-square amplitude, reflection intensity, and amplitude variation rate [3].

Reservoir genesis analysis indicates that fractures and fracture zones play a critical role in forming large caves and dissolution cavity systems. Fracture zones themselves, however, rarely serve as major oil and gas reservoirs. They function primarily as conduits for fluid migration and significantly enhance the connectivity of fracture–cavity networks. As a result, they strongly influence overall reservoir connectivity. This study treats large caves, dense dissolution cavity systems, and their associated fracture zones as an integrated system. To predict these targets systematically, structural gradient tensor attributes and seismic discontinuity attributes are applied. These methods enable comprehensive identification and precise characterization of fracture–cavity unit boundaries.

4.1. Cave Identification Based on Structural Gradient Tensors

Structural gradient tensors offer an effective approach for analyzing and characterizing the structural features within seismic data [17]. The concept of structural tensors was originally developed in the field of image processing to distinguish planar regions, edges, and corners within images. In 2002, Bakker et al. introduced this method to seismic exploration, applying the structural gradient tensor algorithm to three-dimensional seismic interpretation. They also introduced an anisotropy parameter derived from this algorithm, primarily used to characterize faults and predict fractures [18]. Subsequent studies have progressively refined the structural gradient tensor analysis method, which is now widely employed for reservoir identification and prediction.

In the study area, fracture–cavity carbonate reservoirs are predominantly controlled by strike-slip fault systems. Reservoir space develops progressively. Fracture activity initially creates fracture zones. These zones are subsequently enlarged by dissolution, ultimately evolving into caves and dissolution cavities. On seismic profiles, large cave systems exhibit high-amplitude, bead-like reflections. These reflections serve as direct indicators of effective reservoirs (Figure 3). Traditional characterization methods primarily rely on amplitude attributes. For instance, root-mean-square amplitude (RMSA) effectively highlights bead-like reflections. However, RMSA is also sensitive to continuous strong reflection axes. This sensitivity makes it difficult to distinguish bead-like reflections from horizontal reflections, hindering accurate delineation of cave boundaries. The structural gradient tensor method enhances this process by calculating gradients and constructing structural tensors from seismic data. This approach improves recognition of large cave system boundaries. It also evaluates the morphology and connectivity of fracture–cavity units and their relationship with surrounding rock. Thus, it supports accurate reservoir characterization. Drilling data from the study area were subjected to statistical analysis. Blowouts and gas anomalies identified during drilling and logging were mapped onto seismic profiles. Structural gradient tensor values were subsequently extracted at these locations. By integrating the distribution of strong bead-like reflections, a threshold value was established (Figure 4). The results indicate that, in the Tarim S91 well area, structural gradient tensor values exceeding 3000 correspond to effective reservoir space.

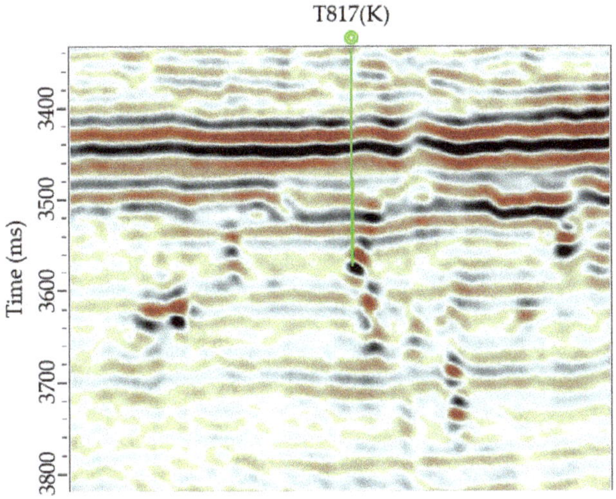

Figure 3. Fracture–cavity bead reflection seismic profile.

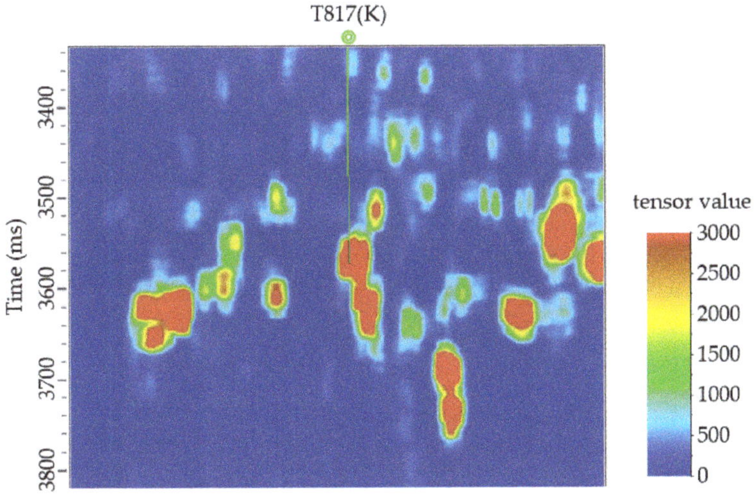

Figure 4. Structural gradient tensor profile.

4.2. Pore Identification Based on Stratigraphic Continuity

In contrast to the strong, bead-like reflections associated with large caves in seismic data, reflections from pore-type reservoirs are typically disordered or discontinuous, making them substantially more challenging to identify. These reservoirs are primarily formed through fluid-driven dissolution processes, with microscopic storage spaces comprising small-scale caves, dissolution pores, and dissolution-related fractures. Through core observations, well-log interpretations, and studies of small-scale pore distribution in outcrops, previous researchers have demonstrated that these pores preferentially develop along large caves, fault zones, or major faults, thereby forming a complex secondary reservoir network surrounding the primary reservoir space [19].

The typically disordered or discontinuous seismic responses of pore-type reservoirs are more effectively captured by the stratigraphic continuity attribute, which provides a more accurate representation of their distribution [6]. This study utilizes the stratigraphic continuity analysis method to detect variations in stratigraphic structure and discontinuity characteristics. This is achieved through the calculation of seismic waveform continuity parameters. The effective range of this attribute for characterizing pore-type reservoirs is calibrated using drilling and seismic data from the study area. The goal is to identify regions with well-developed pores.

Pore-type reservoirs predicted using the stratigraphic continuity attribute are predominantly located at the periphery of strong bead-like seismic reflections. They also occur along same-phase axis fault segments and within the disordered, weak-reflection zones surrounding the fault area (Figure 5). This distribution closely corresponds to the small-scale pore development patterns reported in previous studies. This approach effectively overcomes the limitations of the structural gradient tensor method in detecting small-scale reservoir spaces. It provides critical technical support for the comprehensive characterization of fracture–cavity units. By integrating the identification results from both methods, a more precise characterization of pore locations, sizes, and distribution patterns can be achieved. This integration establishes a solid foundation for subsequent connectivity analysis.

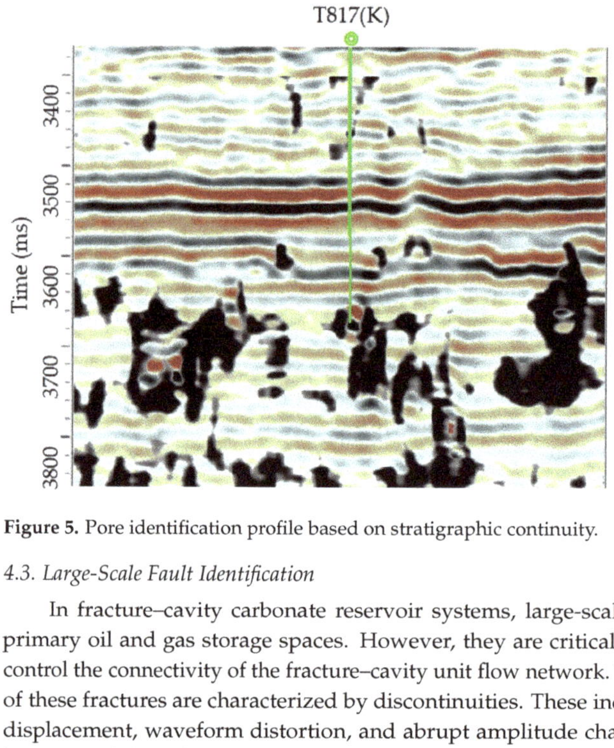

Figure 5. Pore identification profile based on stratigraphic continuity.

4.3. Large-Scale Fault Identification

In fracture–cavity carbonate reservoir systems, large-scale faults do not serve as primary oil and gas storage spaces. However, they are critical structural elements that control the connectivity of the fracture–cavity unit flow network. Typical seismic responses of these fractures are characterized by discontinuities. These include co-directional axial displacement, waveform distortion, and abrupt amplitude changes. Consequently, the locations of these fractures can generally be identified through the analysis of seismic geometric structural attributes. Common techniques include coherence, curvature, and variance attributes.

Research indicates that caves in Unit S91 predominantly occur within fault zones or regions of intensive faulting. Karst features, including caves, are closely associated with faults. Large caves frequently form at the intersections of multiple fault sets. Deep, extensive regional faults also affect cave stratification. In such areas, cave depths deviate from typical layer patterns and are generally greater. Thus, regional faults strongly control both the spatial distribution and depth of cave development. In the study area, major faults are well developed. Conventional single-attribute methods are insufficient for clearly delineating fault morphology in densely faulted zones. To overcome this limitation, the RGB attribute fusion technique is employed to enhance fault recognition.

To achieve a high signal-to-noise ratio suitable for subsequent attribute calculations, the three-dimensional seismic data volume of the study area was processed using structure-oriented filtering. This method, based on the principle of anisotropic compensation, effectively preserves the edge features of geological bodies. It also significantly enhances fracture details and suppresses random seismic noise. Following the evaluation of multiple attributes, variance, coherence, and maximum curvature were selected for RGB attribute fusion. In seismic attribute analysis, the coherence algorithm highlights discontinuities between adjacent seismic traces, such as amplitude and phase variations. It thereby emphasizes boundary information associated with tectonic and geological features (Figure 6a). Low coherence values generally indicate geological anomalies, including faults, fractures, or unconformities. The maximum curvature attribute quantifies the geometric deformation of subsurface layers. Its absolute value is positively correlated with the degree of curvature. Elevated curvature values often correspond to pronounced folds or fault zones (Figure 6b). The variance attribute captures the lateral heterogeneity of strata or rock types by quantifying variations in the similarity of seismic signals between adjacent traces within a local range (Figure 6c). Such heterogeneity may arise from lithological changes, small faults, or variations in pore fluids.

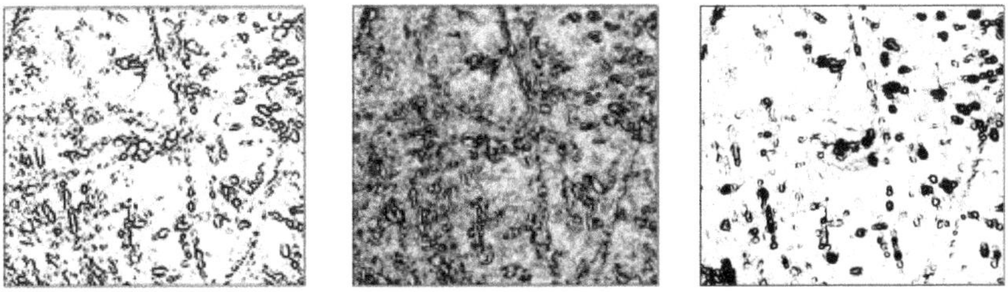

(a) (b) (c)

Figure 6. Geometric construction attribute: (**a**) coherent; (**b**) maximum curvature; (**c**) variance.

By employing a multi-attribute fusion approach, information from various geometric structural attributes can be comprehensively integrated. This allows fracture features to be characterized with greater accuracy. This study conducts a comprehensive attribute analysis using RGB attribute fusion technology. Also known as three-primary-color display technology, RGB attribute fusion maps three distinct types of data onto the primary colors red (R), green (G), and blue (B). Using a defined mapping function, these colors are adjusted to generate an RGB image. Each value in the fused image corresponds to a unique color. The primary advantages of this technology include high fusion speed and strong capability to extract and integrate structural and lithological information from seismic attributes. Furthermore, RGB fusion markedly enhances image integrity by providing richer and more detailed textural information. This greatly improves the discernibility of geological features. Current applications demonstrate that RGB attribute fusion technology provides outstanding performance in key areas, including small fault identification, geological anomaly detection, and reservoir prediction [20–22]. The variance attribute (R), coherence attribute (G), and maximum curvature attribute (B) are combined to form RGB attributes. By adjusting the hue, saturation, and brightness parameters, fault information can be effectively emphasized. Parameter testing indicates that setting the hue to 80, the saturation to 50, and the brightness to 0 enhances the visual presentation of the cross-sectional structure. This adjustment makes its morphological characteristics more distinct (Figure 7a). This fused dataset provides a robust foundation for large-scale fault analysis. Multi-angle attribute analyses—such as cross-sections, along-strike profiles, and slices—are conducted in combination with geological feature constraints. These analyses allow thorough interpretation of the morphology, scale, and distribution of large-scale faults. As a result, detailed characterization of large-scale fault systems is enabled (Figure 7b).

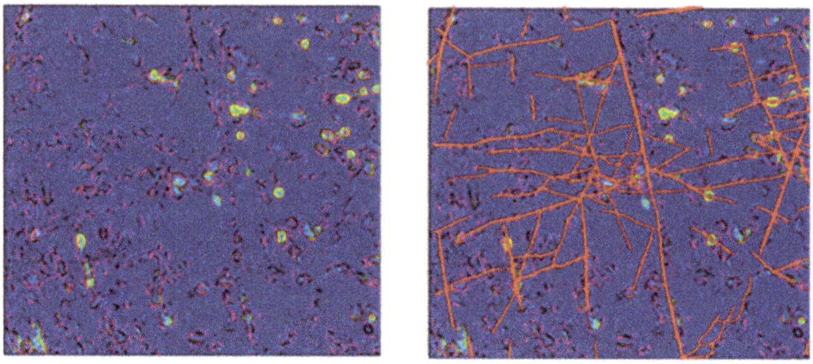

(a) (b)

Figure 7. RGB attribute fusion: (**a**) coherent–curvature–variance RGB fusion attribute; (**b**) large-scale fault identification.

4.4. Fracture Development Zones Identification

In fracture–cavity carbonate reservoirs, accurate characterization of fracture zones is essential for oil exploration and development. These fracture-rich areas not only provide effective storage space but, when combined with caves and pores (e.g., fracture–cave networks), also significantly influence reservoir heterogeneity and effective storage capacity, directly affecting the accuracy of reserve estimates. As critical connecting channels between fracture–cavity units, the spatial distribution of small fractures provides a fundamental basis for identifying connected units and understanding communication pathways between wells. Characterizing these small fractures enables the prediction of dominant flow paths, thereby mitigating the risks of ineffective circulation and water intrusion during injection and production operations, while providing essential support for optimizing well-group injection and replacement parameters.

To accurately determine fracture distribution and density characteristics, ant-tracking and fracture simulation analyses were conducted using the previously extracted geometric structural properties. Geometric structural properties, including coherence, curvature, and variance, not only reveal large-scale fault information but also provide detailed fracture data. Ant bodies derived from geometric structural attributes can rapidly delineate fracture development locations and clearly characterize the distribution of fracture systems. By configuring active ant-tracking parameters (Table 2), the ant bodies generated from different structural attributes were systematically analyzed and compared. By integrating the results of large-scale fault characterization, the ant colony outputs that most effectively identify fracture development zones were ultimately selected. However, because the ant-tracking algorithm relies on point-to-point tracking, its fracture identification results are susceptible to noise interference. Consequently, this method provides only qualitative fracture predictions and does not generate quantitative estimates.

Table 2. Ant-tracking parameters.

Parameters	Value	Parameter Meaning
Initial ant boundary	4	Ant density in seismic data
Ant track deviation	2	Maximum allowable deviation within the effective step size
Ant step size	3	Pixels searched per step
Illegal steps allowed	2	When ants cannot find the boundary value, the distance they continue searching
Legal steps required	2	The maximum distance ants can travel without effective information
Stop criteria (%)	10	Illegal step ratio in the search process

Post-stack seismic data, the most commonly used type, are typically employed to qualitatively characterize fracture development through basic seismic attributes. Chen et al. proposed a quantitative fracture prediction technique based on tensor direction fields, enabling the use of post-stack seismic data for quantitative estimation [23]. The present study applies this technique for quantitative fracture prediction.

The implementation of quantitative fracture prediction comprises four main steps. (1) A high-precision coherence volume is generated from post-stack seismic data using a third-generation eigenvalue-based coherence algorithm. (2) An ant colony tracking algorithm based on directional consistency is applied, followed by post-processing of the coherence volume under ridge-line enhancement constraints to further enhance coherence. (3) Using the simulated point tensor direction field, weighting kernel functions for each direction are calculated within the neighborhood, and an optimized clustering method is applied for weighted fitting to derive fracture attribute values along the principal orientations of the simulated points. (4) The simulated fracture attribute values are used to statistically analyze the relative fracture density within a unit radius and the primary orientations of fracture development, thereby providing reliable information on relative fracture density and fracture orientation.

As shown in Figure 8, the fracture distribution simulated using this method is consistent with the ant-tracking results regarding the overall locations and distribution patterns of fractures. The simulation results provide greater detail, indicating that this method is reliable for fracture prediction in the study area. Based on the simulation results, relative fracture density and fracture orientation can be determined (Figure 9), with short lines representing fracture orientation. This method therefore provides a quantitative foundation for fracture prediction.

Figure 8. Small-scale fracture and crack identification: (**a**) ant tracking; (**b**) fracture simulation.

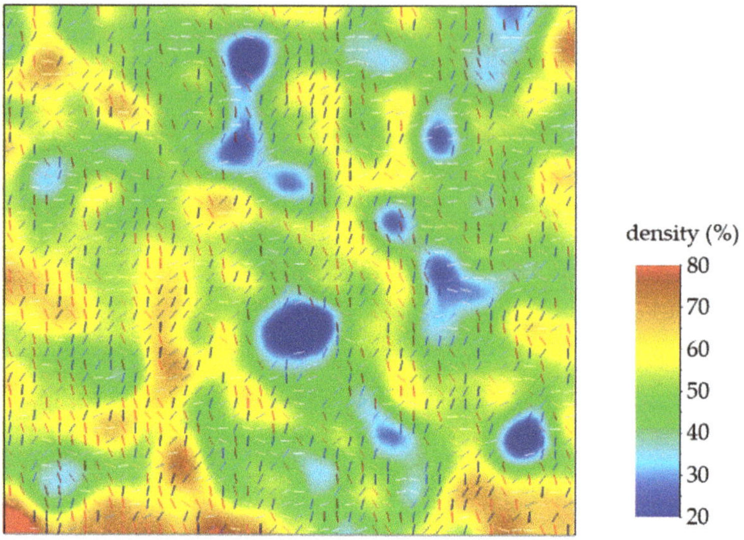

Figure 9. Fracture density and fracture orientation.

5. Fracture–Cavity Unit Connectivity Analysis

5.1. Static Connectivity

Accurately characterizing the reservoir's internal spatial structure requires a detailed analysis. This analysis includes examining the spatial distribution and structural relationships of fracture–cavity units and fracture systems. It also involves assessing connectivity within the fracture–cavity units. Three-dimensional spatial carving is conducted on fracture–cavity units and fractures. Based on the connectivity revealed by these carving results, and in combination with actual oil-field production experience, the study area is partitioned into several connectivity units. The three-dimensional modeling results obtained through this analytical process illustrate the reservoir's spatial structure. They also show the relationships among fracture–cavity systems. This analysis reveals potential interwell connectivity patterns.

The study area demonstrates a static connectivity relationship among wells S91, TK725, and TH10222XK. These wells are classified within the same connected well cluster. TK725 is located approximately 750 m north of TH10222XK and roughly 730 m south of S91. As shown by the three-dimensional carving results of cavities, pores, and fractures (Figure 10), the fracture–cavity units within this well group are distributed discontinuously. Large-scale faults intersect the area, and dense fractures develop throughout. The karst cavities and pores between TH10222X(K) and TK725 are discontinuous. A large-scale north–south-oriented fault intersects both wells. Fracture density in the interwell zone exceeds 50%. A potential connectivity channel exists between TH10222X(K) and TK725 along this fault. Karst development between S91 and TK725 is discontinuous. Large-scale faults do not directly link the two wells. Pores are densely distributed between the wells, with fracture density exceeding 50%. Potential connectivity channels may exist between S91 and TK725 via fractures and pores.

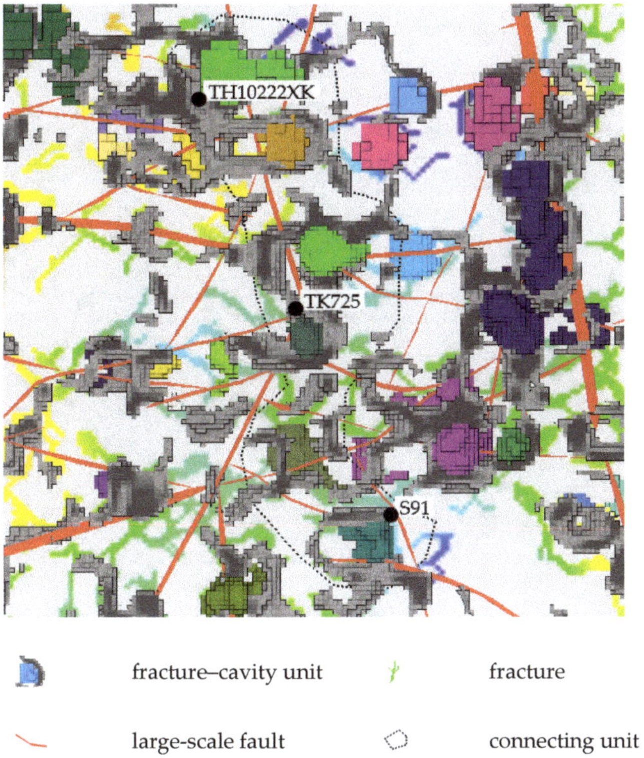

Figure 10. Three-dimensional space carving.

5.2. Dynamic Connectivity

The "potential connectivity" indicated by static models may differ from the "effective connectivity" observed during production. Therefore, validating static model predictions using dynamic production data is essential. This validation enables accurate evaluation of reservoir fluid flow. It also supports optimization of injection and production well networks and the formulation of effective adjustment strategies. This section analyzes dynamic production responses to validate the reliability of the previously discussed static connectivity methods and results. The analysis provides a robust scientific foundation for assessing reservoir connectivity.

During the high-volume water injection at Well TH10222X(K) in 2005, the adjacent Well TK725 exhibited a notable response: well pressure increased by 1.25 MPa, daily oil production rose by 18 tons, and the pressure rise rate reached 0.57 MPa per 10 days. The tracer monitoring peak persisted for 10 days, consistent with the well's dynamic response cycle, thereby indicating clear waterflood connectivity between Wells TH10222X(K) and TK725.

On 20 February 2012, following the resumption of a daily injection rate of 50 m^3 at Well TH10222X(K), Well TK725 exhibited significant dynamic changes over 11 days, during which a cumulative total of 450 m^3 of fluid was injected. By 1 March, oil pressure at TK725 had increased by 1.4 MPa, daily crude oil production had risen by 23 tons, and the water cut had decreased by 34%. The response timing corresponded with the peak observed in the previous tracer monitoring.

In 2015, a composite displacement test confirmed the connectivity between Wells S91 and TK725: from 21 August to 14 September, Well S91 was injected with 800,000 cubic meters of gas, supplemented by 3619 cubic meters of water. Eighteen days later, without any adjustments to oil nozzle parameters or dilution processes, oil pressure at Well TK725 increased by 2.1 MPa, indicating a clear dynamic response. This operation resulted in a cumulative increase in crude oil production of 601 tons.

Comprehensive dynamic monitoring data indicate that the S91–TK725–TH10222XK well group established an effective displacement channel during production. This finding is consistent with the connectivity predicted by the static fracture–cavity reservoir model (Table 3), thereby confirming the reliability of the reservoir connectivity analysis.

Table 3. Comparison of dynamic response and static prediction.

Type	Injection Well	Effect Well	Connectivity Direction	Dynamic Response	Static Prediction
water injection displacement	TH10222X(K)	TK725	North → South	Increasing pressure	Connected along large-scale faults
gas injection displacement	S91	TK725	South → North	Increasing pressure	Connected along fractures and pores

5.3. Connecting Well Group Division

The results of the static and dynamic connectivity analyses for the verification well groups are consistent. This consistency demonstrates that the characterization method developed for fracture–cavity reservoirs is reliable, effective, and well-suited for evaluating interwell connectivity. Building on this conclusion, additional interwell static connectivity analyses were conducted for wells in areas lacking production dynamic data. These wells were subsequently identified and classified into multiple interconnected well groups. These findings provide a solid geological basis for predicting the distribution of remaining oil. They also support optimization of injection and production well networks and the formulation of strategies to enhance recovery rates.

The comprehensive analysis conducted in this study reveals two typical connectivity patterns, which are primarily controlled by large-scale faults and karst–fracture complexes (Figure 11).

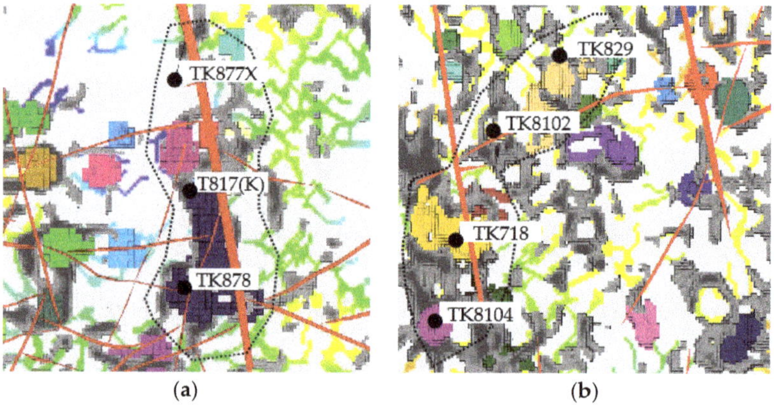

Figure 11. Connecting well group: (**a**) Well Group 1; (**b**) Well Group 2.

Pattern 1: Fault-dominated linear connectivity

This pattern is characterized by wells arranged linearly along major faults. Both Wells T817(K) intersect the same karst cavity system, confirming their direct connectivity. This cavity system is connected to a major fault structure. The trajectory of the inclined Well

TK877X directly intersects this fault. Consequently, this nearly north–south-trending fault serves as the primary pathway linking the TK877X and T817(K) well groups.

Pattern 2: Composite Connectivity Governed by the Fracture–Cavity Network

This pattern is characterized by discontinuously distributed karst cavities. These cavities are interconnected through a dense network of pores and fractures. Wells TK829 and TK8102 intersect the same karst cavity, indicating direct connectivity. Wells TK718 and TK8104 intersect distinct karst cavities. However, the abundant pores distributed between them provide effective connectivity pathways. Furthermore, two intersecting faults and a high-density fracture zone (55%) are present between Wells TK8102 and TK718. This fault–fracture system constitutes an additional potential pathway for enhanced connectivity.

6. Discussion

The S91 unit of the Tarim Oilfield represents a typical shallow-covered karst fracture–cavity reservoir. It exhibits pronounced heterogeneity and complex connectivity. These characteristics are key geological factors limiting the efficient exploitation of residual oil. This study systematically investigates the reservoir connectivity mechanisms. An integrated approach combining multi-scale static and dynamic methods is employed.

Well-logging data provide information only at specific well locations. In contrast, the three-dimensional seismic data employed in this study allow comprehensive identification of carbonate fracture–cavity reservoirs across a regional scale. This approach offers more extensive geological insights. Nevertheless, due to the resolution limitations of seismic data, the method is limited in detecting small-scale geological structures. Consequently, seismic techniques cannot accurately predict the distribution of karst cavities of all sizes. They reliably identify only larger-scale cavities that generate distinct seismic responses. Spectral analysis and property testing, combined with comparisons to logging, image logging, and production data, confirm that the structural gradient tensor and stratigraphic continuity attributes achieve sufficient accuracy in identifying karst cavities and pores to meet the practical requirements of current oilfield production.

In characterizing karst cavities and pores, multiple seismic attributes were evaluated. Among them, amplitude-based attributes—particularly root-mean-square (RMS) amplitude—are effective in highlighting the bead-like responses associated with karst cavities. However, the target interval is overlain by a strong reflection layer (3400–3500 ms, Figure 3), whose lower boundary lies close to the underlying karst cavities. RMS amplitude emphasizes strong bead-like reflections, but it is also highly sensitive to continuous strong reflection events. This sensitivity can exaggerate the influence of the overlying layer, obscure cave boundaries near the interface, and hinder accurate delineation of cave geometry. When seismic waves propagate through fluid-bearing reservoirs, they undergo low-frequency attenuation, which reduces the dominant frequency. The resulting "low-frequency shadow" beneath fracture–cavity bodies serves as an important seismic indicator of hydrocarbon accumulation. Frequency-based attributes can therefore assist in identifying fluid-bearing fracture–cavity bodies. However, these attributes are highly susceptible to noise interference. The presence of thin interlayers may introduce tuning effects that cause misleading frequency variations. These limitations reduce the accuracy of frequency-based identification in the study area. In contrast, the application of structural gradient tensor and formation continuity attributes enables more precise characterization of cave and pore boundaries. This approach effectively suppresses interference from strong reflective layers and allows a clearer delineation of the spatial distribution of fracture–cavity units. The feasibility of this method was validated by integrating cavity and pore identification results from logging, image logging, and related datasets, with the achieved accuracy meeting practical application requirements.

RGB multi-attribute fusion technology offers a significant advantage by generating synergistic information through intuitive color mapping. This reduces the interpretational ambiguity inherent in single-attribute analyses. The seismic attributes of variance, coherence, and maximum curvature are assigned to the red, green, and blue channels, respectively, to generate a superimposed display. Compared with single-attribute analysis, which reveals only one aspect of the subsurface, this method delineates the boundary contours of geological bodies more clearly. The technique leverages the human eye's sensitivity to color differences to emphasize subtle geological features. It also markedly enhances both the efficiency and accuracy of large-scale fault interpretation. Applied to the study area, this method significantly enhanced interpretation accuracy in zones with dense faulting.

Common seismic methods for fracture prediction include post-stack seismic attributes, pre-stack wide-angle data, and rock physics modeling. Among these, post-stack attributes are generally employed solely for qualitative characterization of fractures. In this study, a quantitative fracture prediction method based on tensor direction fields is employed. This method enables estimation of relative fracture density and orientation from post-stack seismic data. Fracture prediction using post-stack data is sensitive to factors such as signal-to-noise ratio and migration imaging, which can reduce modeling accuracy. Tests in the S91 unit indicate that the predicted fracture distributions generally align with ant-tracking results. They also provide more detailed structural information. Accordingly, this technique demonstrates strong applicability in the study area.

Dynamic production data provide direct evidence of connectivity and associated dynamic parameters. However, their application is constrained by high costs and time delays in data acquisition. In contrast, static connectivity analysis based on 3D seismic data offers broader spatial coverage. It also enables the identification of potential connectivity pathways. Integrating dynamic and static approaches allows simultaneous consideration of real-time production responses and spatial structural information. This combined approach enhances the accuracy and reliability of regional connectivity evaluations. Large-scale faults and dense fracture networks are the primary controls on the static connectivity of fracture–cavity units. These networks interconnect isolated units, forming potential pathways for fluid migration. Dynamic production data from the S91–TK725–TH10222XK well group confirm interwell fluid flow. This demonstrates that the fracture network functions as an effective displacement channel, in agreement with the static connectivity model. The consistency between dynamic production results and the static geological model validates the reliability of 3D spatial carving technology for delineating fracture–cavity systems and connectivity channels. The dynamic connectivity results align closely with the static geological model, confirming the reliability of the 3D spatial carving technique for inferring fracture–cavity relationships and connectivity pathways.

Despite the discrete spatial distribution of the reservoir, the highly heterogeneous connectivity network—controlled by the fracture–fracture system—can form locally efficient flow units. Mapping the distribution patterns of these connectivity pathways provides a critical geological basis for the targeted recovery of residual oil. These findings offer valuable guidance for optimizing injection–production strategies and enhancing overall recovery rates. In reservoir management, identifying dominant flow paths enables precise monitoring and control. This helps prevent premature breakthrough and inefficient circulation of injected fluids. Consequently, displacement efficiency and sweep volume in water- and gas-injection projects are enhanced. Based on the spatial distribution of fracture–connectivity units, the existing well network can be optimized. Directional or horizontal wells can target high-permeability channels. Water-blocking and profile-control measures can be implemented in low-efficiency zones. These measures support targeted recovery of residual oil and enhance overall recovery rates.

The connectivity evaluation framework developed in this study integrates static and dynamic data. It provides a valuable reference for similar highly heterogeneous carbonate reservoirs. This framework can inform injection–production optimization, measurement strategies, and enhanced recovery practices. Nevertheless, certain limitations exist. Seismic resolution constraints create blind spots in identifying small-scale fracture–cavity pairs. This may result in overlooked reservoir space. Fracture prediction outcomes depend on seismic data quality, with increased uncertainty in areas exhibiting low signal-to-noise ratios. Although dynamic production data were employed for validation, their representativeness is limited and influenced by production regimes. Current research integrates geological, logging, seismic, and production data. This integration improves identification accuracy and reduces multi-solution ambiguity. Future studies could incorporate higher-frequency seismic, wellbore seismic, or electromagnetic data. They could also develop multi-scale data fusion algorithms. In addition, sensitivity and risk analyses of connectivity on development dynamics could be conducted through numerical simulation. Such efforts would further enhance the robustness and predictive capability of connectivity evaluation results.

7. Conclusions

The S91 unit of the Tarim Oilfield is a fracture–cavity reservoir developed through karstification in a shallowly covered area. The reservoir exhibits high heterogeneity and complex connectivity among fracture–cavity systems. Despite high utilization and recovery levels, considerable residual oil potential remains.

Structural gradient tensor and stratigraphic continuity attributes are integrated to delineate cave–pore boundaries. They also reveal the spatial distribution of fracture–cavity units. RGB fusion is applied to identify large fractures. This approach integrates coherence, maximum curvature, and variance attributes. It substantially enhances the clarity and continuity of fracture imaging. In the study area, fracture–cavity units exhibit discontinuous spatial patterns. Connectivity is governed by large-scale faults and dense fracture networks. Ant-tracking and fracture simulation techniques are employed to identify fracture development zones. They are also used to characterize the connecting channels between fracture–cavity units.

Three-dimensional spatial carving technology is employed to visualize fracture–cavity units and associated fractures. This method effectively reveals the internal spatial structure of the reservoir. It also shows the organization of fracture–cavity systems. In the S91–TK725–TH10222XK well group, fracture–cavity units exhibit discontinuous distributions. They are intersected by extensive fracture systems and dense fractures. These fractures form a network that interconnects the fracture–cavity units. This network establishes potential static connectivity channels.

Production dynamic data from the S91–TK725–TH10222XK well group confirm interconnectivity. They also validate effective displacement pathways during production. These dynamic responses corroborate the static characterization of potential connectivity within the fracture–cavity reservoir. The consistency between dynamic and static results further reinforces the reliability of the connectivity analysis.

This study conducts a regional connectivity analysis across the study area. The area is divided into multiple connected well groups. This approach reduces the cost of acquiring production data from individual wells. It also provides a geological basis for optimizing fracturing and injection–production operations. Additionally, it supports prediction of the distribution of remaining oil.

Author Contributions: Conceptualization, Y.X. and X.P.; Methodology, Y.X. and X.P.; Software, Y.X.; Validation, Y.X.; Formal analysis, Y.X.; Investigation, Y.X.; Writing – original draft, Y.X.; Writing—review & editing, Y.G. and X.P.; Visualization, Y.X. and X.P.; Supervision, Y.G. and X.P.; Project administration, X.P.; Funding acquisition, Y.G. and X.P. All authors have read and agreed to the published version of the manuscript.

Funding: This research received no external funding.

Data Availability Statement: The data presented in this study are available on request from the corresponding author. The data are available only to the collaborating scientists due to relevant data protection law.

Conflicts of Interest: Author Yinghan Gao was employed by the company Sinopec Group Northwest Petroleum Branch Oil Production Plant No. 3. The remaining authors declare that the research was conducted in the absence of any commercial or financial relationships that could be construed as a potential conflict of interest.

References

1. Chen, Z.; Xu, S.; Xu, C. Study on the connectivity of Ordovician karstic dissolution bodies in the Yue Man block of the Halahatang Oilfield. *J. Gansu Sci.* **2023**, *35*, 22–30.
2. Yang, M. Study on interwell connectivity of karstic fracture-cavity type carbonate reservoirs in Area 4 of Tarim Oilfield. *Xinjiang Geol.* **2004**, *22*, 196–199.
3. Wu, M.; Chai, X.; Zhou, B.; Li, H.; Yan, N.; Peng, P. Characterization and application of fracture-cavity type carbonate reservoir connectivity. *Xinjiang Pet. Geol.* **2022**, *43*, 188–193.
4. Geng, T.; Lv, Y.; Wu, B.; Zhang, X.; Wen, H. Fracture-Cavity Reservoir Reserve Evaluation Methods and Development Strategies. *Spec. Oil Gas Reserv.* **2021**, *28*, 129–136.
5. Xie, X.; Yan, C.; Lai, S.; He, G.; Luo, J. Study on interwell connectivity of fracture-cavity type carbonate reservoirs in the Tarim River Area 6. *Pet. Geol. Eng.* **2013**, *27*, 103–105.
6. Liu, J.; Liao, M.; Wang, Y.; Gong, W.; Huang, C.; Zha, M. Evaluation of Interwell Connectivity of Reservoir Bodies Controlled by Fault Zones in the Middle Segment of the Shunbei 4 Fault Zone, Shunbei Oilfield. *Xinjiang Pet. Geol.* **2023**, *44*, 456–464.
7. Zhu, G.; Wu, J.; Ge, X.; Miao, X. Logging evaluation technology for fracture-cavity carbonate reservoirs. *Well Logging Technol.* **2014**, *38*, 695–699. [CrossRef]
8. Du, H. Study on Logging Evaluation Methods for Fracture-Cavity Reservoirs in Ordovician Carbonate Rocks in the Tarim Oilfield. Master's Thesis, China University of Geosciences, Beijing, China, 2018.
9. Lin, X.; Xiao, H. Simulation and application of acoustic remote detection logging response in fracture-cavity reservoirs: The case of carbonate fracture-cavity reservoirs in Tarim Oilfield. *J. Jilin Univ. Earth Sci. Ed.* **2024**, *55*, 312–327. [CrossRef]
10. Chen, K.; Liu, S.; Wang, Z.; Zhang, Y. Study on the identification of fracture-cavity carbonate reservoirs based on curve reconstruction: The case of the Pz layer in the Karabulak oilfield, South Turge Basin. *Adv. Earth Sci.* **2018**, *33*, 1154–1160.
11. Li, H.; Liu, J.; Gong, W.; Zhang, R. Seismic prediction of medium- to small-scale fracture-controlled fracture-cavity bodies in the Shunbei area. *Xinjiang Pet. Geol.* **2025**, *46*, 240–245.
12. Li, H.; Dou, Z.; Wang, S.; Tang, J.; Zhu, T. Seismic multi-attribute identification of "weak reflection" characteristics of carbonate fracture-cavity reservoirs. *Geophys. Prospect. Pet.* **2014**, *53*, 713–719.
13. Xie, W.; Bi, C.; Hu, H.; Ma, Y.; Ma, L.; Yao, M. Identification and application of fracture-cavity reservoirs in carbonate rocks based on diffraction waves. *Sci. Technol. Eng.* **2021**, *21*, 453–457.
14. Zhang, X.; Li, L. Tracer monitoring of interwell connectivity in fracture-cavity type reservoirs in carbonate rocks: The case of the T402 injection-production well group in the Tarim Oilfield. *J. Xi'an Shiyou Univ. (Nat. Sci. Ed.)* **2019**, *34*, 55–59.
15. Rong, Y. Study on the Identification and Characterization of Interwell Connectivity Structures in Fracture-Cavity Type Reservoirs Based on Tracer Technology. Doctoral Thesis, Southwest Petroleum University, Chengdu, China, 2016.
16. Wang, P.; Pan, W.; Li, S.; Liu, Z.; He, J.; Ma, H. Identification of multiple fracture-cavity bodies in fracture-cavity type carbonate reservoirs using single-well dynamics: The case of the Ha 6 well area in the Halahatang oilfield. *Xinjiang Pet. Geol.* **2017**, *38*, 363–368.
17. Zhang, J.; Dong, M.; Zhou, Z.; Tan, M.; Feng, D. Study application of the coherent body method based on GST. *Nat. Gas Ind.* **2007**, *27*, 381–383.
18. Bakker, P. Image Structure Analysis for Seismic Interpretation. Doctoral Thesis, Delft University of Technology, Delft, The Netherlands, 2002.

19. Lü, X.; Li, H.; Wei, H.; Si, C.; Wu, X.; Bu, C.; Kang, Z.; Sun, J. Multi-scale classification and characterization of fracture-cavity bodies in carbonate reservoirs: The example of the Ordovician reservoir in the S80 unit of the Tarim Oilfield. *Oil Gas Geol.* **2017**, *38*, 813–821.
20. Zhang, L.; Meng, Y.; Qi, L.; Adilijiang, A.; Dai, J.; Li, A.; Zhang, L. Research progress on fault identification techniques using seismic data. *Prog. Geophys.* **2024**, *40*, 208–219.
21. Zhao, X.; Li, L.; Xue, G.; Zhang, Z.; Yuan, X.; Chai, Y.; Yang, P.; Xu, Y. Fine Characterization of Shallow-Water Turbidite Sand Bodies Using RGB Attribute Fusion and FCM Clustering Algorithm: A Case Study of the X Gas Field in the Yinggehai Basin. *Mar. Geol. Front.* **2025**, *41*, 40–54. [CrossRef]
22. Wang, J.; Hu, H.; Wang, H.; Fu, X.; Hu, S.; Gao, F. Identification and prediction methods for concealed faults associated with extension differences in the Bohai X Block. *Geophys. Prospect. Pet.* **2025**, *64*, 553–562.
23. Chen, Y.; Yang, H.; Zhao, H.; Wang, C.; Jin, G. Quantitative prediction technology for fractures based on post-stack seismic data. In Proceedings of the 2019 Geophysical Exploration Technology Seminar of the Petroleum Society of China, Chengdu, China, 9 September 2019.

Disclaimer/Publisher's Note: The statements, opinions and data contained in all publications are solely those of the individual author(s) and contributor(s) and not of MDPI and/or the editor(s). MDPI and/or the editor(s) disclaim responsibility for any injury to people or property resulting from any ideas, methods, instructions or products referred to in the content.

applied sciences

MDPI

Article

Efficient Three-Dimensional Marine Controlled-Source Electromagnetic Modeling Using Coordinate Transformations and Adaptive High-Order Finite Elements

Feiyan Wang * and Song Cheng

College of Civil Engineering, Xiangtan University, Xiangtan 411105, China
* Correspondence: feiyanwang@xtu.edu.cn

Abstract: Efficient and accurate forward modeling of electromagnetic fields is essential for advancing geophysical exploration in complex marine environments. However, realistic survey conditions characterized by low-frequency spectra, fine sedimentary strata, irregular bathymetry, and anisotropic materials pose significant challenges for conventional numerical methods. To address these issues, this work presents a parallel modeling framework that combines coordinate transformations with an adaptive high-order finite-element approach for 3D marine controlled-source electromagnetic (MCSEM) simulations. The algorithm exploits the form invariance of Maxwell's equations to map the original boundary value problem over the physical domain to one defined over a computationally favorable domain filled with anisotropic media. The transformed model is then discretized and solved using a parallel high-order finite-element scheme enhanced with a goal-oriented adaptive mesh refinement strategy. We examine the performance of the proposed framework using both synthetic models and the realistic Marlim R3D benchmark dataset. The results demonstrate that the proposed approach can effectively reduce computational costs while maintaining high accuracy across a wide frequency range and varying water depths. These findings highlight the framework's potential for large-scale, high-resolution CSEM exploration of offshore resources.

Keywords: marine controlled-source electromagnetic method; coordinate transformation; high-order finite-element method; 3D forward modeling; unstructured grid

1. Introduction

Marine controlled-source electromagnetic (MCSEM) methods have gained wide recognition as a powerful tool for imaging offshore geological structures and identifying submarine resources [1]. These applications rely on towed electric sources and arrays of seafloor or streamer-based receivers to infer the electrical conductivity distribution of the subsurface [2]. Modern acquisition systems are able to conduct large-scale 3D MCSEM surveys [3]. The rich data coverage processed at a few discrete low frequencies helps improve the model resolution at different scales. During the interpretation process, electrical anisotropy is often considered according to the support from well log data. A common case is the vertical transverse isotropy (VTI) due to sedimentary layering [4,5]. However, geological activities such as folding and faulting often disrupt the idealized layered models, resulting in complex anisotropy patterns [6]. Recent studies have demonstrated the significant impact of complex electrical anisotropy, such as the tilted transverse isotropy (TTI), on MCSEM responses [7–9], highlighting the need for more advanced and reliable interpretation tools.

Accurate and efficient forward modeling is fundamental to the success of MCSEM data interpretation, particularly in environments characterized by strong anisotropy and geometrical complexity. To this end, a variety of numerical approaches have been proposed, including finite-difference [10], integral-equation [11], finite-volume [12], and finite-element methods [13–16]. Among these, finite-element techniques are especially attractive due to their flexibility in accommodating irregular topography, heterogeneous media, and complex boundary conditions. In particular, high-order finite-element methods have drawn growing interest owing to their faster convergence rates and superior accuracy on coarser meshes [17,18]. Recent developments in high-order finite-element applications include adaptive mesh refinement techniques [17,19,20] and efficient solvers for the increasingly dense and large-scale equation systems [15]. Nevertheless, most finite-element implementations with unstructured grids continue to operate directly on complex physical domains, which could lead to excessive meshing overhead and degraded computational performance in certain marine settings.

These computational challenges associated with high-order finite-element discretization are further amplified in practical MCSEM surveys, where the low-frequency nature of the signals and the presence of conductive seawater impose additional requirements on the modeling domain. To detect deeply buried hydrocarbon reservoirs, MCSEM signals are typically processed at low frequencies, ranging from 0.1 to 10 Hz, to enhance penetration through conductive seawater and sediments. Accurate simulation of these low-frequency electromagnetic responses requires the computational domain to extend sufficiently far from the survey area to properly represent open-boundary conditions. This becomes even more challenging when seawater conductivity varies with depth. As a result, large-scale domains with sequences of thin, conductive layers are generated, particularly in shallow-water environments or long-offset surveys. Although unstructured mesh discretization offers geometrical flexibility, it is generally inefficient in such scenarios. The presence of strong scale disparity can either cause element distortion or excessively refined meshes to maintain numerical stability [21].

Several techniques have been proposed to mitigate the computational burden associated with large and highly heterogeneous domains. Rigorous model reduction methods, such as hybrid finite-element/boundary-element approaches [22], can provide accurate results over a wide frequency range. However, these methods typically lead to dense equation systems that become expensive to store and solve as the complexity of geological heterogeneity increases. A computationally less involved alternative refers to transformation-based strategies. One representative example is the perfectly matched layer (PML), which surrounds the computational domain with artificially designed absorbing layers using coordinate stretching to absorb outgoing wave reflections [23]. Various formulations of PML have proven effective in frequency-domain CSEM simulations [24–26] and time-domain EM applications [27]. Nonetheless, the performance of PML depends sensitively on its controlling parameters, including the layer thickness, the working frequencies and the physical properties, which should be carefully determined for each modeling scenario [27].

Beyond boundary-condition treatments such as PML, a more general strategy can be derived from the form invariance of Maxwell's equations under coordinate transformations (CT) [28]. This invariance guarantees that the interaction between the electromagnetic fields and the underlying media remains consistent regardless of the selected coordinate system. Based on this, field behavior can be controlled by properly designing the geometry and material properties without altering the fundamental physics. This concept forms the foundation of transformation optics. In contrast to the widespread applications in computational electromagnetics [29,30], this property has not been fully explored in geo-electromagnetic community. One notable exception is the work of Kamm et al. [31], which employed CT to flatten complex topography such that rectilinear grids can be readily employed to discretize irregular geological structures. More recently, Chen et al. [32] adopted the geometrical mapping to construct an exponentially decaying absorbing boundary condition in MCSEM modeling.

This work explores the application of transformation electromagnetics beyond conventional boundary-condition treatments and aims to establish a unified framework for high-resolution 3D MCSEM modeling. The approach integrates coordinate transformations to overcome geometrical limitations with high-order finite elements and goal-oriented mesh adaptivity to ensure numerical accuracy, resulting in an efficient, fully distributed parallel modeling scheme. The framework is suitable for large-scale surveys in both deep and ultra-shallow marine settings, where traditional methods may be inefficient under challenging conditions.

The remainder of this paper is organized as follows. First, we revisit the form-invariant principle of Maxwell's equations under coordinate transformations and derive the transformed boundary value formulation. Next, a parallel adaptive high-order finite-element method tailored to solve the transformed system is introduced. Finally, two representative transformation scenarios are presented to demonstrate the effectiveness of the proposed algorithm.

2. Methodology

2.1. MCSEM Boundary Value Problem Under Coordinate Transformations

Under the quasi-static approximation and assuming a time-harmonic dependence $e^{i\omega t}$, Faraday's law and Ampere's law due to an impressed electric source can be expressed as follows:

$$\nabla\times \mathbf{E}(\mathbf{r}) = -i\omega\mu(\mathbf{r})\mathbf{H}(\mathbf{r}), \tag{1a}$$

$$\nabla\times \mathbf{H}(\mathbf{r}) = \sigma(\mathbf{r})\mathbf{E}(\mathbf{r}) + \mathbf{J}_s(\mathbf{r}), \tag{1b}$$

where $\mathbf{r} \in \mathbb{R}^3$ denotes an arbitrary position vector in 3D Cartesian coordinates, $\mathbf{E}$ denotes the electric field, $\mathbf{H}$ denotes the magnetic field, $\mathbf{J}_s$ denotes the impressed current density, ω denotes the angular frequency, and μ and σ are the magnetic permeability and electrical conductivity, respectively. Without loss of generality, we assume that the subsurface magnetic permeability μ is equal to that of free space. σ is generally anisotropic and noted as a symmetric, positive-definite second-rank tensor [33]. To solve the electromagnetic fields with numerical methods, the unbounded space needs to be truncated into a finite domain Ω (physical domain). Meanwhile, proper boundary conditions should be enforced on the outer boundaries to ensure the well-posedness of the formulation. With this implementation, we considered the Neumann boundary condition:

$$\mathbf{n} \times (\boldsymbol{\mu}^{-1}(\mathbf{r})\nabla\times \mathbf{E}(\mathbf{r})) = \mathbf{0} \text{ on } \partial\Omega, \tag{2}$$

where $\mathbf{n}$ denotes the outward normal vector to $\partial\Omega$.

Assume a real-valued mapping $\mathbf{r} \mapsto \tilde{\mathbf{r}}$, consequently, $\Omega \to \tilde{\Omega}$, and let $\mathbf{J}$ denote the Jacobian of the corresponding coordinate transformation:

$$\mathbf{J} = \begin{pmatrix} \partial\tilde{x}/\partial x & \partial\tilde{x}/\partial y & \partial\tilde{x}/\partial z \\ \partial\tilde{y}/\partial x & \partial\tilde{y}/\partial y & \partial\tilde{y}/\partial z \\ \partial\tilde{z}/\partial x & \partial\tilde{z}/\partial y & \partial\tilde{z}/\partial z \end{pmatrix}. \tag{3}$$

It can be proven that Equations (1a), (1b) and (2) remain form-invariant in the transformed space [34]:

$$\tilde{\nabla} \times \tilde{\mathbf{E}}(\tilde{\mathbf{r}}) = -i\omega\tilde{\mu}(\tilde{\mathbf{r}})\tilde{\mathbf{H}}(\tilde{\mathbf{r}}) \quad \text{in } \tilde{\Omega}, \tag{4a}$$

$$\tilde{\nabla} \times \tilde{\mathbf{H}}(\tilde{\mathbf{r}}) = \tilde{\sigma}(\tilde{\mathbf{r}})\tilde{\mathbf{E}}(\tilde{\mathbf{r}}) + \tilde{\mathbf{J}}_s(\tilde{\mathbf{r}}) \quad \text{in } \tilde{\Omega}, \tag{4b}$$

$$\tilde{\mathbf{n}} \times (\tilde{\mu}^{-1}(\tilde{\mathbf{r}})\tilde{\nabla} \times \tilde{\mathbf{E}}(\tilde{\mathbf{r}})) = \mathbf{0} \quad \text{on } \partial\tilde{\Omega}, \tag{4c}$$

here, $\tilde{\mathbf{n}}$ represents the normal vector to $\partial\tilde{\Omega}$. The new fields and parameters are related to the values in the original coordinates via [35]

$$\tilde{\mu}(\tilde{\mathbf{r}}) = \frac{\mathbf{J}\mu(\mathbf{r})\mathbf{J}^T}{\det(\mathbf{J})}, \tag{5a}$$

$$\tilde{\sigma}(\tilde{\mathbf{r}}) = \frac{\mathbf{J}\sigma(\mathbf{r})\mathbf{J}^T}{\det(\mathbf{J})}, \tag{5b}$$

$$\tilde{\mathbf{J}}_s(\tilde{\mathbf{r}}) = \frac{\mathbf{J}}{\det(\mathbf{J})}\mathbf{J}_s(\mathbf{r}), \tag{5c}$$

$$\tilde{\mathbf{E}}(\tilde{\mathbf{r}}) = \mathbf{J}^{-T}\mathbf{E}(\mathbf{r}), \tag{5d}$$

$$\tilde{\mathbf{H}}(\tilde{\mathbf{r}}) = \mathbf{J}^{-T}\mathbf{H}(\mathbf{r}). \tag{5e}$$

In the new form, $\tilde{\nabla}$ denotes the gradient operator with respect to the transformed coordinates $\tilde{\mathbf{r}} = (\tilde{x}, \tilde{y}, \tilde{z})$, det(.) means the determinant operator, and the subscript (T) stands for the transpose operator. $\tilde{\mu}(\tilde{\mathbf{r}})$ and $\tilde{\sigma}(\tilde{\mathbf{r}})$ are the new constitutive parameters defined in the transformed domain. It is readily seen that the transformed electromagnetic parameters remain real-valued, symmetric, and positive-definite as long as the transformation is real-valued and differentiable, so that the Jacobian is non-singular throughout the entire computational domain to ensure the solution stability [31].

Once the transformation Jacobian in Equation (3) is obtained, the resulting material parameters $\tilde{\mu}$ and $\tilde{\sigma}$ and the impressed source term $\tilde{\mathbf{J}}_s$ in the transformed domain $\tilde{\Omega}$ can be computed according to Equation (5a)–(5c). Following the so-called material interpretation [35], we solve the transformed problem in Equation (4a)–(4c) defined on $\tilde{\Omega}$ as if it is formulated in the original Cartesian coordinates. The only difference is that we use the new anisotropic properties listed in Equation (5a), (5b) to represent the transformed media [36]. For clarity in notation, we omit the tilde from the gradient operator and normal vector, while retaining it for field and parameter variables to distinguish them from the original physical values. Accordingly, substituting Equation (4a) into Equation (4b) to eliminate the magnetic field and applying the homogeneous Neumann boundary condition from Equation (4c), we derive the transformed boundary value problem expressed in terms of the transformed electric field:

$$\nabla \times (\tilde{\mu}^{-1}\nabla \times \tilde{\mathbf{E}}) + i\omega\tilde{\sigma}\tilde{\mathbf{E}} = -i\omega\tilde{\mathbf{J}}_s \text{ in } \tilde{\Omega}, \tag{6a}$$

$$\mathbf{n} \times (\tilde{\mu}^{-1}\nabla \times \tilde{\mathbf{E}}) = \mathbf{0} \text{ on } \partial\tilde{\Omega}. \tag{6b}$$

2.2. Adaptive High-Order Finite-Element Discretization

We choose to discretize the transformed problem in Equation (6a), (6b) using high-order finite elements. While high-order finite-element methods are considered more appropriate for simulating strongly heterogeneous and anisotropic media, increasing the order of basis functions incurs significant computational cost [15,37]. Our preliminary studies indicate that upgrading from the lowest order (P1) to the second order (P2) greatly improves the convergence rate with manageable computational efforts. However, further increasing the polynomial order leads to a significantly denser system matrix, which in turn induces a sharp rise in computational cost that may not be justified by the marginal gain in solution accuracy. To balance numerical efficiency and accuracy, we opted for a P2 discretization.

To illustrate the finite-element procedure, we first define the underlying function space:

$$H(curl,\tilde{\Omega}) = \{\mathbf{w} \in (L^2(\tilde{\Omega}))^3 \mid \nabla\times \mathbf{w} \in (L^2(\tilde{\Omega}))^3\}. \tag{7}$$

Based on the Galerkin finite-element approach, the weak form of the problem in Equation (6a), (6b) reads: find $\tilde{\mathbf{E}} \in H(curl,\tilde{\Omega})$, such that

$$B(\tilde{\mathbf{E}},\mathbf{N}) = F(\mathbf{N}), \;\; \forall \mathbf{N} \in H(curl,\tilde{\Omega}), \tag{8}$$

where

$$B(\tilde{\mathbf{E}},\mathbf{N}) = \iiint_{\tilde{\Omega}} \nabla\times \mathbf{N}\cdot(\tilde{\boldsymbol{\mu}}^{-1}\nabla\times\tilde{\mathbf{E}})\mathrm{d}\tilde{\Omega} + i\omega \iiint_{\tilde{\Omega}} \mathbf{N}\cdot(\tilde{\sigma}\tilde{\mathbf{E}})\mathrm{d}\tilde{\Omega}, \tag{9a}$$

$$F(\mathbf{N}) = -\,i\omega \iiint_{\tilde{\Omega}} \mathbf{N}\cdot\tilde{\mathbf{J}}_s\mathrm{d}\tilde{\Omega}. \tag{9b}$$

To solve the weak form presented above, we first partition the transformed computational domain into a set of non-overlapping tetrahedral cells, denoted by $\mathcal{T}_h$. Over this mesh, we employ the second-order curl-conforming finite-element function space $H(curl,\mathcal{T}_h)$ to approximate the continuous electric field. After assembling the local element matrices [24], the discrete boundary value problem corresponding to Equation (6a), (6b) reduces to the following system of linear equations:

$$\mathbf{Kx} = \mathbf{F}, \tag{10}$$

where $\mathbf{x}$ denotes the solution vector, $\mathbf{K}$ denotes the system matrix, which is complex and symmetric with the element entries given by

$$K_{ij} = \iiint_{\tilde{\Omega}} \nabla\times \mathbf{N}_i\cdot(\tilde{\boldsymbol{\mu}}^{-1}\nabla\times\mathbf{N}_j)\mathrm{d}\tilde{\Omega} + i\omega \iiint_{\tilde{\Omega}} \mathbf{N}_i\cdot(\tilde{\sigma}\mathbf{N}_j)\mathrm{d}\tilde{\Omega}, \; \forall \mathbf{N}_i,\mathbf{N}_j \in H(curl,\mathcal{T}_h), \tag{11}$$

and $\mathbf{F}$ is the source vector with

$$F_i = -i\omega \iiint_{\tilde{\Omega}} \mathbf{N}_i\cdot\tilde{\mathbf{J}}_s\mathrm{d}\Omega, \; \forall \mathbf{N}_i \in H(curl,\mathcal{T}_h). \tag{12}$$

Once the solution vector $\mathbf{x}$ is obtained, we can compute the transformed fields at measuring points through finite-element interpolation. After that, inverse transformation via Equation (5d), (5e) must be performed to obtain the true field values.

Before conducting the finite-element discretization, it is important to address a few critical considerations. First of all, the differentiability of the mapping function is essential to ensure that the Jacobian matrix is well defined, thereby guaranteeing the validity of the coordinate transformation. Even when the mapping is differentiable, excessively rapid variations within a single finite element may still give rise to numerical instabilities. Such effects may manifest in several ways, including poor conditioning of the system matrix, reduced accuracy of numerical quadrature, or the need for significant mesh refinement to adequately capture local variations. These observations highlight that the quality of the mesh plays a decisive role in balancing accuracy and stability. In particular, the mesh must be sufficiently refined in regions with strong field variations while avoiding unnecessary refinement elsewhere.

To achieve this, we employ a goal-oriented adaptivity strategy coupled with a residual-based error estimator. This estimator accounts for the volume residual as well as the numerical discontinuities in the normal component of the electrical current density and the tangential component of the magnetic field, yielding

$$\begin{aligned}\eta_{K,u}^2 =& h_K^2 c_{\mu^{-1}} \|\nabla\times(\tilde{\boldsymbol{\mu}}^{-1}\nabla\times\mathbf{U}) + i\omega\tilde{\sigma}\mathbf{U}\|_{\Omega_k}^2 \\ &+ \frac{1}{2}\sum_{\mathcal{F}\in\mathcal{E}_K^{\mathrm{int}}} \frac{h_{\mathcal{F}}}{\omega\overline{\sigma}} \|[\mathbf{n}_{\mathcal{F}}\cdot(i\omega\tilde{\sigma}\mathbf{U})]\|_{\mathcal{F}}^2 \\ &+ \frac{1}{2}\sum_{\mathcal{F}\in\mathcal{E}_K^{\mathrm{int}}} \frac{h_{\mathcal{F}}}{\overline{\mu^{-1}}} \|[\mathbf{n}_{\mathcal{F}}\times(\tilde{\boldsymbol{\mu}}^{-1}\nabla\times\mathbf{U})]\|_{\mathcal{F}}^2,\end{aligned} \tag{13}$$

where $[\cdot]_{\mathcal{F}}$ denotes the jump across face $\mathcal{F}$, $\mathcal{E}_K^{\mathrm{int}}$ denotes the set of interior faces of tetrahedron K, $\mathbf{n}_{\mathcal{F}}$ denotes the face normal vector, h_K and $h_{\mathcal{F}}$ denote the size of element K and face $\mathcal{F}$, respectively, $c_{\mu^{-1}}$, $\overline{\mu^{-1}}$, and $\overline{\sigma}$ are three scaling coefficients to address the heterogeneity of physical properties. $c_{\mu^{-1}}$ is chosen to be the inverse of the maximum eigenvalue of $\tilde{\boldsymbol{\mu}}^{-1}$, while $\overline{\mu^{-1}}$and $\overline{\sigma}$ the averaged $\tilde{\boldsymbol{\mu}}^{-1}$ and $\tilde{\sigma}$ at each inner face, respectively.

Following Ren et al. [38], we derive the goal-oriented error estimation by first constructing a linear functional that quantifies the data errors at the receiver locations. We then solve the corresponding auxiliary (dual) problem, which yields the influence function $\mathbf{W}$. After obtaining the element-wise error indicators from Equation (13) for both the primal ($\eta_{K,e}$) and dual ($\eta_{K,w}$) problems, the goal-oriented error indicator, which estimates the error in the quantity of interest, can be approximated as follows:

$$\eta = \sum_K \eta_K = \sum_K \eta_{K,e}\eta_{K,w}. \tag{14}$$

We mark a subset of the elements with the largest error indicators and refine them by reducing their volumes to the half of the original value. This mesh refinement process is repeated iteratively until either the total number of degrees of freedom (DoFs) exceeds 10^7, or the root mean squared relative error between two consecutive steps is less than 0.5%.

3. Numerical Experiments and Analysis

In this section, we present two sets of examples to evaluate the performance of the adaptive finite-element solver based on the coordinate transformation method (AFEM-CT). The algorithm was implemented based on the scalable finite-element library MFEM [39]. The domain partitioning and inter-processor communication are handled automatically within MFEM's parallel framework, ensuring both scalability and robustness. Mesh generation and refinement were carried out using the TetGen software package [40]. The resulting finite-element systems were solved using the parallel direct solver MUMPS [41] for moderate-size problems. For large-scale models, the FGMRES preconditioned with the auxiliary-space Maxwell solver (AMS) [15] was adopted to relieve memory usage. The convergence tolerance was set to 10^{-9} to ensure an accurate solution. All computations were executed on the High Performance Computing Platform at Xiangtan University, comprising 85 Intel® Xeon® 2.6 GHz CPU nodes, each equipped with 64 cores.

3.1. Application to Domain Compression

We begin by demonstrating how the coordinate transformation approach functions as an effective model reduction strategy. The design of this domain compression is illustrated in Figure 1, where Figure 1a shows the original physical domain, and Figure 1b depicts the transformed computational model. The entire physical model Ω is partitioned into two subdomains—the inner domain Ω^{int} and the exterior domain Ω^{ext}—such that $\Omega = \Omega^{\mathrm{int}} \cup \Omega^{\mathrm{ext}}$. Without loss of generality, the inner domain Ω^{int} is constructed as a brick-shaped region based on geological data. Its lateral extent fully encloses the survey area, while its vertical extent covers, from top to bottom, a portion of the air space, the seawater column, and the subsurface region to be investigated. The distance between the outer boundary of Ω^{ext} ($\partial\Omega^{\text{ext}}$, marked by solid black lines) and the outer boundary of Ω^{int} ($\partial\Omega^{\text{int}}$, marked by solid white lines) is chosen to be several times the skin depth, aiming to minimize boundary effects. We then design a uniaxial real-valued coordinate transformation that preserves the geometry of Ω^{int} while compressing the exterior domain Ω^{ext} along the x, y, and z directions into a much thinner cubic shell $\tilde{\Omega}^{\text{ext}}$ (see Figure 1b). Inspired by the compact expression for perfectly matched layers given by Schwarzbach [24], the geometrical relation between the fictitious computational domain $\tilde{\Omega}$ and the original physical model Ω can be described by the following mapping function $\mathbf{r} = \mathbf{r}(\tilde{\mathbf{r}})$:

$$\mathbf{r} = \sum_{i=x,y,z} a_i(\omega,\sigma)\hat{H}((\tilde{\mathbf{r}} - \mathbf{r}_i^a)^T\mathbf{n}_i)\left(\frac{(\tilde{\mathbf{r}} - \mathbf{r}_i^a)^T\mathbf{n}_i}{L_i}\right)^m \mathbf{n}_i + \tilde{\mathbf{r}},\ \tilde{\mathbf{r}} \in \tilde{\Omega}, \tag{15}$$

where $\hat{H}$ is a Heaviside-like function:

$$\hat{H}(x) = \begin{cases} 1 & x > 0 \\ 0 & x \leq 0 \end{cases}. \tag{16}$$

As shown in Figure 1, $\mathbf{r}_i^a$ $(i = x, y, z)$ denotes the intersections of the corresponding axes with the inner boundary $\partial\Omega^{\text{int}}$. L_i $(i = x, y, z)$ represents the thickness of transformed boundary layers along the three axes. $a_i(\omega, \sigma)$ $(i = x, y, z)$ characterizes the amounts of expansion toward the original outer boundary, which will be determined according to the skin depth of electromagnetic signals on the specific geological model. The rate of spatial variation is controlled by a scaling factor m.

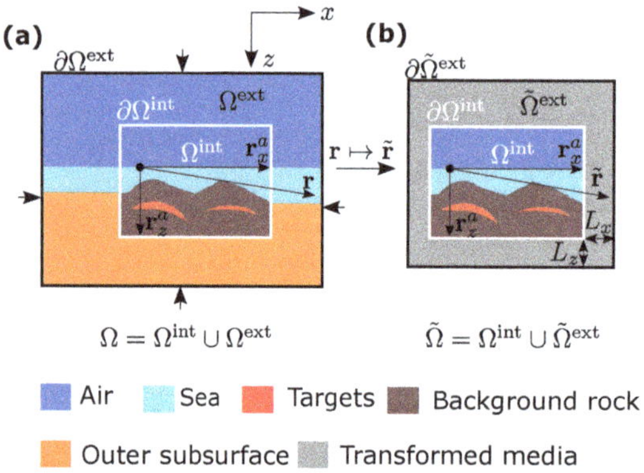

Figure 1. Illustration of the geometrical relation between (**a**) the physical model $\Omega = \Omega^{\text{int}} \cup \Omega^{\text{ext}}$ and (**b**) the compressed model $\tilde{\Omega} = \Omega^{\text{int}} \cup \tilde{\Omega}^{\text{ext}}$. Here, $\tilde{\Omega}^{\text{ext}}$ denotes the new boundary domain compressed from the original boundary region Ω^{ext}.

Taking partial derivatives on both sides of Equation (15) with respect to the transformed coordinates $(\tilde{x}, \tilde{y}, \tilde{z})$, we obtain the explicit expression of the inverse of Jacobian in Equation (3):

$$\mathbf{J}^{-1} = \begin{pmatrix} \Lambda_{x\tilde{x}} & 0 & 0 \\ 0 & \Lambda_{y\tilde{y}} & 0 \\ 0 & 0 & \Lambda_{z\tilde{z}} \end{pmatrix} \tag{17}$$

with

$$
\begin{aligned}
\Lambda_{v\tilde{v}} &= \frac{\partial v}{\partial \tilde{v}} \\
&= \frac{a_v m}{L_v} \hat{H}((\tilde{\mathbf{r}} - \mathbf{r}_v^a)^T \mathbf{n}_v) \left(\frac{(\tilde{\mathbf{r}} - \mathbf{r}_v^a)^T \mathbf{n}_v}{L_v} \right)^{m-1} + 1,\ v = x, y, z.
\end{aligned}
\tag{18}
$$

It is readily seen that the Jacobian is non-singular throughout the entire computational domain, which will be reduced to an identity matrix in the interior region. Therefore, the transformed problem shown in Equation (4a)–(4c) is well defined over $\tilde{\Omega}$.

Before performing numerical simulations, it is useful to investigate the spatial variation in the equivalent material properties so as to predict the need of proper grid density. To draw a rough yet general conclusion for different model situations, we examine the following transformation tensor:

$$
\mathbf{T} = \frac{\mathbf{J}\mathbf{J}^T}{\det(\mathbf{J})}. \tag{19}
$$

The underlying meaning of **T** becomes clear if we assume, without loss of generality, that the original electromagnetic coefficients are isotropic. Under this assumption, **T** directly reflects the heterogeneity and anisotropy that the coordinate transformation introduces in the transformed media. As an example, we show in Figure 2 the diagonal components of **T** at the slice $y = 0\,\mathrm{m}$ on the model with dimensions $[-2\,\mathrm{km}, 14\,\mathrm{km}] \times [-3\,\mathrm{km}, 3\,\mathrm{km}] \times [-6\,\mathrm{km}, 4\,\mathrm{km}]$. The interior domain is enclosed with dashed black lines and outside is the compressed boundary region. In the transformation function, m was set to 3 and scaling $a_i = 25L_i, i = \pm x, \pm y, \pm z$ has been performed. This amounts to an original model with dimensions $[-52\,\mathrm{km}, 64\,\mathrm{km}] \times [-53\,\mathrm{km}, 53\,\mathrm{km}] \times [-81\,\mathrm{km}, 29\,\mathrm{km}]$. From this figure, it is clear to see that a smaller computational domain comes at a cost of a stronger heterogeneous parameter distribution. Specifically, a_i is directly related to the original scale of the computational domain. The factor $a_i m / L_i$ controls the parameter contrast while the factor m determines the spatial variation pattern. For the sake of numerical stability, the scaling factors a_i / L_i and m should not be too large. We found with many tests that $m = 3$ is suitable for MCSEM applications and therefore we used this value for all the experiments presented in this section.

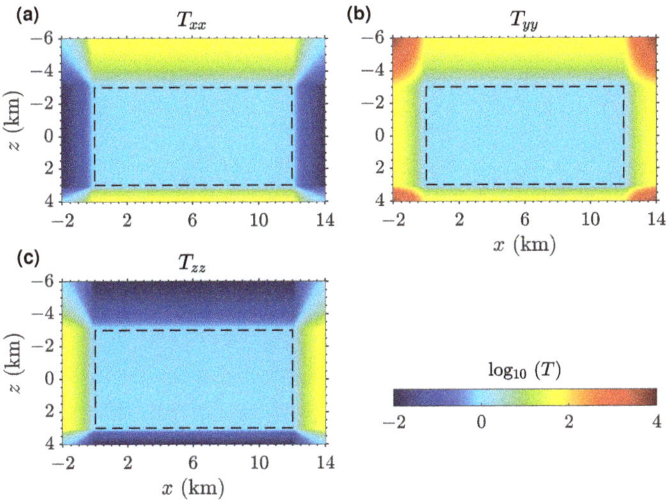

Figure 2. The diagonal component (**a**) T_{xx}, (**b**) T_{yy} and (**c**) T_{zz} of the transformation matrix **T** along $y = 0\,\mathrm{m}$ defined in Equation (19). Dashed lines enclose the inner region.

3.1.1. 1D Marine Hydrocarbon Scenario

First, we validate the accuracy and practicability of our algorithm on 1D VTI models. The template consists of an air layer of 1×10^{-7} S/m, a seawater column of 3.2 S/m, a 1000 m thick sediment with a horizontal conductivity of 1 S/m and a vertical conductivity of 0.2 S/m, and a 100 m thick isotropic layer of 0.01 S/m, followed by a deep basement with a horizontal conductivity of 0.5 S/m and a vertical conductivity of 0.1 S/m. We consider two model variants that are a 1000 m deep-sea environment (Model I) and a 100 m shallow-water case (Model II), respectively. As an example, Figure 3a displays the geometry of Model I and the survey setup, and Figure 3b–d show the final refined meshes at frequency 0.25 Hz. For both models, the interior computational domain laterally encloses the survey region and vertically spans a 3000 m thick air layer and a 3000 m thick section from the sea surface to the deep subsurface, resulting in a total domain of size $[0\,\text{km}, 12\,\text{km}] \times [-1\,\text{km}, 1\,\text{km}] \times [-3\,\text{km}, 3\,\text{km}]$. On both models, a horizontal electric dipole oriented with a 4.76° azimuth from the x-axis is located 50 m above the seabed. 24 inline receivers are positioned 0.05 m above the seafloor and along the source polarization direction with a lateral spacing of about 502 m to record the six electric and magnetic field components at frequencies 0.1, 0.25, 1, and 10 Hz, respectively. For this survey geometry, the vertical component of the magnetic field B_z becomes trivial. We used the same scaling coefficients $a_i = 25L_i$ with $L_{\pm x} = 2\,\text{km}$, $L_{\pm y} = 2\,\text{km}$, $L_{+z} = 1\,\text{km}$, and $L_{-z} = 3\,\text{km}$ in both experiments. This gives rise to a final model size of dimensions $[-2\,\text{km}, 14\,\text{km}] \times [-3\,\text{km}, 3\,\text{km}] \times [-6\,\text{km}, 4\,\text{km}]$.

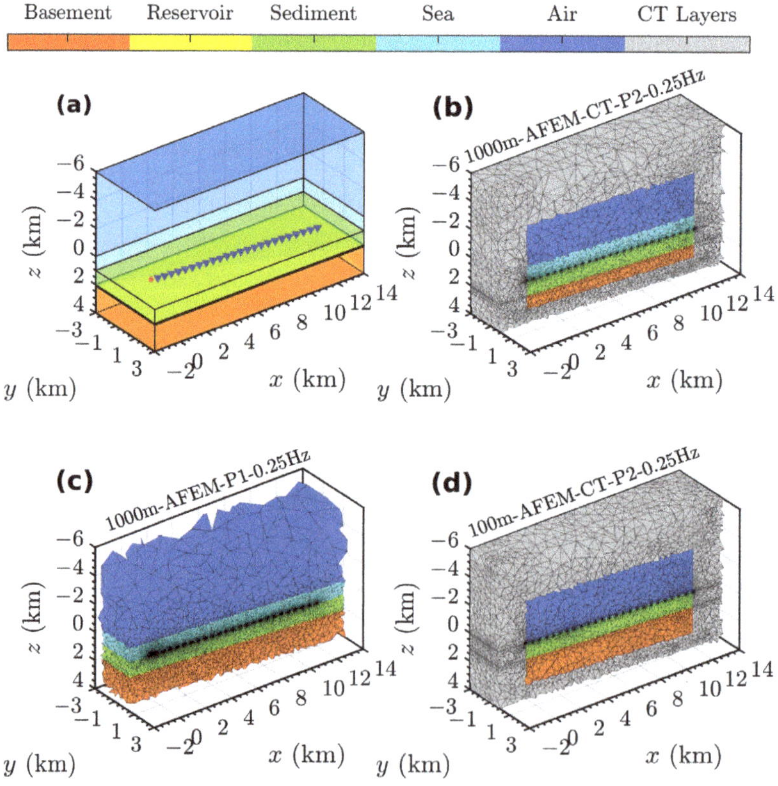

Figure 3. (**a**) Sketch of the 1D layered MCSEM model; (**b**) converged AFEM-CT-P2 mesh at 0.25 Hz for Model I; (**c**) converged AFEM-P1 mesh at 0.25 Hz for Model I; (**d**) converged AFEM-CT-P2 mesh at 0.25 Hz for Model II.

We executed the AFEM-CT-P2 algorithm for Model I at each frequency separately. Figure 4 depicts the comparison of the converged AFEM-CT-P2 results against the semi-analytical solutions [42]. For the sake of illustration, the errors for data below the noise floor (5×10^{-17} V/Am^2 for the electric field and 5×10^{-20} T/Am for the magnetic field) have been excluded from the error curves. It is evident from the figure that our AFEM-CT-P2 results agree well with the references. The maximum relative amplitude error among all data points is about 1.6% and the maximum phase difference is 0.8°. Although not shown, we also computed the original forward problems on the physical model of the same scale as above. These results confirmed that the current model extent was sufficient for the 10 Hz data due to the rapid attenuation of high-frequency signals. However, at the three lower frequencies, significant boundary effects were observed, resulting in relative errors greater than 10% as the measuring points approach the outer boundary.

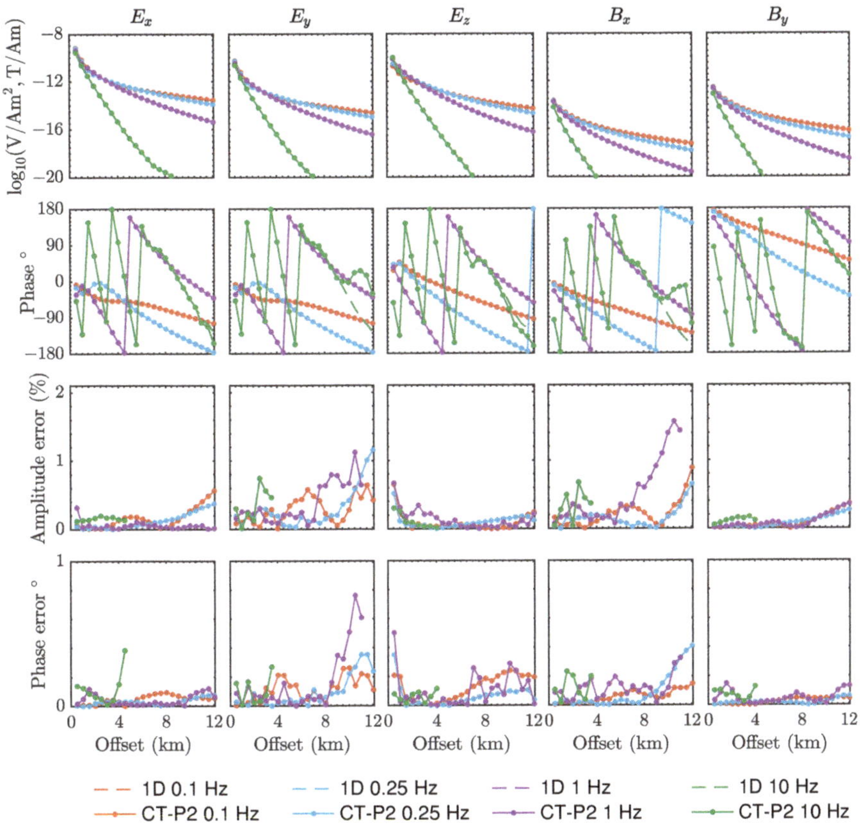

Figure 4. AFEM-CT-P2 results on the deep-sea Model I in comparison with the semi-analytical solutions.

To demonstrate the effectiveness of the presented algorithm, we have prepared a larger computational domain without the CT functionality. To make a justified comparison, we have run a few tests and picked the smallest-scale model which is able to render reasonable accuracy. The selected proper dimensions are $[-20\,\text{km}, 20\,\text{km},] \times [-20\,\text{km}, 20\,\text{km}] \times [-70\,\text{km}, 40\,\text{km}]$. The initial model discretization gives 1,787,181 tetrahedra and 11,323,722 DoFs with the second-order finite-element approximation. Apparently, the computational costs on the initial mesh with the traditional P2 discretization already exceed those on the final meshes with the AFEM-CT-P2 algorithm (Table 1). Therefore, further comparison was not executed. Instead, we switch to the lowest-order (P1) finite-element approximation. These converged AFEM-

P1 results in comparison with the 1D solutions at each frequency are illustrated in Figure 5. We list the computational efforts for AFEM-CT-P2, AFEM-P2 and AFEM-P1 in Table 1. According to these statistics, it is evident that AFEM-CT-P2 shows decisive computational superiority. It reduces the number of DoFs by an order of magnitude compared to the conventional AFEM-P2 approach, while achieving a 63–76% reduction in computing time and a 14–47% reduction in memory usage versus the AFEM-P1 method.

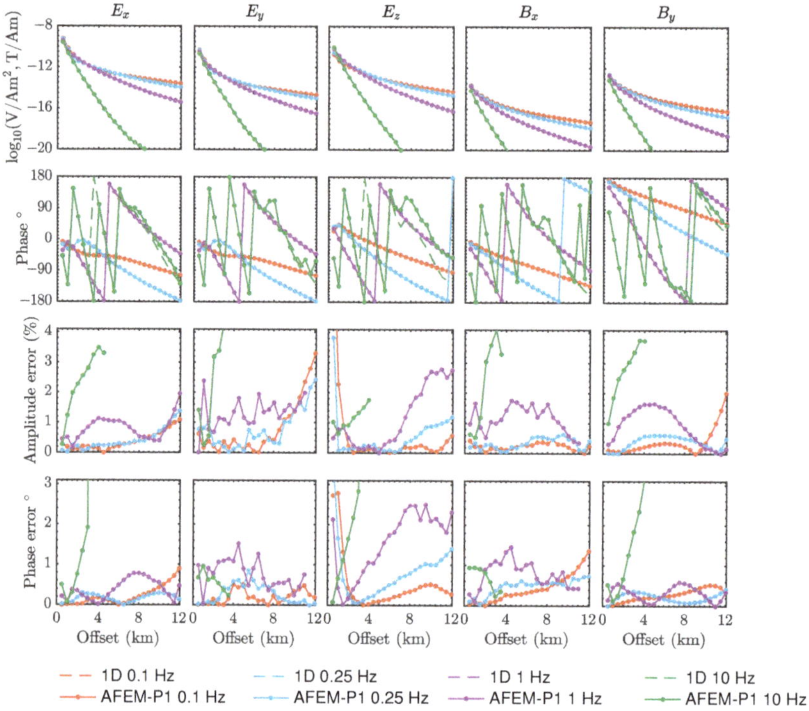

Figure 5. AFEM-P1 results on the deep-sea Model I without transformation in comparison with the semi-analytical solutions.

Table 1. Comparison of computational costs for Model I using AFEM-CT-P2 versus AFEM-P2 and AFEM-P1. All experiments were conducted using 14 MPI processes with 4 OpenMP threads per process. '–' denotes the process is not further executed.

	AFEM-CT-P2 **16 × 6 × 10 km³**				**AFEM-P1** **40 × 40 × 110 km³**				**AFEM-P2** **40 × 40 × 110 km³**			
Frequency (Hz)	0.1	0.25	1	10	0.1	0.25	1	10	0.1	0.25	1	10
AFEM steps	6	7	6	8	13	13	13	14	0	0	0	0
Elements	373,568	375,050	218,112	192,130	3,298,259	2,909,945	2,655,371	2,674,748	1,787,181	1,787,181	1,787,181	1,787,181
DoFs	2,384,808	2,391,656	1,393,634	1,225,292	3,830,210	3,381,016	3,086,587	3,109,081	11,323,722	11,323,722	11,323,722	11,323,722
CPU time (s)	512	576	364	447	1665	1560	1495	1597	–	–	–	–
Memory (GB)	82	81	43	36	118	94	75	68	–	–	–	–

Following the same procedure, we examine the algorithm on the shallow-water Model II. In this experiment, we set the noise levels for the electric field and magnetic field to 5×10^{-16} V/Am^2 and 5×10^{-19} T/Am, respectively. Figure 6 shows the non-trivial electromagnetic components at the four frequencies compared to the analytical solutions. Among all data points, we observe a maximum amplitude discrepancy of 4.8% in E_z at 10 Hz at the farthest measuring location, where the signal amplitude is already very close to the noise floor. Apart from this data point, the relative amplitude deviations all remains below 1.2%. The maximum phase error is about 1.7°. Based on the computational statistics given in Table 2, we highlight the capability of the CT technique in dealing with shallow-water models at low frequencies. Mesh accuracy at all acquisition frequencies can be ensured by about 350,000 to 490,000 tetrahedra, which correspond to 2 to 3 million DoFs for the second-order finite-element approximations. In contrast, this model becomes extremely challenging for the conventional FEM approach without the CT technique. By constraining the maximum aspect ratio to 1.4, the model with dimensions $[-30\,\text{km}, 30\,\text{km}] \times [-30\,\text{km}, 30\,\text{km}] \times [-70\,\text{km}, 40\,\text{km}]$ is discretized into 7,549,120 elements, yielding 8,772,487 DoFs for the P1 formulation and 47,775,376 DoFs for the P2 case. It is apparent that both alternatives are inferior to the AFEM-CT-P2 approach, for which detailed comparisons are not conducted.

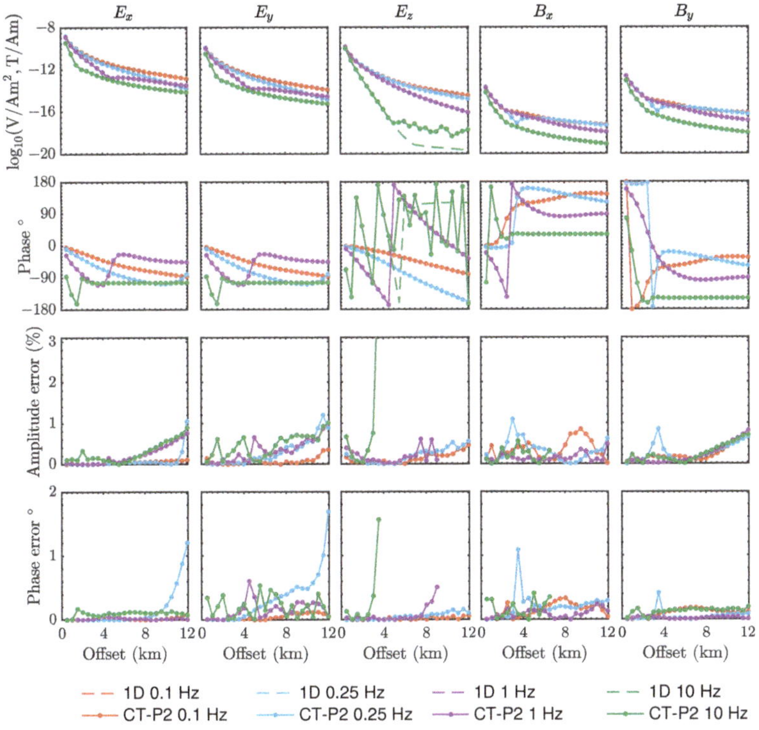

Figure 6. AFEM-CT-P2 results on the 100-m shallow-water Model II compared to the semi-analytical solutions.

Table 2. The computational costs using AFEM-CT-P2 for Model II versus AFEM-P1 and AFEM-P2. All experiments were finished on 14 MPI processes with 4 OpenMP threads each process. '–' denotes the process is not further executed.

	AFEM-CT-P2 $16 \times 6 \times 10$ **km**^{3}				**AFEM-P1** $60 \times 60 \times 110$ **km**^{3}				**AFEM-P2** $60 \times 60 \times 110$ **km**^{3}			
Frequency (Hz)	0.1	0.25	1	10	0.1	0.25	1	10	0.1	0.25	1	10
AFEM steps	5	5	5	6	0	0	0	0	0	0	0	0
Elements	485,915	449,031	370,881	347,303	7,549,120	7,549,120	7,549,120	7,549,120	7,549,120	7,549,120	7,549,120	7,549,120
DoFs	3,100,168	2,864,300	2,366,002	2,214,098	8,772,487	8,772,487	8,772,487	8,772,487	47,775,376	47,775,376	47,775,376	47,775,376
CPU time (s)	708	667	605	677	–	–	–	–	–	–	–	–
Memory (GB)	121	104	84	72	–	–	–	–	–	–	–	–

3.1.2. 3D Anticline Trap with Bathymetry

Based on the above validation, we now utilize this algorithm to analyze a 3D bathymetric model with a local anticline trap [7,43]. While comprehensive studies are essential to fully understand the impact of complex electrical anisotropy, this work focuses on the anticline scenario to demonstrate the capability of our approach in handling generally anisotropic conductivity within complex geological settings. Figure 7 displays the geometry and initial mesh of the computational domain. The solid black lines mark the boundaries of the interior subdomain. Outside the black lines are the transformed regions. In the target domain, the water depth varies from about 600 m to about 1200 m. A 3000 m thick air layer of 1×10^{-7} S/m was appended on top of the sea surface. The hydrocarbon reservoir was modeled as a local 3D resistive body of 0.01 S/m and dimensions 5 km × 5 km × 0.1 km, trapped in a local symmetric anticline structure of $|\beta| = 20^\circ$ dip formed by two limbs of dimensions 4 km × 6 km × 1.5 km. The rest of the rugged sediment is assumed to be vertically transversely isotropic with a horizontal conductivity of 1 S/m and a vertical conductivity of 0.25 S/m. A VTI basement is underlain with a horizontal conductivity of 1 S/m and a vertical conductivity of 0.2 S/m. In the transformed domain, the deep basement was compressed to 1 km thickness. The boundary effect along the other directions was taken into account by 2 km thick layers surrounding the vertical boundaries and a 3 km thick layer on top of the air space with equivalent materials. Again, the transformation scaling $a_i = 25L_i$ was utilized.

We deployed a dense acquisition array consisting of 207 receivers arranged in a 23 × 9 grid with each station spaced 500 m apart (see Figure 7b) to record the EM responses. Due to the bathymetry, the vertical positions of the receivers vary from about 644 m to about 1197 m. An x-directed horizontal electric dipole is situated at $(-4600\,\text{m}, 0, 880\,\text{m})$, roughly 50 m above the seabottom. The transmission frequencies were set to 0.1, 0.25, 1, and 10 Hz.

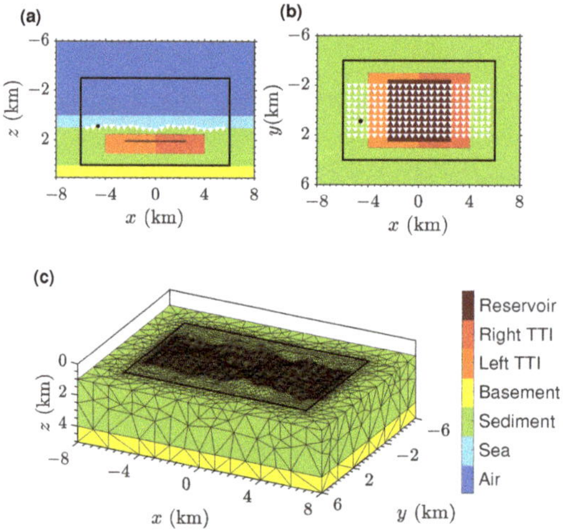

Figure 7. Panels (**a**,**b**) show the plane view of the 3D bathymetric model at $y = 0$ m and $z = 2050$ m, respectively; (**c**) the initial discretization of the subsurface. The black dot and white triangles in (**a**,**b**) mark the transmitter's and receivers' positions, respectively.

We apply the AFEM-CT-P2 procedure to compute the TTI model (referred to as TTIA) responses at the four frequencies separately. To analyze the impact of electrical anisotropy, the background model excluding the anticline and reservoir bodies (named by VTIB) as well as the model without the anticline (named by VTIA) were also computed. Given field measurements are often contaminated with different sources of errors, we conduct a realistic sensitivity analysis according to Brown et al. [5]:

$$S_{F_i} = \frac{|F_i| - |F^{\mathrm{VTIB}}|}{F_i^{\mathrm{noise}}}, \ i = x, y, \tag{20}$$

where F_i^{noise} is the estimated data uncertainty for each horizontal component, reading

$$F_{x,y}^{\mathrm{noise}} = \sqrt{(\alpha^2 + (\cos\phi - 1)^2)|F_{x,y}|^2 + \sin^2\phi|F_{y,x}|^2 + e^2}. \tag{21}$$

As described in Equation (21), this formulation accounts for three error sources: the relative error in the measurements (α), the rotation uncertainty (ϕ), and the noise floor (e). In this experiment, we set $\alpha = 0.03$, $\phi = 2°$, the noise floor for the electric and magnetic field to 1×10^{-16} V/Am^2 and 1×10^{-19} T/Am, respectively. Figures 8 and 9 show the resolution of the anomalous signals with respect to the data uncertainty for each horizontal data component on model VTIA and TTIA, respectively. It is clearly seen that the effective sensitivity of the composite responses of the anticline and reservoir is much weaker than that in the VTI case. Given the weak responses, an accurate forward operator becomes essential for imaging the challenging anticline traps.

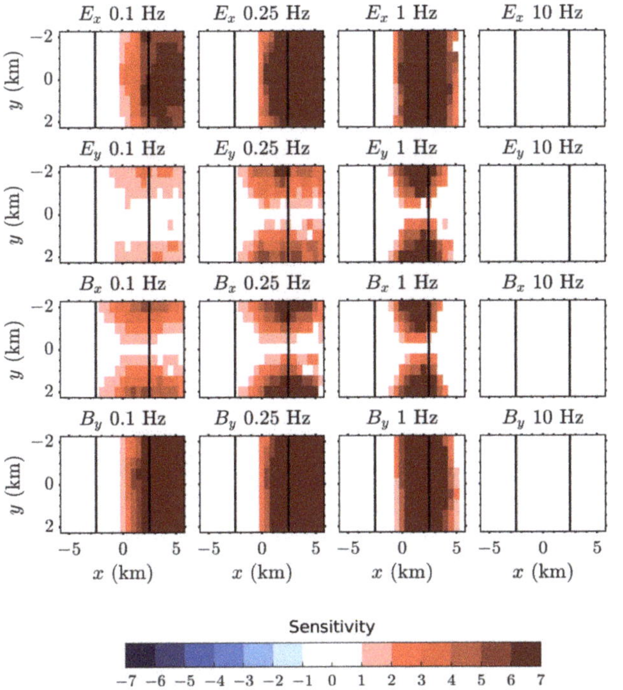

Figure 8. The resolution of anomalous signals on model VTIA with respect to the data uncertainty. The black lines outline the horizontal extent of the oil reservoir.

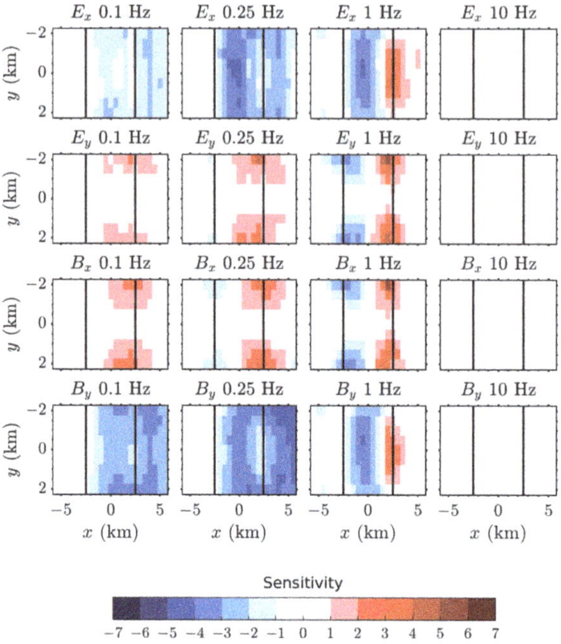

Figure 9. The resolution of anomalous signals on model TTIA with respect to the data uncertainty.

3.2. Application to High-Resolution Models

3.2.1. Depth-Varying Seawater Conductivity Profile

Next, we examine the potential of the CT technique in handling complex, high-resolution models. Due to the combined effects of temperature, salinity, and pressure, seawater conductivity can vary with depth [44]. Given the limited spatial extent of MCSEM surveys, such vertical variations in seawater conductivity may influence electromagnetic responses, thereby highlighting the need for further investigation. To explore this effect, we design a simplified seawater conductivity profile following Zheng et al. [44], where conductivity decays exponentially from about 5.6 S/m at the sea surface to about 3.4 S/m at the seabottom of 1000 m. The depth-varying conductivity profile is discretized into 50 m thick layers. The conductivity of each layer is calculated as the average conductivity across the layer's thickness (see Figure 10). The model setup of the subsurface is identical to that in the 1D benchmark described in the last section.

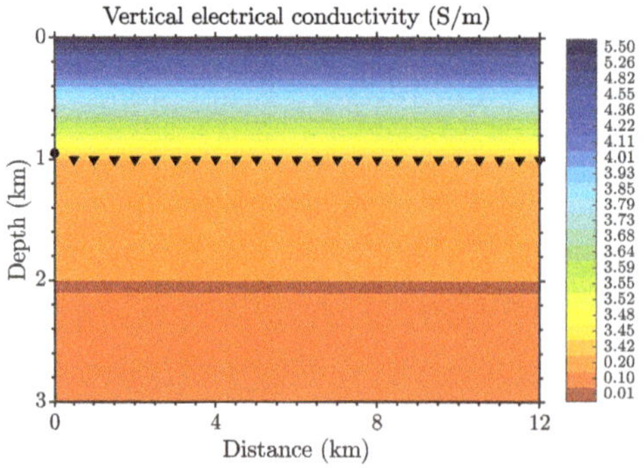

Figure 10. Sketch of the depth-varying seawater conductivity profile. The black dot and triangles mark the transmitter's and receivers' positions, respectively.

We apply a composite coordinate transformation to this model. First, a stretching function is designed to expand the vertical spacing between layers beneath the sea surface by a factor of six. Then, a compression function, similar to Equation (15) but with $L_{\pm x} = 3\,\text{km}$ and $L_{\pm y} = 3\,\text{km}$, is composited to squeeze the boundary subdomain. This yields a transformed model with dimensions $18\,\text{km} \times 8\,\text{km} \times 30\,\text{km}$. Figure 11 depicts the numerical results in comparison with the semi-analytical solutions. It is seen that the AFEM-CT-P2 results match well with the references with a maximum amplitude error less than 1.6% and phase error 0.5°. However, modeling such continuously fine layers using standard finite-element schemes with unstructured grids becomes computationally intensive. For example, an initial tetrahedralization of this profile shaped into a 3D computational domain of size $16\,\text{km} \times 6\,\text{km} \times 11\,\text{km}$ yields over 3.2 million cells. In contrast, the AFEM-CT-P2 scheme is designed to mitigate the geometrical disparity between structures of different scales. The procedure converges to accurate solutions at all working frequencies with no more than 525,379 cells and 3,346,976 DoFs, which corresponds to only 16.4% of the elements needed in the conventional methods.

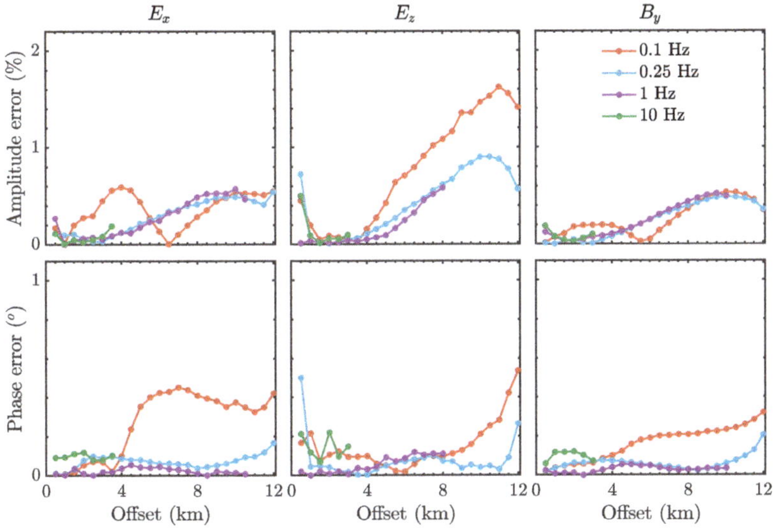

Figure 11. AFEM-CT-P2 results on the seawater conductivity profile in comparison with the semi-analytical solutions.

To evaluate the impact of depth-varying seawater conductivity on the electromagnetic responses, we compare the fields with the responses of a constant seawater column of 3.42 S/m, namely the conductivity measured at receiver positions. To account for realistic environmental conditions, an optimistic noise floor $1 \times 10^{-15}\,\text{V/Am}^2$ and $1 \times 10^{-18}\,\text{T/Am}$ have been assumed for the electric and magnetic field, respectively. It is seen from Figure 12 that the vertical variation in seawater conductivity most strongly affects the electromagnetic field at 0.1 Hz. A moderate impact is observed at 0.25 Hz within a short transmitter–receiver offset less than 4 km. In contrast, trivial influence is found at frequencies higher than 1 Hz. This frequency-dependent behavior seems reasonable since lower-frequency signals penetrate deeper and are more sensitive to the spatial variation in the conductivity profile. Whereas higher-frequency signals are primarily influenced by the local environment near the source and receivers. Among the EM field components, E_x shows strong sensitivity to changes in the seawater conductivity, as it is affected by broad, horizontally distributed current systems. In contrast, E_z is less affected, as it is primarily governed by localized vertical current paths.

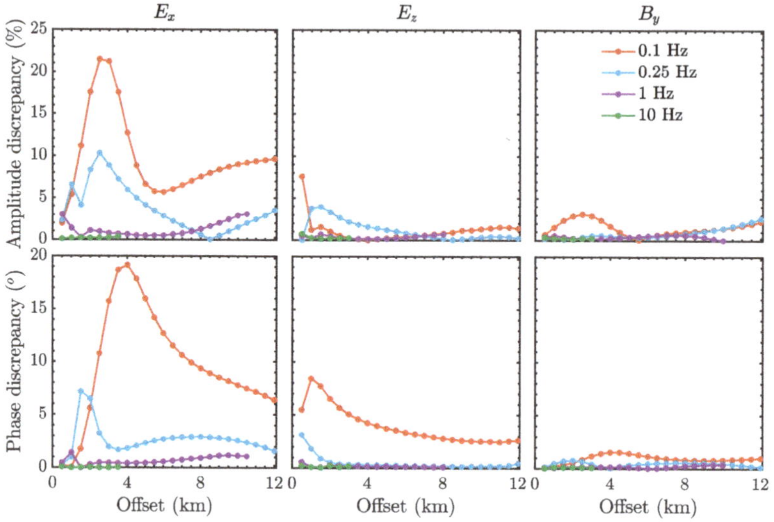

Figure 12. The difference in field responses between the depth-varying seawater conductivity profile and the constant conductivity seawater column.

3.2.2. Marlim R3D

Finally, we apply the CT approach to a real-world MCSEM model. Marlim R3D (MR3D) is an open-source and realistic geoelectric model developed for simulating CSEM responses of postsalt turbiditic reservoirs along the Brazilian offshore margin [45]. The MR3D resistivity model is initially defined on a uniform hexahedral mesh with the target region comprising nearly 90 million cells. Reconstructing such a high-resolution conductivity distribution using numerical methods based on unstructured mesh discretization is computationally demanding. Rochlitz et al. [37] and Werthmüller et al. [46] conducted a comprehensive study on this model with a robust mesh generation and resistivity interpolation workflow tailored to tetrahedral meshes. We utilize this preprocessing procedure to construct the discretized conductivity model. Prior to that, the CT approach is applied to vertically stretch the subdomain beneath the sea surface by a factor of six. A gradual global refinement is then performed over the target region, generating three successively finer unstructured meshes. Throughout this process, the initial triangulation of known geological interfaces is preserved to ensure geometrical consistency. The coarsest mesh comprises 1,060,715 elements, the intermediate mesh comprises 1,182,126 elements, and the finest mesh comprises 3,120,950 elements. For each mesh, the conductivity distribution is re-interpolated accordingly (see Figure 13). Each model is individually solved using the FEM-CT-P2 approach. We note that although a much coarser discretization similar to that in Rochlitz et al. [37] and Werthmüller et al. [46] could have been reused, we chose to construct new meshes in order to better represent the complexity of the MR3D model, and, furthermore, to analyze the sensitivity of the electromagnetic fields to finely resolved geological bodies. We employ the AMS-preconditioned FGMRES method to solve the three models. Despite the strong heterogeneity and anisotropy, which typically increase the condition number of the system matrix [47], the solver consistently converges within 17–20 iterations to a tolerance of 10^{-9} across all experiments in the frequency range of

0.125–1.25 Hz. Its performance is comparable to the non-transformed cases with different meshes, which converge in 17–19 iterations.

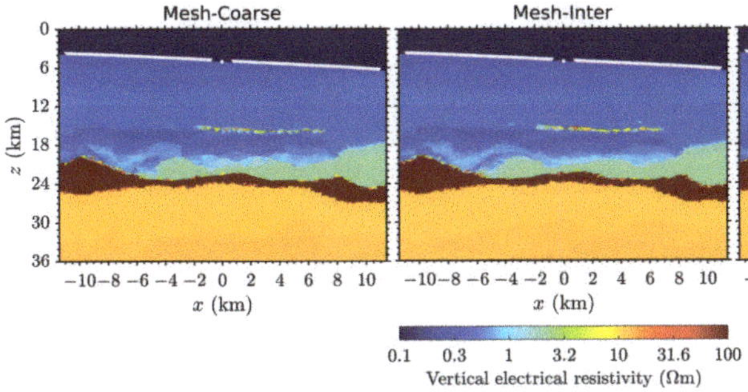

Figure 13. Slices of the interpolated transformed MR3D model at different levels of mesh refinement. White dot and triangles indicate the source location and measurement positions.

To illustrate the results, Figures 14 and 15 depict the comparison of the computed inline and broadside components on the three meshes at frequencies of 0.25 and 1 Hz, respectively, with references on the hexahedral mesh [45]. For additional verification, the results from Rochlitz et al. [37] are also included to help identify the source of any discrepancies. We have excluded the errors for data below the noise floor (1×10^{-16} V/Am^2 for the electric field) from the error curves. The results reveal that the coarsest mesh deviates most significantly from the reference solution. This may result from insufficient mesh density to ensure the numerical accuracy for the equivalent materials. In contrast, the moderate and finest meshes produce results with similar levels of accuracy, indicating that additional mesh refinement, or accordingly, a further increase in model resolution yields marginal improvements. Nevertheless, a residual discrepancy of approximately 10% persists [46]. The observed non-negligible difference is likely due to deviations in how irregular geometries are represented and how boundary domain extensions are treated in different implementations.

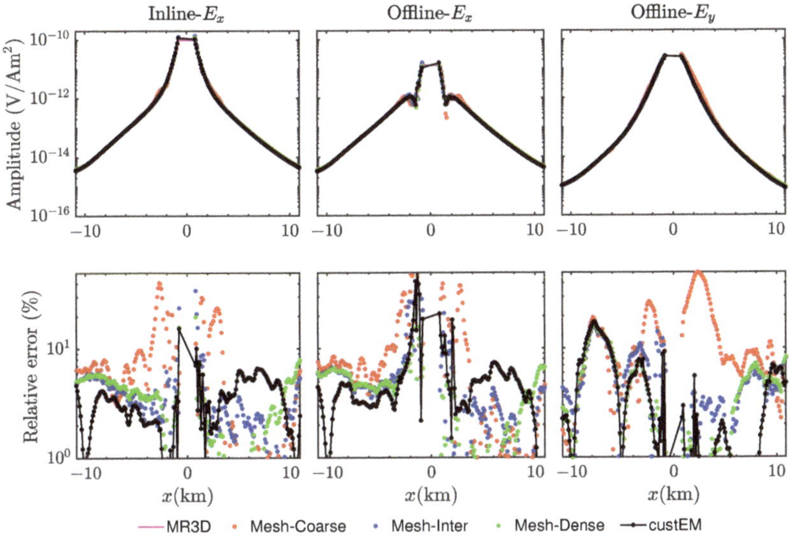

Figure 14. Comparison of the computed electric field components on the three meshes at 0.25 Hz with the reference 'MR3D' [45].

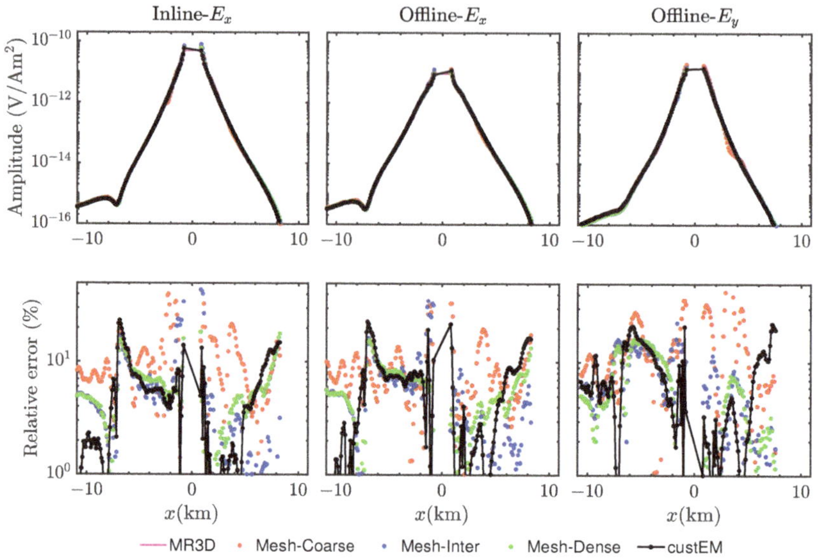

Figure 15. Comparison of the computed electric field components on the three meshes at 1 Hz with the reference 'MR3D' [45].

4. Conclusions

We have developed a parallel adaptive high-order finite-element framework for large-scale 3D marine controlled-source electromagnetic modeling in the frequency domain. Although high-order finite-element methods offer high accuracy and robustness, their application to 3D MCSEM problems remains constrained for models involving thin layers or regional multi-scale heterogeneities. To address this challenge, this work introduced a coordinate transformation approach based on the form invariance of Maxwell's equations. By designing a problem-specific spatial mapping, the structural complexity and scale contrasts among different structures can be effectively reduced. As a trade-off, the transformation yields a new formulation characterized by equivalent anisotropic materials, while still preserving the original mathematical structure of Maxwell's equations. This allows the transformed problem to be solved using standard high-order finite-element solvers. Since the new electromagnetic parameters become spatially varying and anisotropic, we employed a goal-oriented mesh refinement strategy to ensure accurate modeling of the complex field behavior in target regions.

The 1D benchmark tests demonstrate that our solver performs competitively in deep-sea environments and offers clear advantages over the traditional approach in shallow-water settings and scenarios with depth-dependent conductivity variations. The practicality of our approach was further illustrated using a TTI model with rugged seafloor topography. The resulting sensitivity analysis highlights the critical importance of using an accurate forward solver to resolve the weak electromagnetic responses associated with anticline traps. Finally, we revisited the MR3D model. The convergence study inferred that high-resolution models may not be fully resolved due to the diffusive behavior of low-frequency electromagnetic fields. This suggests that MCSEM inversion workflows could benefit from adaptive model parameterization to balance the resolution and computational efficiency.

Although only MCSEM applications were considered here, this method can be naturally applied to plane wave EM modeling. However, it is worth pointing out that the CT approach does not inherently improve the simulation accuracy. Rather, it alleviates geometrical complexity at the cost of increased contrast in material properties. This trade-off could present further challenges for iterative solvers. Nevertheless, by employing a robust Maxwell solver such as the AMS-preconditioned FGMRES, we observe no significant difficulties in solving the resulting system of equations.

Author Contributions: Conceptualization, methodology, validation, F.W.; software, F.W. and S.C.; writing—original draft preparation, F.W. and S.C.; writing—review and editing, F.W. and S.C.; visualization, F.W. and S.C.; funding acquisition, F.W. All authors have read and agreed to the published version of the manuscript.

Funding: This research was funded by the Natural Science Foundation of Hunan Province (grant no. 2023JJ30584), the National Natural Science Foundation of China (grant no. 42004066) and the fund support from Xiangtan University (grant no. 23QDZ10).

Institutional Review Board Statement: Not applicable.

Informed Consent Statement: Not applicable.

Data Availability Statement: The data from the experiments are available on request from the corresponding author.

Conflicts of Interest: The authors declare no conflicts of interest.

References

1. Key, K. Marine electromagnetic studies of seafloor resources and tectonics. *Surv. Geophys.* **2012**, *33*, 135–167. [CrossRef]
2. Zhdanov, M.S.; Endo, M.; Yoon, D.; Čuma, M.; Mattsson, J.; Midgley, J. Anisotropic 3D inversion of towed-streamer electromagnetic data: Case study from the Troll West Oil Province. *Interpretation* **2014**, *2*, SH97–SH113. [CrossRef]
3. Zhdanov, M.S.; Endo, M.; Cox, L.H.; Čuma, M.; Linfoot, J.; Anderson, C.; Black, N.; Gribenko, A.V. Three-dimensional inversion of towed streamer electromagnetic data. *Geophys. Prospect.* **2014**, *62*, 552–572. [CrossRef]
4. Newman, G.A.; Commer, M.; Carazzone, J.J. Imaging CSEM data in the presence of electrical anisotropy. *Geophysics* **2010**, *75*, F51–F61. [CrossRef]
5. Brown, V.; Hoversten, M.; Key, K.; Chen, J. Resolution of reservoir scale electrical anisotropy from marine CSEM data. *Geophysics* **2012**, *77*, E147–E158. [CrossRef]
6. Cook, A.E.; Anderson, B.I.; Malinverno, A.; Mrozewski, S.; Goldberg, D.S. Electrical anisotropy due to gas hydrate-filled fractures. *Geophysics* **2010**, *75*, F173–F185. [CrossRef]
7. Jaysaval, P.; Shantsev, D.V.; de la Kethulle de Ryhove, S.; Bratteland, T. Fully anisotropic 3-D EM modelling on a Lebedev grid with a multigrid pre-conditioner. *Geophys. J. Int.* **2016**, *207*, 1554–1572. [CrossRef]
8. Li, J.; Liu, Y.; Han, B.; Lu, J. Impacts of complex geological structures on marine controlled-source electromagnetic responses. *J. Appl. Geophys.* **2020**, *180*, 104126. [CrossRef]
9. Peng, R.; Hu, X.; Li, J.; Liu, Y. Finite element simulation of 3-D marine controlled source electromagnetic fields in anisotropic media with unstructured tetrahedral grids. *Pure Appl. Geophys.* **2020**, *177*, 4871–4882. [CrossRef]
10. Jaysaval, P.; Shantsev, D.; de la Kethulle de Ryhove, S. Fast multimodel finite-difference controlled-source electromagnetic simulations based on a Schur complement approach. *Geophysics* **2014**, *79*, E315–E327. [CrossRef]
11. Zhdanov, M.; Lee, S.; Yoshioka, K. Integral equation method for 3D modeling of electromagnetic fields in complex structures with inhomogeneous background conductivity. *Geophysics* **2006**, *71*, G333–G345. [CrossRef]
12. Jahandari, H.; Farquharson, C. Finite-volume modelling of geophysical electromagnetic data on unstructured grids using potentials. *Geophys. J. Int.* **2015**, *202*, 1859–1876. [CrossRef]
13. Cai, H.; Xiong, B.; Han, M.; Zhdanov, M. 3D controlled-source electromagnetic modeling in anisotropic medium using edge-based finite element method. *Comput. Geosci.* **2014**, *73*, 164–176. [CrossRef]
14. Ren, Z.; Kalscheuer, T.; Greenhalgh, S.; Maurer, H. A finite-element-based domain-decomposition approach for plane wave 3D electromagnetic modeling. *Geophysics* **2014**, *79*, E255–E268. [CrossRef]
15. Grayver, A.V.; Kolev, T.V. Large-scale 3D geoelectromagnetic modeling using parallel adaptive high-order finite element method. *Geophysics* **2015**, *80*, E277–E291. [CrossRef]
16. Zhou, F.; Chen, H.; Tang, J.; Zhang, Z.; Yuan, Y.; Wu, Q. A comparison of A-phi formulae for three-dimensional geo-electromagnetic induction problems. *J. Geophys. Eng.* **2022**, *19*, 630–649. [CrossRef]
17. Schwarzbach, C.; Börner, R.U.; Spitzer, K. Three-dimensional adaptive higher order finite element simulation for geo-electromagnetics—A marine CSEM example. *Geophys. J. Int.* **2011**, *187*, 63–74. [CrossRef]

18. Qin, C.; Li, H.; Li, H.; Zhao, N. High-order finite-element forward-modeling algorithm for the 3D borehole-to-surface electromagnetic method using octree meshes. *Geophysics* **2024**, *89*, F61–F75. [CrossRef]
19. Grayver, A.V.; Bürg, M. Robust and scalable 3-D geo-electromagnetic modelling approach using the finite element method. *Geophys. J. Int.* **2014**, *198*, 110–125. [CrossRef]
20. Castillo-Reyes, O.; de la Puente, J.; García-Castillo, L.E.; Cela, J.M. Parallel 3-D marine controlled-source electromagnetic modelling using high-order tetrahedral Nédélec elements. *Geophys. J. Int.* **2019**, *219*, 39–65. [CrossRef]
21. Ren, Z. New Developments in Numerical Modeling of Broadband Geo-Electromagnetic Fields. Ph.D. Thesis, ETH Zurich, Zurich, Switzerland, 2012.
22. Ren, Z.; Kalscheuer, T.; Greenhalgh, S.; Maurer, H. A hybrid boundary element-finite element approach to modeling plane wave 3D electromagnetic induction responses in the Earth. *J. Comput. Phys.* **2014**, *258*, 705–717. [CrossRef]
23. Bérenger, J.P. *Perfectly Matched Layer (PML) for Computational Electromagnetics*, 1st ed.; Synthesis Lectures on Computational Electromagnetics; Springer: Cham, Switzerland, 2007; pp. 1–117.
24. Schwarzbach, C. Stability of Finite Element Solutions to Maxwell's Equations in Frequency Domain. Ph.D. Thesis, Technische Universität Bergakademie Freiberg, Freiberg, Germany, 2009.
25. Li, G.; Li, Y.; Han, B.; Liu, Z. Application of the perfectly matched layer in 3-D marine controlled-source electromagnetic modelling. *Geophys. J. Int.* **2018**, *212*, 333–344. [CrossRef]
26. Liu, J.; Xiao, X.; Tang, J.; Ren, Z.; Huang, X.; Zhang, J. Fast 3-D controlled-source electromagnetic modeling combining UPML and rational Krylov method. *IEEE Geosci. Remote Sens. Lett.* **2022**, *19*, 3006505. [CrossRef]
27. de la Kethulle de Ryhove, S.; Mittet, R. 3D marine magnetotelluric modeling and inversion with the finite-difference time-domain method. *Geophysics* **2014**, *79*, E269–E286. [CrossRef]
28. Pendry, J.B.; Schurig, D.; Smith, D.R. Controlling electromagnetic fields. *Science* **2006**, *312*, 1780–1782. [CrossRef] [PubMed]
29. Ozgun, O.; Kuzuoglu, M. Form invariance of Maxwell's equations: The pathway to novel metamaterial specifications for electromagnetic reshaping. *IEEE Antennas Propag. Mag.* **2010**, *52*, 51–65. [CrossRef]
30. Shyroki, D.M.; Ivinskaya, A.M.; Lavrinenko, A.V. Free-space squeezing assists perfectly matched layers in simulations on a tight domain. *IEEE Antennas Wirel. Propag. Lett.* **2010**, *9*, 389–392. [CrossRef]
31. Kamm, J.; Becken, M.; Abreu, R. Electromagnetic modelling with topography on regular grids with equivalent materials. *Geophys. J. Int.* **2020**, *220*, 2021–2038. [CrossRef]
32. Chen, H.; Xiong, B.; Lu, Y.; Jiang, Q.; Huang, L.; Zhou, S. Three-dimensional modeling of marine controlled source electromagnetic using a new high-order finite element method with an absorption boundary condition. *IEEE Trans. Geosci. Remote Sens.* **2023**, *61*, 2001909. [CrossRef]
33. Weidelt, P. 3D conductivity models: Implications of electrical anisotropy. In *Three-Dimensional Electromagnetics*; Oristaglio, M., Spies, B., Eds.; Society of Exploration Geophysicists: Houston, TX, USA, 1999.
34. Kottke, C.; Farjadpour, A.; Johnson, S.G. Perturbation theory for anisotropic dielectric interfaces, and application to subpixel smoothing of discretized numerical methods. *Phys. Rev. E* **2008**, *77*, 036611. [CrossRef]
35. Schurig, D.; Pendry, J.B.; Smith, D.R. Calculation of material properties and ray tracing in transformation media. *Opt. Express* **2006**, *14*, 9794–9804. [CrossRef] [PubMed]
36. Li, J.; Huang, Y.; Yang, W. An adaptive edge finite element method for electromagnetic cloaking simulation. *J. Comput. Phys.* **2013**, *249*, 216–232. [CrossRef]
37. Rochlitz, R.; Seidel, M.; Börner, R.U. Evaluation of three approaches for simulating 3-D time-domain electromagnetic data. *Geophys. J. Int.* **2021**, *227*, 1980–1995. [CrossRef]
38. Ren, Z.; Kalscheuer, T.; Greenhalgh, S.; Maurer, H. A goal-oriented adaptive finite-element approach for plane wave 3-D electromagnetic modelling. *Geophys. J. Int.* **2013**, *194*, 700–718. [CrossRef]
39. Anderson, R.; Andrej, J.; Barker, A.; Bramwell, J.; Camier, J.S.; Cerveny, J.; Dobrev, V.; Dudouit, Y.; Fisher, A.; Kolev, T.; et al. MFEM: A modular finite element methods library. *Comput. Math. Appl.* **2021**, *81*, 42–74. [CrossRef]
40. Si, H. TetGen, a Delaunay-based quality tetrahedral mesh generator. *ACM Trans. Math. Softw.* **2015**, *41*, 1–36. [CrossRef]
41. Amestoy, P.R.; de la Kethulle de Ryhove, S.; L'Excellent, J.Y.; Moreau, G.; Shantsev, D.V. Efficient use of sparsity by direct solvers applied to 3D controlled-source EM problems. *Comput. Geosci.* **2019**, *23*, 1237–1258. [CrossRef]
42. Streich, R.; Becken, M. Sensitivity of controlled-source electromagnetic fields in planarly layered media. *Geophys. J. Int.* **2011**, *187*, 705–728. [CrossRef]
43. Davydycheva, S.; Frenkel, M.A. The impact of 3D tilted resistivity anisotropy on marine CSEM measurements. *Lead. Edge* **2013**, *32*, 1374–1381. [CrossRef]
44. Zheng, Z.; Fu, Y.; Liu, K.; Xiao, R.; Wang, X.; Shi, H. Three-stage vertical distribution of seawater conductivity. *Sci. Rep.* **2018**, *8*, 9916. [CrossRef]
45. Correa, J.L.; Menezes, P.T.L. Marlim R3D: A realistic model for controlled-source electromagnetic simulations—Phase 2: The controlled-source electromagnetic data set. *Geophysics* **2019**, *84*, E293–E299. [CrossRef]

46. Werthmüller, D.; Rochlitz, R.; Castillo-Reyes, O.; Heagy, L. Towards an open-source landscape for 3-D CSEM modelling. *Geophys. J. Int.* **2021**, *227*, 644–659. [CrossRef]
47. Liu, Z.; Yao, H.; Wang, F. Performance investigations of auxiliary-space Maxwell solver preconditioned iterative algorithm for controlled-source electromagnetic induction problems with electrical anisotropy. *Geophys. Prospect.* **2024**, *72*, 2861–2879. [CrossRef]

Disclaimer/Publisher's Note: The statements, opinions and data contained in all publications are solely those of the individual author(s) and contributor(s) and not of MDPI and/or the editor(s). MDPI and/or the editor(s) disclaim responsibility for any injury to people or property resulting from any ideas, methods, instructions or products referred to in the content.

applied sciences

MDPI

Article

Kalman–FIR Fusion Filtering for High-Dynamic Airborne Gravimetry: Implementation and Noise Suppression on the GIPS-1A System

Guanxin Wang [1,2,3], Shengqing Xiong [2,*], Fang Yan [4,*], Feng Luo [2], Linfei Wang [2] and Xihua Zhou [2]

1 School of Geophysics and Information Technology, China University of Geosciences (Beijing), Beijing 100083, China; 3010230033@email.cugb.edu.cn
2 China Aero Geophysical Survey and Remote Sensing Center for Natural Resources, Beijing 100083, China; luofeng80@126.com (F.L.); wlfei4103@126.com (L.W.); xihua_wt@163.com (X.Z.)
3 Key Laboratory of Airborne Geophysics and Remote Sensing Geology Ministry of Natural Resources, Beijing 100083, China
4 Beijing Institute of Automation Control Equipment, Beijing 100074, China
* Correspondence: xsqagrs@126.com (S.X.); y15601390126@163.com (F.Y.)

Abstract: High-dynamic airborne gravimetry faces critical challenges from platform-induced noise contamination. Conventional filtering methods exhibit inherent limitations in simultaneously achieving dynamic tracking capability and spectral fidelity. To overcome these constraints, this study proposes a Kalman–FIR fusion filtering (K-F) method, which is validated through engineering implementation on the GIPS-1A airborne gravimeter platform. The proposed framework employs a dual-stage strategy: (1) An adaptive state-space framework employing calibration coefficients (S_x, S_y, S_z) continuously estimates triaxial acceleration errors to compensate for gravity anomaly signals. This approach resolves aliasing artifacts induced by non-stationary noise while preserving low-frequency gravity components that are traditionally attenuated by conventional FIR filters. (2) A window-optimized FIR post-filter explicitly regulates cutoff frequencies to ensure spectral compatibility with downstream processing workflows, including terrain correction. Flight experiments demonstrate that the K-F method achieves a repeat-line internal consistency of 0.558 mGal at 0.01 Hz—a 65.3% accuracy improvement over standalone FIR filtering (1.606 mGal at 0.01 Hz). Concurrently, it enhances spatial resolution to 2.5 km (half-wavelength), enabling the recovery of data segments corrupted by airflow disturbances that were previously unusable. Implemented on the GIPS-1A system, K-F enables precision mineral exploration and establishes a noise-suppressed paradigm for extreme-dynamic gravimetry.

Keywords: airborne gravity; finite impulse response filter; Kalman filter; fusion filter

1. Introduction

Airborne gravimetry, as a pivotal technology for efficient and environmentally sustainable acquisition of Earth's gravity field, has been extensively deployed in challenging terrains (e.g., mountains, swamps, and oceans) inaccessible to ground surveys or requiring rapid reconnaissance. It plays a vital role in mineral resource exploration and geological structure research, and related domains [1]. This technique employs aircraft or unmanned aerial vehicles (UAVs) as platforms, integrating dual-channel measurements from airborne gravimeters and differential GNSS (Global Navigation Satellite System). During survey flights, the output signal of the gravimeter's vertical sensor contains composite information comprising true gravitational acceleration, motion-induced acceleration, and multi-source noise. The core objective of airborne gravity data processing is to extract gravity anomalies from this composite signal. However, the acquired gravity data exhibit severely degraded signal-to-noise ratios (SNR < 10^{-5}–10^{-6}) due to multi-source interference, including airflow disturbances (e.g., wind shear, wingtip vortices), platform vibrations, and sensor noise [2]. Consequently, achieving effective noise suppression while preserving the spectral integrity of weak gravity anomalies remains a critical challenge for improving measurement accuracy.

Widely adopted approaches include low-pass filtering [3–5] and Kalman filtering [6–12]. The finite impulse response (FIR) filter is widely utilized in airborne gravimetric data processing due to its deterministic frequency-domain selectivity. By optimizing its design parameters, the FIR filter effectively suppresses high-frequency interference components while retaining the primary gravity signal content. However, despite the predominant spectral concentration of gravity anomalies within targeted bands, the absence of a distinct transition between signal and noise spectra limits the efficacy of conventional FIR implementations. Static filter configurations fail to adapt to the time-varying noise profiles inherent to airborne platforms, resulting in incomplete suppression of non-stationary disturbances and insufficient separation of spectrally overlapping signal–noise mixtures. In contrast, Kalman filtering (a recursive time-domain estimation method) excels in dynamic environments by constructing state-space models to decouple gravity anomalies from noise, with demonstrated efficacy in mitigating platform-induced disturbances [6]. Nevertheless, its lack of explicit frequency-domain control hinders compatibility with downstream geophysical workflows (e.g., terrain correction requiring frequency-matched filtering), thereby limiting its engineering adoption.

To synergize the complementary strengths of FIR and Kalman filters, Zou et al. [13] proposed a cascaded filtering framework for strapdown airborne gravimeters. Their method sequentially applies Kalman filtering and FIR low-pass filtering (fc = 0.00625 Hz) to raw gravity data, achieving a 17.2% improvement in repeat-line internal consistency accuracy (1.2 mGal vs. 1.45 mGal for standalone FIR). While innovative, their Markov-model-based state-space formulation assumes stationary noise statistics, rendering it ineffective against time-varying interference spectra encountered in actual flights. Despite its conceptual merit, the achieved precision (1.2 mGal) remains insufficient for operational requirements. Alternative methods, such as wavelet transform [14], improved function filter [15], and minimum curvature techniques [16], have also been explored but face limitations in dynamic adaptability or computational practicality, preventing widespread application. For a comprehensive methodological comparison, see Table 1. Addressing these limitations, this study introduces a novel Kalman–FIR fusion filtering (K-F) algorithm tailored for airborne gravimetry systems. The K-F framework combines adaptive Kalman filtering for dynamic noise suppression with deterministic FIR post-filtering to enforce explicit spectral constraints. Validated through flight tests on the GIPS-1A airborne gravimeter, the method demonstrates dual capabilities: preserving Kalman's noise suppression robustness in dynamic environments and providing FIR's frequency-domain transparency. Operational deployment on the GIPS-1A system has enabled high-precision gravity surveys in complex terrains, establishing a practical engineering solution that harmonizes noise attenuation and spectral fidelity.

In airborne gravity measurements, gravity anomaly signals are intrinsically coupled with aircraft motion-induced accelerations. When the platform encounters in-flight disturbances (e.g., turbulence, maneuvers), horizontal accelerations become partially coupled into the gravimeter's vertical sensor, significantly degrading measurement accuracy. Existing studies generally lack separate consideration for such interference induced by horizontal motion coupling, primarily relying on increasing the filtering bandwidth of vertical signals to suppress such interference. This approach compromises the half-wavelength spatial resolution and attenuates genuine gravity anomaly information. This persistent limitation stems from a disconnect between instrument developers and operational end-users: developers often overlook application-specific interference profiles, while operators rarely delve into underlying instrument technologies.

Table 1. Comparison of airborne gravity data processing methods.

Algorithm	Core Principle	Physical Significance and Characteristics for Gravity Anomaly Extraction	Engineering Application Status
FIR filtering [3]	Digital filter design	• Clear spatial resolution • Poor dynamic adaptability • Resolution-accuracy tradeoff under disturbances	• Longest-developed and most widely applied method • Achieves up to 3 km half-wavelength resolution in stable flights
Kalman filtering [9]	State-space estimation	• Unclear spatial resolution • Moderate dynamic adaptability • Focused on vertical sensor signals	• Rapidly developed in recent decades • Controversial in industrial applications • Primarily used for FIR result verification
Wavelet transform [14]	Time–frequency multiscale decomposition	• Unclear spatial resolution • Suitable for stable flight operations	• Theoretically innovative • Limited practical value • No widespread adoption
Improved function filter [15]	Digital filter design, controlling the cutoff frequency and transition band length of the filter	• Clear spatial resolution • Suitable for stable flight operations	• Theoretically innovative • Limited practical value • No widespread adoption • Achieves 3.5 km half-wavelength resolution
Minimum curvature techniques [16]	Surface interpolation fitting	• Unclear spatial resolution • Suitable for stable flight operations	• Theoretically innovative • Limited practical value • No widespread adoption
Kalman–FIR fusion filtering	Digital filter design and state-space estimation	• Explicit spatial resolution • Maximizes retention of broadband information • Balances resolution and high accuracy • Integrates vertical and horizontal sensor data • Superior dynamic adaptability	• Equipped with GIPS-1A gravimeter • Widely deployed in marine and high-altitude engineering operations • Achieves up to 2.5 km half-wavelength resolution in high dynamic flight

Based on an in-depth understanding of the design and measurement principles of platform-type airborne gravimeters, this study fully utilizes the unique attitude isolation and stability maintenance capabilities of platform gravimeters. For the first time, it systematically constructs and incorporates a horizontal motion coupling interference error model, achieving effective estimation and compensation for interference signals introduced by motion accelerations in both horizontal directions.

The innovative contributions of this study in airborne gravity data processing and its engineering applications are summarized as follows:

(1) Based on the working principles and hardware characteristics of platform-type triaxial gravimeters, short-term platform misalignment angle linearized state equations are established. This model introduces horizontal motion acceleration information into the gravity anomaly estimation state equation, expanding the state space of existing Kalman filtering methods to directly compensate for horizontal motion interference. This state estimation equation effectively improves gravity measurement accuracy in high-dynamic flight regions without sacrificing half-wavelength resolution.
(2) A deterministic finite impulse response (FIR) filter is cascaded to form the innovative Kalman–FIR fusion (K-F) framework. This design ensures clear mapping of physical meaning (spatial resolution) in gravity anomaly data processing; explicit spectral control guarantees compatibility with downstream processes like terrain correction, while eliminating application barriers caused by spectral opacity of Kalman filtering.

2. Methodology

The extraction of airborne gravity anomalies comprises three sequential steps: multi-source data synchronization, comprehensive gravity corrections, and noise elimination. The procedural workflow is illustrated in Figure 1. Specifically, it includes the following steps:

(1) Unify sampling rates and synchronize time signals between the gravity sensor channel data and differential satellite navigation data.
(2) Apply corrections:
   - Gravity sensor data: Zero-drift correction and reference point calibration;
   - Differential satellite data: Eccentricity correction, Eötvös correction, and normal field correction;
   - Integrate all corrected results to generate the raw gravity anomaly.
(3) Filter the raw anomaly to extract the gravity anomaly signal.

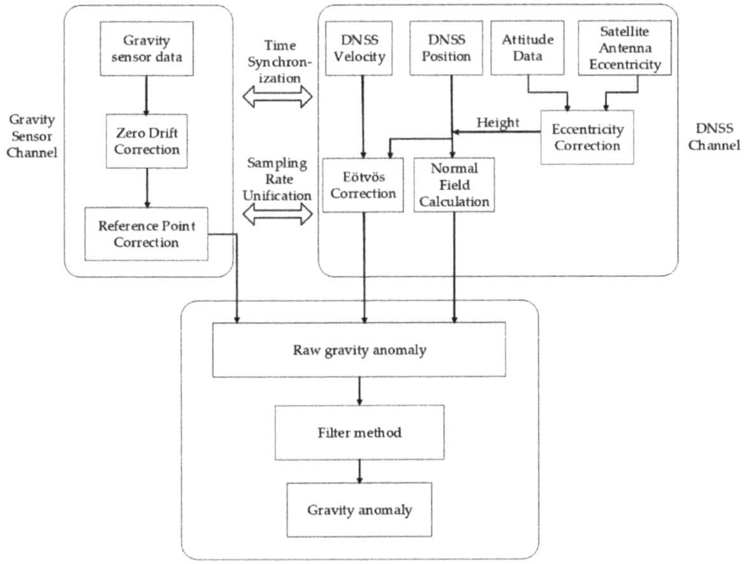

Figure 1. Flowchart of gravity anomaly extraction.

2.1. Data Synchronization

Airborne gravimetry comprises two parallel measurement systems: (1) Differential satellite data measurement acquires the gravimeter's velocity and position; (2) gravimeter sensors measure triaxial accelerations (east, north, vertical) and attitude data. For the GIPS-1A system, the gravimeter outputs data at 100 Hz while differential GNSS provides data at 2 Hz. Adhering to the Nyquist sampling theorem, this study downsampled the gravimeter outputs and performed temporal alignment using timestamp interpolation.

2.2. Comprehensive Gravity Corrections

The idealized measurement model assumes a perfectly leveled instrument platform:

$$\Delta g = f_u - \dot{v}_u + \Delta G_E - G(\varphi, h) \tag{1}$$

where

Δg: Gravity anomaly;

f_u: Observed gravity under platform-leveled conditions;

ΔG_E: Eötvös correction;

$G(\varphi, h)$: Normal gravity at the computation point.

Actual flight dynamics induce platform tilting, requiring revision to

$$\Delta g = f_z - \dot{v}_u + \Delta G_E - G(\varphi, h) \tag{2}$$

where f_z denotes the vertical component of f_u (true sensor measurement).

As airborne gravimetry constitutes relative measurement, pre-flight static observations at airport base stations establish datum ties:

$$\Delta g = g_b + (f_z - f_{z0}) - \dot{v}_u + \Delta G_E - G(\varphi, h) + \delta a_K + \delta a_I \tag{3}$$

where g_b is the reference gravity at the airport, f_z and f_{z0} denote the observed vertical gravity component and its initial value, δa_K represents the zero-drift correction, and δa_I encompasses random noise from both the gravimeter and differential GNSS.

2.3. FIR Low-Pass Filtering

The FIR filter is an all-zero system with guaranteed stability and straightforward design. The output signal $y(n)$ of an N-th order FIR filter can be expressed as the convolution of the input signal $x(n)$ and the filter's impulse response $h(k)$:

$$y(n) = \sum_{k=0}^{N-1} h(k)x(n-k) \tag{4}$$

where $h(k)$ denotes the filter coefficients.

To approximate the infinite impulse response of an ideal filter, FIR design employs finite-length window functions $w(n)$ to truncate the primary components of $h_d(n)$:

$$h(n) = h_d(n).\omega(n), n = 0, 1, 2, \ldots, N-1 \tag{5}$$

Common window functions include the Hanning, Hamming, and Blackman windows. Among these, the Hanning window is frequently regarded as better suited to the extraction of airborne gravity anomalies; therefore, this study employs the Hanning window as the window function. Its mathematical form is as follows:

$$\omega(n) = \begin{cases} \frac{1}{2}(1 - \cos\frac{2\pi n}{N-1}), 0 \leq n \leq N-1 \\ 0, n \text{ are others} \end{cases} \quad (6)$$

2.4. Kalman Filtering

Unlike frequency-domain filtering methods that treat noise as undesirable signals to be removed, the Kalman filter leverages statistical properties of system noise and measurement noise to estimate useful signals—fundamentally representing an optimal estimation theory known as recursive minimum linear variance estimation. As a time-domain filtering technique applicable to multidimensional and non-stationary stochastic processes, the Kalman filter operates as follows:

Let the system state $\mathbf{X}_k$ at time k be driven by system noise $\mathbf{W}_k$ and deterministic inputs $\mathbf{U}_k$. The state-space representation is

$$\mathbf{X}_{k+1} = \mathbf{\Phi}_k\mathbf{X}_k + \mathbf{B}_k\mathbf{U}_k + \mathbf{\Gamma}_k\mathbf{W}_k \quad (7)$$

$$\mathbf{Z}_k = \mathbf{H}_k\mathbf{X}_k + \mathbf{V}_k \quad (8)$$

where

$\mathbf{\Phi}_k$, $\mathbf{H}_k$: State transition and measurement matrices;

$\mathbf{B}_k$: Control input matrix;

$\mathbf{\Gamma}_k$: System noise driving matrix;

$\mathbf{W}_k$, $\mathbf{V}_k$: Zero-mean white noise with covariances $\mathbf{Q}_k$ and $\mathbf{R}_k$, respectively, satisfying

$$E[\mathbf{W}_k\mathbf{W}_j^T] = \mathbf{Q}_k\delta_{kj}, E[\mathbf{V}_k\mathbf{V}_j^T] = \mathbf{R}_k\delta_{kj} \quad (9)$$

Here, $\mathbf{Q}_k$ is non-negative definite, and $\mathbf{R}_k$ is positive definite [17].

The discrete fundamental equations of Kalman filtering (10)–(14) enable recursive solutions to both the state (7) and measurement Equation (8).

State One-Step Prediction:

$$\mathbf{X}_{k,k-1} = \mathbf{\Phi}_{k-1}\mathbf{X}_{k-1} + \mathbf{B}_{k-1}\mathbf{U}_{k-1} \quad (10)$$

State Estimation:

$$\mathbf{X}_k = \mathbf{X}_{k,k-1} + \mathbf{K}_k(\mathbf{Z}_k - \mathbf{H}_k\mathbf{X}_{k,k-1}) \quad (11)$$

Filter Gain:

$$\mathbf{K}_k = \mathbf{P}_{k,k-1}\mathbf{H}_k^T(\mathbf{H}_k\mathbf{P}_{k,k-1}\mathbf{H}_k^T + \mathbf{R}_k)^{-1} \quad (12)$$

One-Step Prediction Mean Square Error:

$$\mathbf{P}_{k,k-1} = \mathbf{\Phi}_{k-1}\mathbf{P}_{k-1}\mathbf{\Phi}_{k,k-1}^T + \mathbf{\Gamma}\mathbf{Q}\mathbf{\Gamma}^T \quad (13)$$

Estimated Mean Square Error:

$$\mathbf{P}_k = (\mathbf{I} - \mathbf{K}_k\mathbf{H}_k)\mathbf{P}_{k,k-1}(\mathbf{I} - \mathbf{K}_k\mathbf{H}_k)^T + \mathbf{K}_k\mathbf{R}_k\mathbf{K}_k^T \quad (14)$$

2.5. Kalman–FIR Fusion Filtering (K-F Filtering)

In airborne gravimetric practice, the mathematical model for gravity anomalies is formulated based on Newton's second law by analyzing the forces acting on the measurement platform and its flight dynamics:

$$\Delta g = f_\Sigma - \dot{v}_u + q_\Sigma \tag{15}$$

where

Δg: Gravity anomaly;

f_Σ: Corrected gravity value incorporating normal field, elevation, Eötvös, drift, and base station corrections;

$\dot{v}_u$: Vertical acceleration of the aircraft;

q_Σ: Aggregate noise from multi-source disturbances.

Notably, f_Σ and $\dot{v}_u$ can be derived from differential GNSS signals and raw gravimetric measurements. Their computational methodologies are beyond the scope of this paper and are treated as known quantities [18].

To enable dynamic compensation of airborne gravity anomalies, the aggregate error q_Σ is decomposed into four components: (1) X-axis gravitational-acceleration-induced error; (2) Y-axis gravitational-acceleration-induced error; (3) Z-axis gravitational-acceleration-induced error; (4) residual noise.

Three calibration coefficients (S_x, S_y, S_z) are introduced into the gravity anomaly model to independently regulate errors caused by gravitational accelerations (f_x, f_y, f_z) along each axis:

$$\Delta g = f_\Sigma - \dot{v}_u + S_x f_x + S_y f_y + S_z f_z + q_{others} \tag{16}$$

It is noteworthy that the effective operation and optimal estimation performance of Equation (16) critically depend on one prerequisite: the inertial stabilized platform must maintain real-time attitude tracking and stabilization capability. Under this condition, S_x and S_y physically represent the platform misalignment angles over short time intervals.

Furthermore, within short time scales, S_x and S_y are predominantly attributed to the null drift of horizontal-axis gyroscopes in the gravimeter. Given the minute magnitude of these angles, they can be linearized as first-order constants, with their time derivatives quantitatively characterizing the gyroscopic drift rate.

Equation (7) is rewritten into a Kalman filter state-space representation:

$$\begin{aligned} &\dot{h} = v_u \\ &\dot{v}_u = f_\Sigma + S_x f_x + S_y f_y + S_z f_z - \Delta g + q_{others} \\ &\dot{S}_x = q_{S_x} \\ &\dot{S}_y = q_{S_y} \\ &\dot{S}_z = q_{S_z} \\ &h' = H_f X_f + \delta h \end{aligned} \tag{17}$$

where

h': Measured reference ellipsoid height;

$\delta\, h$: Measurement error of h'.

The parameterization adopts

$$\mathbf{X}_f = \begin{pmatrix} h & v_u & S_z & S_y & S_x \end{pmatrix}^T, \mathbf{H}_f = \begin{pmatrix} 1 & 0 & 0 & 0 & 0 \end{pmatrix}^T$$

The gravity anomaly is modeled as a random process using a forming filter:

$$\begin{aligned}\dot{\mathbf{X}}_g &= \mathbf{A}_g\mathbf{X}_g + \mathbf{\Gamma}_g\mathbf{q}_g \\ \Delta g &= \mathbf{H}_g\mathbf{X}_g\end{aligned} \tag{18}$$

where

$\mathbf{X}_g$: State vector to be estimated;

$\mathbf{A}_g$, $\mathbf{\Gamma}_g$: Constant system matrices;

$\mathbf{q}_g$: White noise with intensity $\mathbf{Q}_g$.

The parameterization adopts

$$\mathbf{X}_g = \begin{pmatrix} \Delta g \\ \Delta \dot{g} \end{pmatrix}, \mathbf{A}_g = \begin{pmatrix} 0 & 1 \\ 0 & 0 \end{pmatrix}, \\ \mathbf{\Gamma}_g = \begin{pmatrix} 0 \\ 1 \end{pmatrix}, \mathbf{H}_g = \begin{pmatrix} 1 & 0 \end{pmatrix}.$$

By combining Equations (17) and (18), the unified state-space model is derived.

To synergize the dynamic adaptation capability of Kalman filtering with the well-defined cutoff frequency of FIR low-pass filtering, the proposed K-F algorithm operates as follows:

(1) Kalman Filtering Stage:

Estimate the calibration coefficients (S_x, S_y, S_z) via Kalman filtering to dynamically adjust axis-specific errors.

(2) Error Compensation:

Subtract the calibrated errors (S_x, S_y, S_z) from the corrected gravity field f_Σ using Equation (7), yielding a noise-contaminated gravity anomaly Δg_{noisy}.

$$\Delta g_{noisy} = f_\Sigma - \dot{v}_u - (S_x f_x + S_y f_y + S_z f_z) \tag{19}$$

(3) FIR Post-Filtering:

Apply a low-pass FIR filter with a predefined cutoff frequency f_c to suppress residual noise in Δg_{noisy}, producing the refined gravity anomaly Δg final.

$$\Delta g_{final} = \mathrm{FIR}(\Delta g_{noisy}, f_c) \tag{20}$$

The workflow is illustrated in Figure 2.

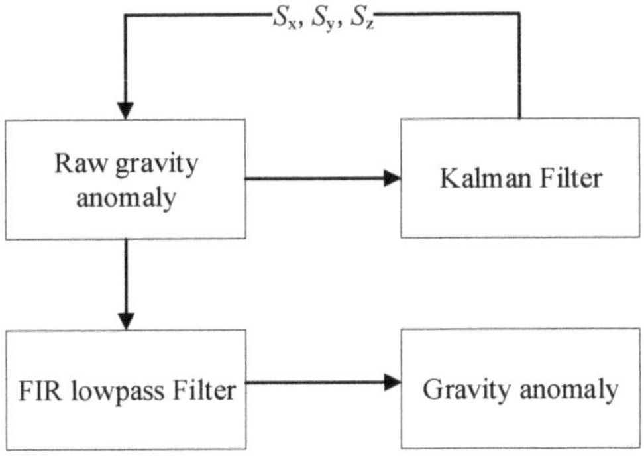

Figure 2. Flowchart of K-F filtering algorithm.

2.6. Quality Evaluation Method

The internal consistency accuracy of airborne gravimetry is evaluated using repeat-line test data to assess the dynamic precision of repeated measurements. This metric quantifies the agreement between multiple repeat-line datasets relative to their averaged gravity field, reflecting the system's dynamic performance [19,20].

For each repeat line, the root mean square internal consistency accuracy is calculated as

$$\sigma_j = \sqrt{\frac{\sum_{i=1}^{n} \left(\Delta g_{ij} - \overline{\Delta g_i}\right)^2}{n}}, (j = 1, 2, \ldots, m) \tag{21}$$

Δg_{ij}: Gravity anomaly at point i on repeat line j;
$\overline{\Delta g_i} = \frac{1}{m}\sum_{i=1}^{m} \Delta g_{ij}$: Mean gravity anomaly at point i across all repeat lines;
m: Number of repeat lines;
n: Number of data points in the common segment of repeat lines.

The overall internal consistency accuracy across all repeat lines is given by

$$\sigma_j = \sqrt{\frac{\sum\limits_{j=1}^{m} \sum\limits_{i=1}^{n} \left(\Delta g_{ij} - \overline{\Delta g_i}\right)^2}{m \times n}} \tag{22}$$

3. Field Data Validation

3.1. Survey Overview

This study validates the K-F filtering using actual measurement data. The test employed the GIPS-1A platform-type airborne gravimeter, jointly developed by China Aero Geophysical Survey & Remote Sensing Center and Beijing Institute of Automation Control Equipment, implementing a "three-axis stabilized platform + quartz-flexure accelerometer" design.

The test was conducted aboard a Cessna 208B (Figure 3) fixed-wing aircraft over a maritime survey line (Line 2010, ~60 km north–south trending, GPS altitude: 600 m, ground speed: 220 km/h). Due to airspace and weather constraints, the measurements were conducted at midday when airflow disturbances were more complex compared to other periods.

(**a**) Cessna 208 aircraft

(**b**) GIPS-A gravimeter

Figure 3. Installation of the GIPS-A gravimeter in an amphibious Cessna 208 aircraft.

The flight path is illustrated in Figure 4. The aircraft took off from the airport, reached the survey area, and executed two round-trip flights along the east–west survey lines, resulting in four repeated east–west survey lines (two pairs). A single north–south round-trip flight was also conducted, yielding a pair of repeated north–south survey lines. Due to the limited number of north–south lines, statistical analysis was only performed on the east–west lines. For these four repeated east–west lines, raw gravity anomalies ranged from −31,715.506 to 27,754.607 mGal and showed an internal consistency accuracy of 4349.245 mGal (Figure 5). This indicates complete signal masking by noise and an absence of repeat-line coherence.

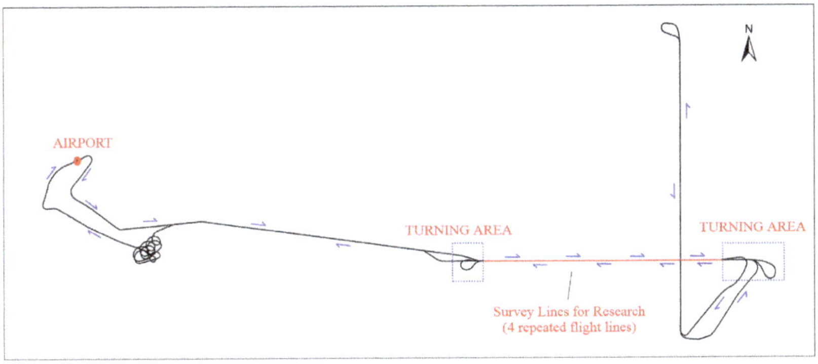

Figure 4. Test flight track map.

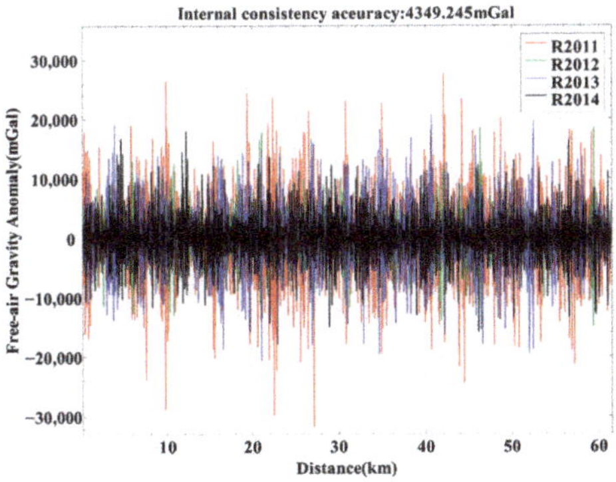

Figure 5. Raw gravity anomaly map K-F filtering algorithm flowchart.

3.2. Repeated Line Data Processing Results

In airborne gravity surveys, the determination of filtering parameters and geospatial half-wavelength resolution adheres to stringent physical constraints: the cutoff frequency (f_c) of low-pass filtering or time window length (T filter) combined with flight velocity (v) dictates the minimum resolvable half-wavelength ($\lambda = v/(2f_c)$) for geological structures. While the selection of filtering parameters and their impact on resolution are relatively straightforward, the core challenge lies in balancing high-frequency noise suppression (e.g., >10 mGal spikes induced by airflow disturbances) and short-wavelength signal preservation (essential for resolving geological features like faults or salt dome boundaries).

In operational airborne engineering applications, a 100 s low-pass filter (cutoff frequency: 0.01 Hz) is typically applied, with an internal consistency accuracy threshold of 0.8 mGal deemed acceptable for production-grade data (half-wavelength spatial resolution: ~3 km) [20]. For comparative analysis, both FIR and K-F filters were tested at three cutoff settings: 80 s (0.0125 Hz), 100 s (0.01 Hz), and 120 s (0.00833 Hz). The gravity anomaly along the line exhibits an inverted "S" morphology with amplitudes of 50–60 mGal, transitioning smoothly to background values at the edges. Results are summarized in Figures 6–8 and Table 2.

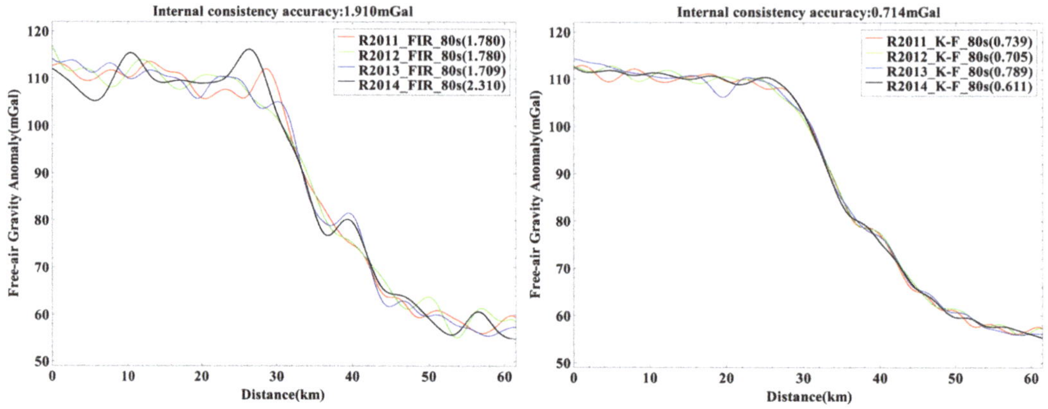

(a) (b)

Figure 6. Processing results of FIR (**a**) and K-F filter (**b**) with 80 s filtering time.

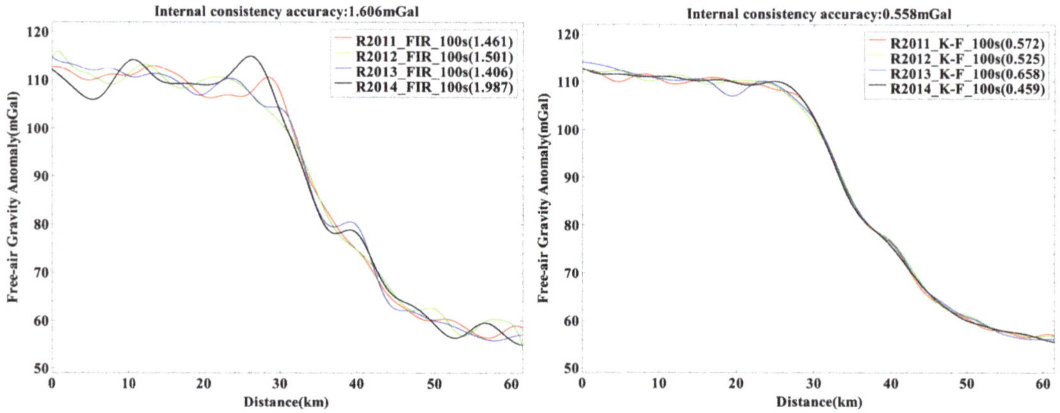

(a) (b)

Figure 7. Processing results of FIR (**a**) and K-F filter (**b**) with 100 s filtering time.

Table 2. Comparison of processing results of FIR Filter and K-F Filter.

Filter Time (s)	Half-Wavelength (km)	FIR Filter (mGal)	K-F Filter (mGal)	Improvement Percentage
80	2.44	1.910	0.714	62.62%
100	3.06	1.606	0.558	65.26%
120	3.67	1.414	0.475	66.41%

Table 2 shows that the FIR-filtered data exhibited internal consistency accuracies of 1.910 mGal (80 s), 1.606 mGal (100 s), and 1.414 mGal (120 s), with persistent artifacts and false anomalies attributed to low-frequency airflow-disturbance-induced noise unattenuated by conventional filtering. Extending the filtering window to 120 s marginally improved repeatability but degraded spatial resolution (half-wavelength > 3 km), rendering the dataset operationally unusable. In contrast, the K-F algorithm achieved significantly higher precision, with internal consistency accuracies of 0.714 mGal (80 s), 0.558 mGal

(100 s), and 0.475 mGal (120 s), corresponding to improvements of 62.6%, 65.3%, and 66.4% over FIR filtering, respectively. The K-F framework demonstrated superior airflow disturbance resistance, suppressing low-frequency noise while preserving spectral integrity and achieving a half-wavelength spatial resolution of ~2.5 km.

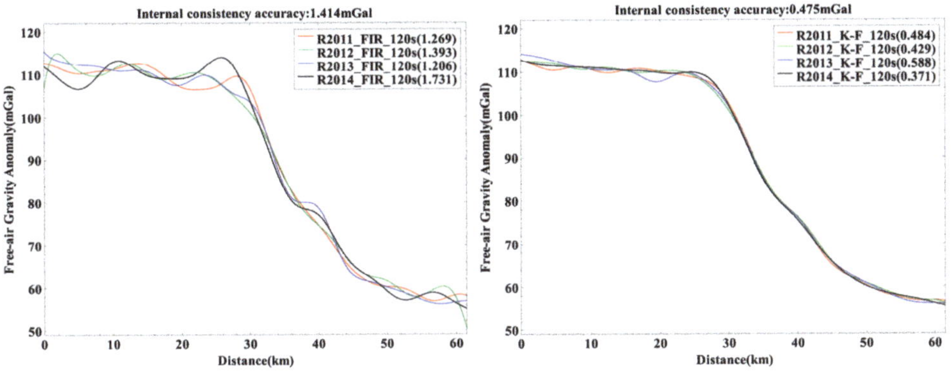

(**a**) (**b**)

Figure 8. Processing results of FIR (**a**) and K-F filter (**b**) with 120 s filtering time.

This validation confirms the K-F algorithm's engineering viability for high-dynamic airborne gravimetry, particularly in challenging survey environments. It should be emphasized that while the K-F filter demonstrated a significant improvement in data processing results compared to the FIR filter in this particular experiment, this advantage is not consistently substantial across all flight tests. When the aircraft is in steady-state flight, the processing results achieved by the K-F filter and the FIR filter show comparable performance.

Figure 9 presents the gravity anomaly extraction results from another flight pass, indicating that the FIR filter achieves an internal coincidence accuracy of 0.603 mGal/100 s, while the K-F filter attains 0.564 mGal/100 s. This confirms that the FIR filter can also deliver satisfactory results under steady flight conditions. However, the K-F filter demonstrates superior stability in more complex flight conditions.

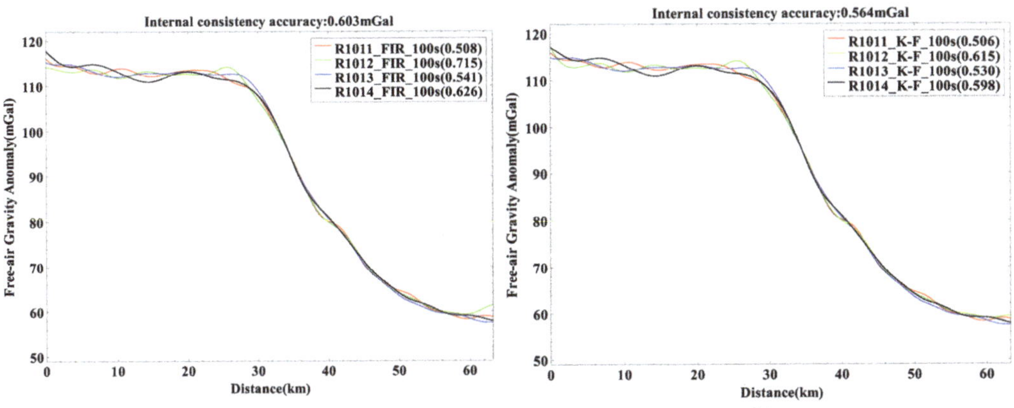

(**a**) (**b**)

Figure 9. Processing results of FIR (**a**) and K-F filter (**b**) with 100 s filtering time.

3.3. Full Fight Data Processing Results

To further evaluate the suppression capability of the algorithm against flight platform motion interference, the entire flight sortie data was analyzed using 100 s filtering results as an example. Figure 10 compares the processing results of three methods: K-F filter, FIR filter, and Kalman filter. The repeat-line measurement areas and turning areas are color-coded in cyan and gray, respectively. Considering that gravity anomaly signals inherently exhibit low-frequency slow variations whose theoretical waveforms should not show drastic fluctuations, while the curves in turning areas display large fluctuations, the abnormal oscillations in the figure can be attributed to motion-induced interference from the platform. By measuring the fluctuation amplitude of these curves, the algorithm's motion interference compensation capability can be characterized.

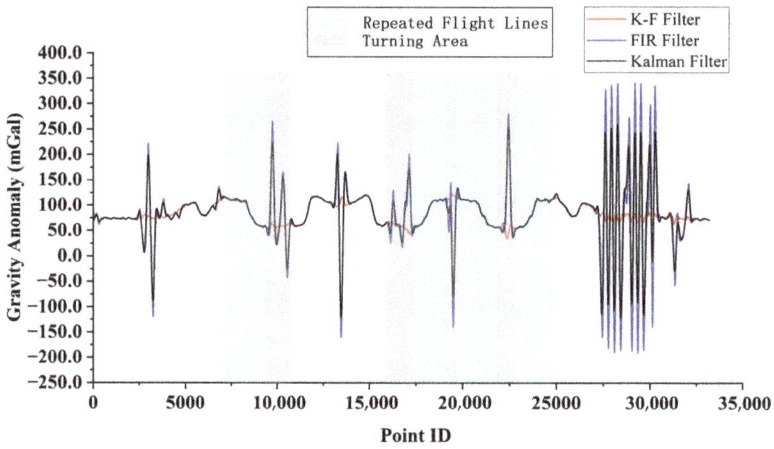

Figure 10. Comparison of processing results: K-F filter vs. FIR filter vs. Kalman filter.

In turning segments (gray zones in Figure 10), fluctuations in the filtered curve of the K-F filter proposed in this paper decreased substantially compared to the other two methods. Statistical results in Table 3 further corroborate this: the K-F filtered output exhibits a peak-to-peak value of only 89.792 mGal (standard deviation: 19.928 mGal), significantly below the FIR filter's peak-to-peak value of 533.342 mGal (standard deviation: 63.036 mGal) and the Kalman filter's peak-to-peak value of 381.775 mGal (standard deviation: 50.390 mGal). Statistical analysis demonstrates that the K-F filter improves motion interference suppression by approximately 5.9 times (vs. FIR) and 4.3 times (vs. Kalman). Figure 11 shows the difference between the K-F filter and FIR filter results, providing a more intuitive visualization of the K-F filter's superior noise attenuation capability compared to the FIR filter.

Table 3. Statistical results of flight sortie data processed by three methods.

	K-F Filter (mGal)	FIR Filter (mGal)	Kalman Filter (mGal)
Maximum Value	124.399	342.779	259.569
Minimum Value	34.607	−190.563	−122.206
Peak-to-Peak Value	89.792	533.342	381.775
Standard Deviation	19.928	63.036	50.390
Maximum Peak in Turning Segments	80.620	342.779	243.214

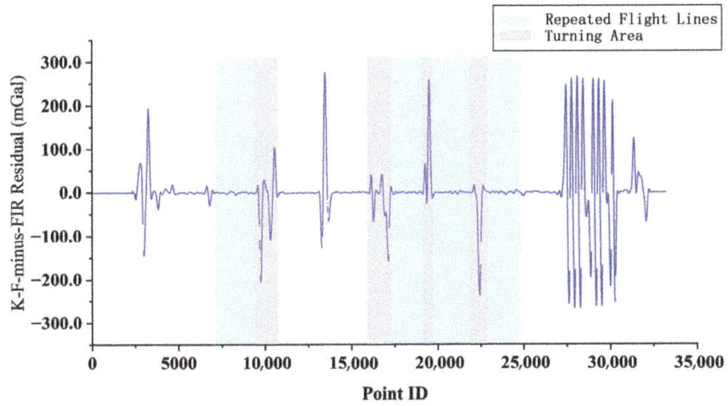

Figure 11. Residual gravity anomaly: K-F filter minus FIR filter.

Figure 12 displays the horizontal motion accelerations f_x and f_y during the flight. Values in turning segments exhibit significantly larger magnitudes compared to those during repeat-line flights. Consequently, accurate estimation of platform attitude errors becomes critical for effective horizontal motion interference compensation, specifically the parameters S_x and S_y in Equation (12). Figure 13 demonstrates stable convergence of S_x and S_y throughout the flight, validating the efficacy of the gravity anomaly estimation state equation.

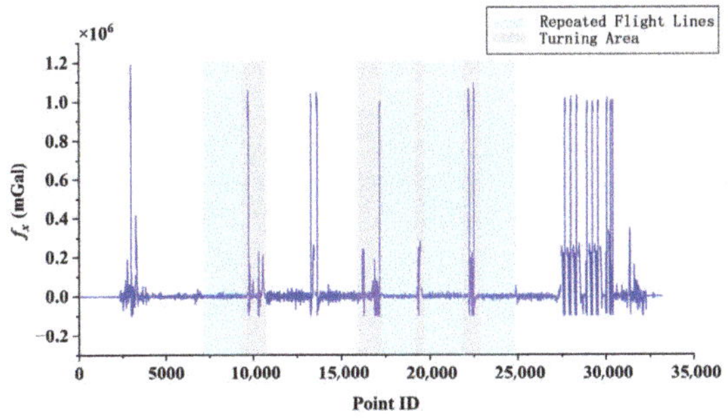

(a) Horizontal accelerations f_x

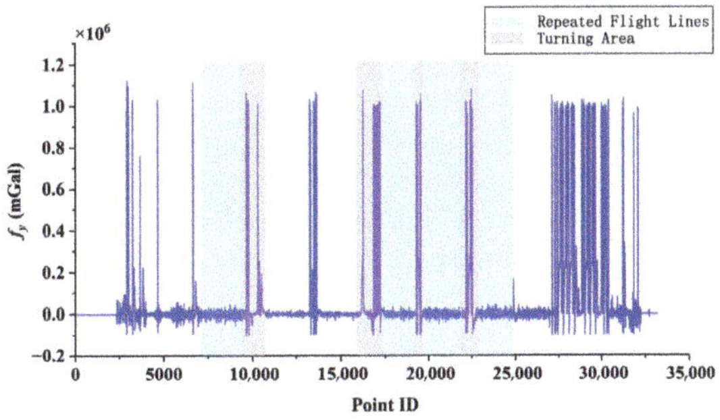

(b) Horizontal accelerations f_y

Figure 12. Raw measurements of horizontal accelerations f_x and f_y.

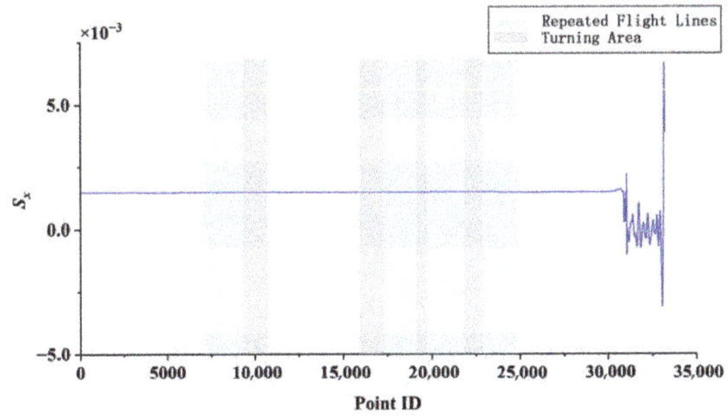

(a)

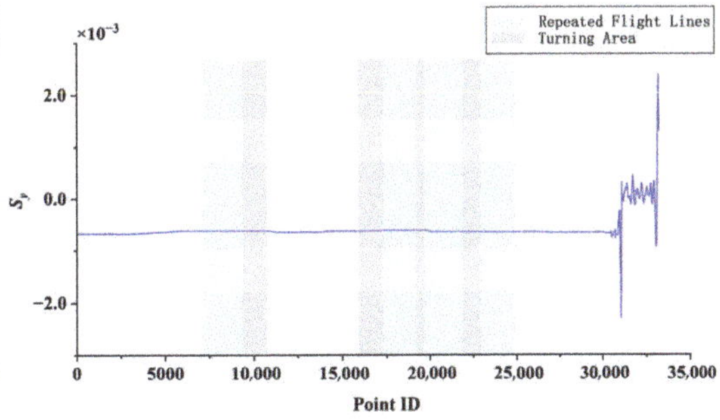

(b)

Figure 13. Processing results of S_x (**a**) and S_y (**b**).

4. Discussion

To address the challenges of dynamic noise suppression and signal fidelity in airborne gravimetry, this study proposes a novel fusion algorithm integrating Kalman filtering with FIR low-pass filtering (K-F filtering). The algorithm leverages inherent physical design advantages of platform-type gravimeters by linearly representing platform attitude errors and embedding them into the gravity anomaly state estimation equation, thereby compensating for horizontal-motion-induced errors. Additionally, cascaded FIR filters provide explicit cutoff frequency control. Operational validation on the GIPS-1A platform gravimeter confirms the algorithm's efficacy.

1. Breakthrough in High-Dynamic Noise Suppression via K-F Filtering

(1) Enhanced Dynamic Noise Suppression

Through Kalman and FIR filtering, a repeat-line accuracy of 0.558 mGal (65.3% improvement over conventional FIR) is attained on the GIPS-1A system (Figure 6), effectively resolving the conflict between noise contamination from external disturbances and gravimetric signal fidelity. Experimental results demonstrate that under flight conditions of 220 km/h airspeed and airflow disturbances, the K-F algorithm improves repeat-line internal consistency accuracy by over 60% compared to standalone FIR filtering, confirming its robustness against high-dynamic noise.

(2) Improved Spectral Transparency

Post-Kalman correction, FIR low-pass filtering (80–120 s filtering windows, fc = 0.0083–0.0125 Hz) preserves the dominant low-frequency gravity anomaly band while enabling explicit spectral control to match terrain correction requirements. The achieved half-wavelength spatial resolution of 2.5 km represents a 16.7% improvement over conventional methods (3 km), satisfying the resolution demands for mid- to short-wavelength gravity signals in mineral exploration.

(3) Extended Engineering Applicability

The constructed joint state-space model quantifies carrier attitude errors by introducing adjustment coefficients (S_x, S_y, S_z), overcoming the stationary state assumption limitation of second-order Markov models. Validation using actual flight mission data demonstrates that this method adapts to noise spectrum variations across different flight missions. Consequently, data previously rendered unusable due to airflow disturbances (FIR filter accuracy: 1.414–1.910 mGal) was recovered to an engineering-applicable level of 0.475–0.714 mGal.

2. Foundation for the Efficacy of the K-F Filtering

Gravimeters represent highly complex instrumentation within geological exploration, creating inherent industry barriers. Manufacturers focus predominantly on hardware development with limited insight into field operational scenarios, while exploration entities lack expertise in instrument principles and design details. Consequently, gravity data acquired in complex environments cannot be fully deciphered to isolate inherent interference and noise, diminishing the value of high-cost measurements. Academic institutions primarily engage in theoretical research, collectively sustaining technical disconnects in gravity surveying.

The GIPS-1A platform gravimeter employed in this study possesses geodetic frame stabilization capability through its three-axis inertially stabilized platform. The system adopts a design scheme comprising a "dynamically tuned gyro + resolver + analog control circuit," with all stages processing analog signals exclusively. This configuration delivers strong dynamic response capability, high control bandwidth, and exceptional platform stability, outperforming other platform gravimeters utilizing digital sensing and control solutions.

Under stable flight conditions with only the vertical acceleration channel considered, differences between gravimeters are minimal. However, during high-dynamic flight conditions requiring precise compensation for dual-axis horizontal accelerations, the GIPS-1A gravimeter provides instrument-level physical capability assurance. This constitutes the fundamental prerequisite for linearized modeling of platform attitude errors and their compensation via optimal estimation methods.

3. Future Research Perspectives and Recommendations

Building upon the contributions of this work, further research can be advanced through the following two dimensions to refine methodological robustness and theoretical foundations:

(1) Construction of a precise gravimetric measurement model:

The current model's noise identification capability originates from explicit parameterization of platform motion errors (S_x, S_y, S_z). Future work must establish a noise–gravity anomaly coupling theory to quantify statistical characteristics of motion interference under varying meteorological conditions (e.g., correlation between kinematic noise power spectrum and Kalman gain $\mathbf{K}_k$), providing physical constraints for model parameters.

(2) Enhancement of adaptive anti-disturbance mechanisms:

To further improve robustness against airframe motion disturbances, adaptive noise covariance estimation (real-time updates of $\mathbf{Q}_k$, $\mathbf{R}_k$, Equation (9)) can be introduced. This adaptive estimation dynamically responds to non-stationary disturbances during flight, eliminating spectral leakage caused by unmodeled high-frequency interference, thereby maintaining stability of gravity anomaly resolution during abrupt variations in aircraft motion.

5. Conclusions

The Kalman–FIR fusion filtering (K-F) method proposed in this study successfully resolves the inherent conflict between noise suppression and signal fidelity in high-dynamic airborne gravimetry through a dual-mechanism approach combining dynamic error compensation and spectral constraints. Its core value manifests in two aspects: (1) The dynamic compensation mechanism overcomes limitations of static filters in adapting to non-stationary noise, maintaining stable system output even in complex meteorological conditions. (2) Explicit spectral control ensures compatibility between gravity anomaly signals and downstream processes like terrain correction, eliminating application barriers caused by the frequency-domain opacity of Kalman filtering.

This achievement marks a technological transition for airborne gravimetry from "single-dimensional noise reduction" to "dynamic-spectral co-optimization". The method has been implemented in the GIPS-1A gravimetry system and deployed across UAVs, helicopters, and fixed-wing aircraft platforms for industrial applications.

Author Contributions: Methodology, G.W. and F.Y.; Software, G.W.; Data curation, F.L.; Writing—original draft, G.W.; Writing—review & editing, G.W., F.Y., S.X. and X.Z.; Funding acquisition, L.W. All authors have read and agreed to the published version of the manuscript.

Funding: This work was funded by the National Key Research and Development Program of China (Grant No. 2023YFB3907703).

Institutional Review Board Statement: Not applicable.

Informed Consent Statement: Not applicable.

Data Availability Statement: The datasets presented in this article are not readily available because the project supporting this paper is currently ongoing; the original gravity data are subject to government-mandated confidentiality requirements and cannot be disclosed to the public.

Conflicts of Interest: The authors declare no conflict of interest.

References

1. Huang, M.T.; Deng, K.L.; Ouyang, Y.Z.; Wu, T.Q.; Zhai, G.J.; Lu, X.P.; Chen, X.; Liu, M.; Wang, X. Research advances in marine-airborne gravity survey and application technology. *Geomat. Inf. Sci. Wuhan Univ.* **2022**, *47*, 1635–1650.
2. Bolotin, Y.V.; Doroshin, D.R. Adaptive filtering of airborne gravimetry data using hidden Markov models. *Mosc. Univ. Mech. Bull.* **2011**, *66*, 63–68. [CrossRef]
3. Cai, S.K.; Wu, M.P.; Zhang, K.D. A comparison of digital lowpass FIR-filters in airborne gravimetry. *Geophys. Geochem. Explor.* **2010**, *34*, 74–78.
4. Lang, J.; Liang, X.; Liu, L.; Wang, G. Research on the Fourier basis pursuit low pass filter for airborne gravity. *Chin. J. Geophys.* **2018**, *61*, 4737–4745.
5. Pan, G.; Wang, M.; Cao, J.; Xiong, Z.; Yu, R.; Cai, S. Improving the Horizontal Components Accuracy of Strapdown Airborne Vector Gravimetry by Yaw Continuous Rotation Modulation. *IEEE Trans. Instrum. Meas.* **2021**, *70*, 8503611. [CrossRef]
6. Bolotin, Y.V.; Fedorov, A.V. Accuracy analysis of airborne gravimeter calibration using repeated flight paths. *Mosc. Univ. Mech. Bull.* **2008**, *63*, 60–67. [CrossRef]
7. Zheng, W.; Zhang, G.B. Application research on adaptive Kalman filtering for airborne gravity anomaly determination. *Chin. J. Geophys.* **2016**, *59*, 1275–1283.
8. Vyazmin, V.; Bolotin, Y.V. Two-Dimensional Kalman Filter Approach to Airborne Vector Gravimetry. *J. Geod. Sci.* **2019**, *9*, 87–96. [CrossRef]
9. Wang, G.X.; Luo, F.; Zhou, X.H.; Yan, F. Research on Kalman filter method for weak signal extraction of airborne gravity. *Geophys. Geochem. Explor.* **2021**, *45*, 76–83. [CrossRef]
10. Cai, T.J.; Fang, K. A New Method for Determining the Gravity Disturbance Vector in Strapdown Airborne Gravimetry. *IEEE Trans. Instrum. Meas.* **2022**, *71*, 9508607. [CrossRef]
11. Jensen, T.E. Spatial resolution of airborne gravity estimates in Kalman filtering. *J. Geod. Sci.* **2022**, *12*, 185–194. [CrossRef]
12. Ke, F.; Cai, T.J. An Algorithm for Strapdown Airborne Gravity Disturbance Vector Measurement Based on High-Precision Navigation and EGM2008. *Sensors* **2024**, *24*, 5899. [CrossRef] [PubMed]
13. Zou, X.L.; Cai, S.K.; Wu, M.P. The airborne gravity signal processing based on Kalman and FIR cascade filter. In Proceedings of the IEEE International Conference on Information and Automation, Ningbo, China, 1–3 August 2016; pp. 1996–2000.
14. Liu, L.T.; Xu, H.Z. Wavelets in airborne gravimetry. *Chin. J. Geophys.* **2004**, *47*, 490–494. [CrossRef]
15. Liu, L.; Liang, X.; Lang, J.; Shi, Z.; Yao, Y.; Hu, H.; Sheng, C. Improved function filter and its application in airborne gravimetry. *Acta Geod. Cartogr. Sin.* **2023**, *52*, 706–713. [CrossRef]
16. Wang, H.; Xiong, S.Q.; He, T.; WanYin, W. Noise suppression of airborne gravity data with minimum curvature model. *Chin. J. Geophys.* **2024**, *67*, 3959–3972. [CrossRef]
17. Kalman, R.E. A new approach to linear filtering and prediction problems. *J. Fluids Eng.* **1960**, *82*, 35–45. [CrossRef]
18. Forsberg, R.; Olesen, A.V. Airborne gravity field determination. In *Sciences of Geodesy*; Springer: Berlin, Germany, 2010; pp. 83–104.
19. Jiang, Z.X.; Zhang, H.; Guo, Z.H. Calculating method of internal coincidence accuracy in airborne gravimetry. *Geophys. Geochem. Explor.* **2010**, *34*, 672–676.
20. *DZ/T0381-2021*; Specification for Airborne Gravity Survey. Ministry of Natural Resources of China: Beijing, China, 2021.

Disclaimer/Publisher's Note: The statements, opinions and data contained in all publications are solely those of the individual author(s) and contributor(s) and not of MDPI and/or the editor(s). MDPI and/or the editor(s) disclaim responsibility for any injury to people or property resulting from any ideas, methods, instructions or products referred to in the content.

applied sciences

MDPI

Article

Analysis of the Genetic Mechanism of Thermal Anomaly in the A'nan Sag, Erlian Basin Based on 3D Magnetotelluric Imaging

Sen Wang [1], Wei Xu [1,*], Tianqi Guo [1], Wentao Duan [2] and Zhaoyun Wang [1]

[1] Institute of Global Environmental Change, Xi'an Jiaotong University, Xi'an 710049, China; senwang@stu.xjtu.edu.cn (S.W.); xjgtq@stu.xjtu.edu.cn (T.G.); wangzhaoyun@stu.xjtu.edu.cn (Z.W.)
[2] School of Remote Sensing and Geomatics Engineering, Nanjing University of Information Science and Technology, Nanjing 210044, China; eilen4011@nuist.edu.cn
* Correspondence: xwswpi@xjtu.edu.cn

Abstract: This study focuses on the genesis mechanism of thermal anomalies in the southwestern part of the Anan Depression in the Erlian Basin. Based on magnetotelluric 3D inversion data, a high-resolution electrical resistivity structure model was constructed, revealing the spatial configuration of deep heat sources and thermal pathways. The main conclusions are as follows: (1) Magnetotelluric 3D imaging reveals an elliptical low-resistivity anomaly (Anomaly C: 20 km × 16 km × 5 km, 0–5 Ωm) at depths of ~10–15 km. This anomaly is interpreted as a hypersaline fluid (approximately 400 °C, ~1.5% volume fraction, 3–5 wt.% NaCl), acting as the primary heat source. (2) Upward migration along F1/F3 fault conduits (10–40 Ωm) establishes a continuous pathway to mid-depth reservoirs D1/D2 (~5 km, 5–10 Ωm) and shallow crust. An overlying high-resistivity caprock (40–100 Ωm) seals thermal energy, forming a convective "source-conduit-reservoir-cap" system. (3) Integrated seismic data reveal that heat from the Abaga volcanic melt supplements Anomaly C via conduction through these conduits, combining with mantle-derived heat to form a composite source. This research delineates the interacting genesis mechanism of "deep low-resistivity heat source—medium-low resistivity fault conduit—shallow low-resistivity reservoir—relatively high-resistivity cap rock" in the southwestern A'nan Sag, providing a scientific basis for optimizing geothermal exploration targets and assessing resource potential.

Keywords: A'nan Sag; magnetotellurics; electrical resistivity; 3D inversion; geothermal resources

1. Introduction

Against the strategic backdrop of global energy transition, geothermal energy, as a crucial clean and renewable resource, has attracted growing strategic attention due to its abundant reserves, stable operation, and significant carbon reduction potential [1–5]. In recent years, China has implemented a series of intensive policies to pro-mote the standardized and scaled development of medium-deep geothermal resources [6–9]. Within this context, the Erlian Basin—distinguished by a prominent regional high-heat background—has emerged as a key target for medium-deep geothermal exploration in China [10–12]. Situated in the eastern segment of the Central Asian Orogenic Belt (CAOB), the Erlian Basin is a typical Mesozoic continental rift basin, with geothermal measurement revealing multiple high-heat-flow anomalies (e.g., 109.5 ± 17 mW/m^2 in the Manite Depression) [10,13]. Geophysical explorations, including Magnetotelluric (MT) surveys, further indicate the presence of deep-seated heat sources and thermal conductive pathways [14–16]. Specifically, in the southwestern A'nan Sag of this basin, measured terrestrial heat flow exceeds 100 mW/m^2, signaling a significant thermal anomaly and substantial potential for geothermal resource exploitation. However, the genetic mechanisms underlying this thermal anomaly remain unclear [10,11,16].

The MT technology, since its breakthrough in data processing and inversion algorithms in the 1980s, has become a pivotal tool for investigating deep electrical structures and key elements of geothermal systems [17–20]. For shallow thermal responses and heat transport mechanisms, Árnason et al. (2010) integrated MT with Transient Electromagnetic (TEM) data to reveal the linkage between shallow mineral alteration zones and deep magmatic heat sources in Iceland, establishing a quantitative model correlating resistivity with altered mineral content [21]; Araya et al. (2019) precisely delineated deep low-resistivity (<10 Ωm) geothermal fluid reservoirs via MT and elucidated high-permeability (>50 mD) fault-controlled thermal conduction-convection mechanisms connecting shallow systems to deep heat sources through anisotropy analysis [22]. Regarding thermal reservoir architecture and heat source localization, Gao et al. (2018) employed 3D MT inversion to construct a resistivity model of the Gonghe Basin, identifying the coupling relationship between heat sources and hydrothermal fluid channels [23]; He et al. (2016) combined AMT-MT joint exploration to comprehensively map the spatial distribution of thermal reservoirs, heat flow pathways, and magmatic heat source boundaries in the Mapamyum geothermal system [24]; Wu et al. (2018) systematically resolved the spatial configuration of the Zhangzhou Basin's thermal structure and its magmatic-hydrothermal genetic coupling using multidimensional MT datasets [25]. In deep thermal state inversion, Hata et al. (2015) derived the mantle wedge temperature field (800–1200 °C) and partial melt fraction (5–15%) in the Kyushu subduction zone by integrating 3D MT electrical models with thermodynamic boundary constraints [26]; Sheng et al. (2020) fused 3D electrical architecture, Hashin–Shtrikman boundary constraints, and Arrhenius rheological equations to reconstruct the upper mantle thermal evolution model in the eastern Gangdese, jointly inverting temperature gradient fields at 30–100 km depths with surface heat flow data [27].

Given the significant resource potential associated with the high-heat-flow anomaly in the southwestern A'nan Sag and the critical need to understand its controlling mechanisms for effective exploration targeting and regional geothermal assessment, this study employs 3D MT inversion technology to obtain a high-resolution sub-surface resistivity structure of the study area, providing critical electrical constraints for thermal anomaly genesis analysis. Based on the derived 3D resistivity model, we identify key structural features controlling thermal anomalies and establish a correlation model of geothermal system elements (integrating heat source, reservoirs, caprocks, and conduits) in the southwestern A'nan Sag, elucidating their spatial configuration. Finally, by integrating resistivity structures with the correlation model, we systematically analyze the nature of deep heat sources and dominant controlling factors in heat transfer processes, ultimately revealing the genetic mechanism of high-heat-flow anomalies. This research provides a geophysical basis for optimizing geothermal exploration targets in the southwestern A'nan Sag and establishes a scientific foundation for assessing the overall geothermal resource potential of the Erlian Basin.

2. Overview of Regional Geology

The Erlian Basin, situated in the north-central Inner Mongolia Plateau, covers a total area of ~10^5 km^2 [28]. This basin developed atop the suture zone between the Asian and Siberian plates, representing a typical NE-trending Cretaceous rift basin [10,28–30]. Its dominant NE-oriented tectonic framework controls the distribution of first-order structural units, forming a fundamental pattern termed "five depressions and one up-lift" [29]. The internal architecture is complex, with 53 sags and 21 uplifts arranged alternately, making it one of China's major onshore sedimentary basins [13,30,31]. The A'nan Sag, located in the northeastern part of the Manite Depression, is adjacent to the Sonid Up-lift to its southeast (Figure 1). This sag exhibits an elongated NE-trending morphology, characterized as a fault-sag superimposed depression with an area of ~2100 km^2 [32]. It ranks among the largest and most hydrocarbon-rich sags within the Erlian Basin [32–35]. Internally, the sag features complex structures, including multiple sub-sags and uplifts [32]. Among these, the A'nan Sub-sag serves as the primary trough, covering an area of ~375 km^2 with a maximum basement depth of ~4700 m [33]. The Late Paleozoic basement in the A'nan Sub-sag consists of arc-related tuffs and serpentinized ultramafic rocks, overlain by thick Lower Cretaceous and Cenozoic sedimentary sequences [34,36–38].

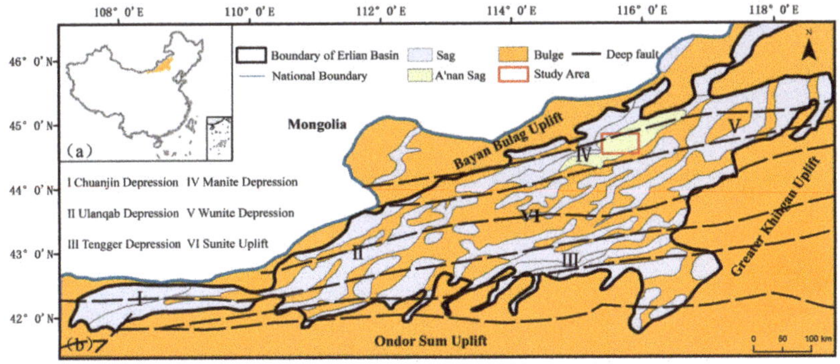

Figure 1. (**a**) The location of the Erlian Basin; (**b**) structural units division map of the Erlian Basin.

The tectonic evolution of the A'nan Sag has been jointly controlled by Yanshanian tectonic movements and the Paleozoic basement, resulting in the present-day structural framework (Figure 2a) [29,32]. This framework comprises three positive structural units (Jing-tewula Fault Zone, Aershan Fault Zone and A'nan Slope Belt) and four negative structural units (A'nan Sub-sag, Shadong Sub-sag, Mengdong Sub-sag and Mengxi Sub-sag) [33]. The evolutionary sequence can be divided into five main stages: (1) the Late Paleozoic–Triassic Basement Uplift and Denudation Phase, which established the crystalline basement framework [29]; (2) the Early Jurassic Initial Rifting Phase, forming fault-bounded lacustrine basins filled with coal-bearing clastic rocks [39]; (3) the Late Jurassic Compression–Thrusting Phase, characterized by basement-involved fold-thrust belts accompanied by basement-involved fold-thrust belts accompanied by predominantly basaltic to andesitic volcanic activity that modified the pre-existing structural configuration [29,40]; (4) the Early Cretaceous Intense Rifting and Subsidence Phase (Main Ba-sin-Forming Period), featuring sustained basement extension that generated half-graben systems with rapid accumulation of thick oil-bearing clastic rocks (Bayanhua Group, K_1b), including sub-stages of fault reactivation controlling secondary unit formation, migration of depocenters, widespread lake transgression, and shallow-water deposits in extensive lacustrine basins [36,41]; and (5) the Saihantala Formation (K_1bs) Decline and Post-rift Modification Phase, marked by regional uplift inducing lake-basin shrinkage transitioning to fluvial–swamp facies, with Late Yanshanian faulting and volcanic eruptions establishing the present-day structural geomorphology of the sag (Figure 2b) [15,29,42,43].

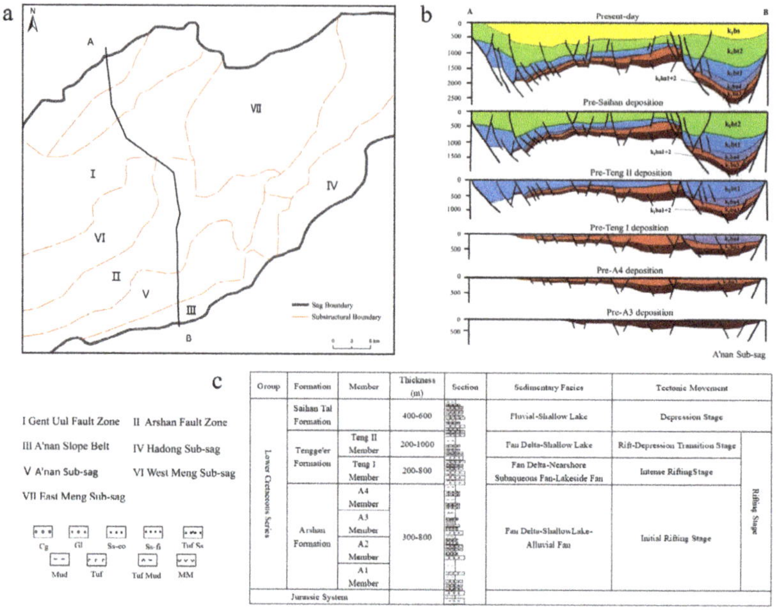

Figure 2. (**a**) Tectonic division map of the A'nan Sag, modified from Zhang et al. (2025) [44]; (**b**) Schematic geological sections across the A'nan Sag, refers to Sun et al. (2017) [45]; (**c**) generalized stratigraphic column of the A'nan Sag, modified from Sun et al. (2001) [46].

The A'nan Sag is a typical Mesozoic continental rift basin lacking Triassic deposits, with a maximum basement burial depth exceeding 3500 m (Figure 2c) [32]. The primary sedimentary cover consists of the Lower Cretaceous Bayanhua Group (K_1b), subdivided ascendingly into the Arshan Formation (K_1ba), Tenggeer Formation (K_1bt), and Saihantala Formation (K_1bs) [41,44]. The Paleozoic (PZ) fold basement comprises metamorphic sequences, tuffs (interbedded Late Paleozoic basic/intermediate volcanic rocks), and localized serpentinites and rhyolites, with intermediate tuffs exhibiting reservoir potential [29,32]. The Jurassic system (J) is dominated by volcaniclastic rocks and coal-bearing sandy mudstones, overlying the basement via an angular unconformity [40]. The Cretaceous Arshan Formation (K_1ba) records initial rift infill: its lower part (K_1ba^1) features conglomerates and variegated sandstones; the middle part (K_1ba^2) consists of grayish-green lacustrine mudstones; and the upper part (K_1ba^3) comprises light-gray breccias and sandy conglomerates interbedded with dark mudstones, serving as the primary oil-bearing and geothermal reservoir unit [13,36,47]. The Tenggeer Formation (K_1bt), deposited during the main syn-rift subsidence phase, includes the Tenggeer Member 1 (K_1bt^1), characterized by strongly reducing deep-to-semi-deep lacustrine facies, with gray sandstone-mudstone intercalations, tuff layers, and dark gray mudstones, constituting both the key hydrocarbon reservoir and regional seal; the Tenggeer Member 2 (K_1bt^2) formed in shallow-water extensive basins, with lower sandy conglomerates to fine sandstones and upper gray mudstones acting as a critical regional caprock [36,37,48]. The Saihantala Formation (K_1bs) reflects rift decline with fluvial-swamp facies, exhibiting interbedded conglomerates, sandstones, siltstones, and mudstones, contacting the underlying Tenggeer Formation via an angular unconformity [32]. The Cenozoic lacks Paleogene and Neogene strata; the Quaternary (Q) comprises grayish-yellow gravelly clays (alluvial-proluvial origin, <35 m thickness) with localized basalt lenses or thin layers, unconformably overlying pre-existing units [29].

3. Data Acquisition and Processing

3.1. MT Data Acquisition

To investigate the deep-seated genetic mechanism of thermal anomalies in the south-western A'nan Sag, this study systematically deployed 16 MT stations within the target area (44°30′–44°50′ N, 115°28′–115°52′ E; coverage 1024 km^2), organized into eight survey profiles (Figure 3a,b). With an average station spacing of ~8 km, a Phoenix Geophysics MTU-5A system was employed to acquire horizontal electric field components (Ex, Ey) and horizontal magnetic field components (Hx, Hy) (Figure 3c). Continuous observation at each station lasted for ≥20 h, with a frequency range spanning 0.001–320 Hz, enabling high-resolution imaging of deep-seated heat sources and heat-controlling structures. Field operations strictly adhered to the following technical protocols: (1) Site Selection and Interference Mitigation: Flat terrains (<10% slope) with homogeneous surficial soils were prioritized; high-voltage lines (>500 m buffer), rail-ways (>200 m buffer), high-population areas, and significant near-surface heterogeneities were strictly avoided. (2) Sensor Deployment and Calibration: Electrodes were arranged in a cross-shaped array, with Ex aligned to magnetic north and electrode spacing optimized between 20–100 m based on ground conditions; horizontal magnetic probes (Hx, Hy) were orthogonally buried at 25–30 cm depth with azimuthal deviations <1°, and all probes were leveled using precision spirit levels. (3) Grounding and Cable Layout: Electrode grounding resistance was maintained <2 KΩ through saline solution infusion and dry soil removal; signal cables were laid in a non-loop "serpentine pattern" to minimize inductive current interference. (4) Real-time Data Quality Control: Pre-acquisition tests (≥15 min) assessed ambient electromagnetic noise and instrument stability; continuous monitoring tracked electrode potential differences, magnetic field amplitudes, and real-time apparent resistivity-phase curves; data col-lection was halted immediately upon detecting anomalies (e.g., strong interference, electrode instability, curve distortion) for site reevaluation or sensor reconfiguration. This protocol, combining ultra-low-frequency signal acquisition (0.001 Hz) and systematic noise suppression measures, yielded high-quality MT datasets with optimal signal-to-noise ratios for deep subsurface probing. These data provide a robust foundation for subsequent 3D inversion to resolve key structures associated with thermal anomalies—such as low-resistivity fluid/melt migration pathways and heat-controlling fault zones—including geometry, spatial distribution, and resistivity attributes.

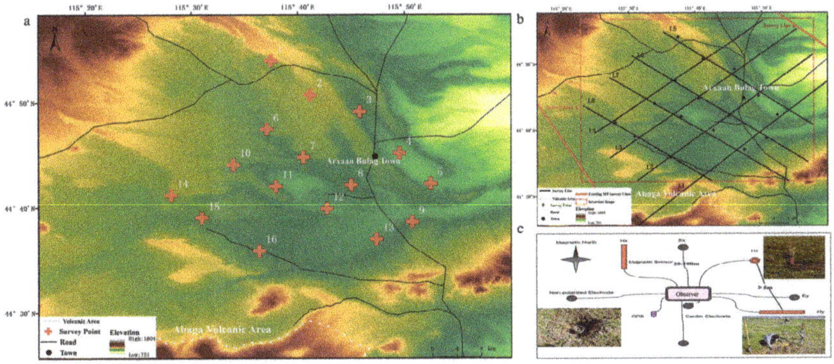

Figure 3. (**a**) Distribution of magnetotelluric stations in the study area; (**b**) layout of magnetotelluric survey lines in the study area; (**c**) schematic diagram of phoenix geophysics MTU-5A system deployment.

3.2. Preprocessing of MT Data

To acquire high-precision MT data for deep thermal structure inversion and thermal anomaly genesis analysis in the southwestern A'nan Sag, this study implemented systematic preprocessing on the original time-series data. First, the data were converted into the frequency domain via Fast Fourier Transform (FFT). Subsequently, the impedance tensor was estimated using the SSMT2000 software, with Robust statistical weighting applied to effectively suppress non-Gaussian noise interference such as electromagnetic pulses and industrial harmonics. Further, the remote reference technique was applied by deploying a reference station in a low-noise background area (west of Abaga Volcano, <200 km from the study area) with synchronized observations. Cross-spectral density analysis between the reference and local magnetic field components was used to reconstruct the impedance tensor, effectively reducing incoherent local noise. This workflow significantly enhanced the data signal-to-noise ratio, reducing scatter in apparent resistivity curves for heavily disturbed sites and decreasing phase curve distortion points by ~70%.

To eliminate outliers and abrupt transitions in apparent resistivity and phase curves induced by residual interference, this study performed spectral editing through manual interactive power spectrum selection in MTEditor software (https://www.phoenixgeophysics.com/, accessed on 10 January 2025). For each frequency point, outlier data points in the raw power spectral estimates were rigorously screened and removed, followed by precise calculation of the impedance tensor via Robust statistical averaging to export high-quality apparent resistivity and phase data. During editing, phase data were assigned higher weighting to ensure physical consistency and smoothness, due to their critical constraint on shallow thermal reservoir boundary in-version. Post-editing analysis demonstrated substantial data quality improvement at representative stations: scatter in apparent resistivity curves decreased by 35%, and phase curve distortion points reduced by ~80%. The integrated noise suppression workflow—combining robust processing, remote reference, and spectral editing—significantly enhanced the reliability and accuracy of the impedance tensor, establishing a robust data foundation for high-resolution characterization of deep heat sources, low-resistivity thermal anomaly conduits, and heat-controlling structures.

4. Electrical Structure Characteristics of the Southwestern A'nan Sag

4.1. 3D Inversion Methods

To resolve the electrical structure of the southwestern A'nan Sag with high fidelity, this study implemented high-resolution 3D inversion of 16 MT stations (42 frequencies spanning 0.000275–80 Hz) using the cloud computing system toPeak [49]. The inversion algorithm synergized dual engines of Occam inversion and Nonlinear Conjugate Gradient (NLCG) method within the ModEM framework. A homogeneous half-space with a background resistivity of 30 Ωm served as the initial model, and a multigrid optimization strategy was implemented to accelerate computation: coarse-grid inversion was first conducted to derive large-scale electrical features as prior constraints; subsequently, topological grid interpolation topologically mapped the prior model to a refined grid ($32 \times 32 \times 83$ cells, total 83,968) for fine-scale inversion. After 200 iterations, a high-precision 3D resistivity model with RMS = 1.27 was obtained (Figure 4).

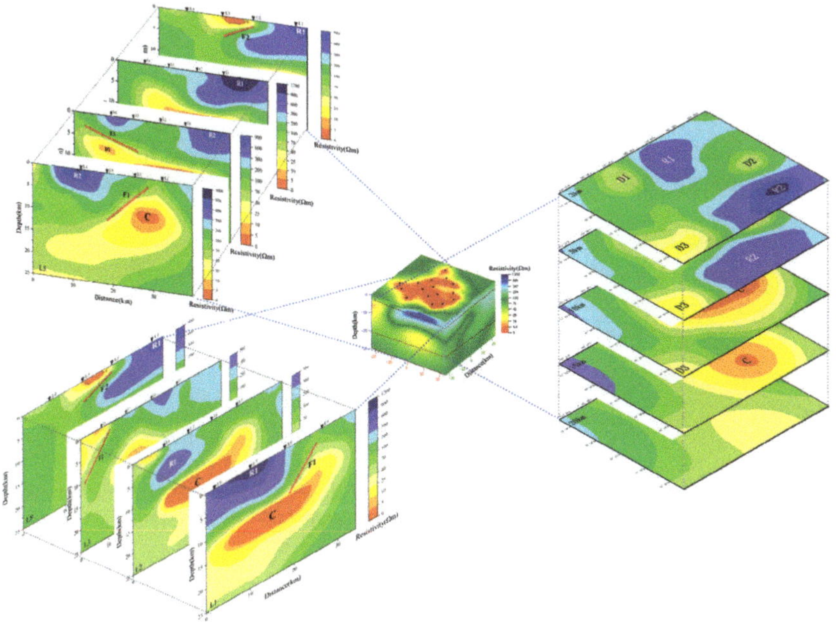

Figure 4. 3D Inverted electrical resistivity structure model of the study.

For the shallow subsurface (0.5–1 km depth), a spurious low-resistivity layer—attributed to precipitation-induced pore fluid saturation (distribution spatially correlated with stations, absent elsewhere)—was identified based on anomaly patterns, corroborated by regional meteorological data and resistivity spatial discontinuities. To minimize shallow interference artifacts in deep thermal structure inversion, spatial interpolation of reliable deep resistivity data reconstructed shallow electrical parameters, ensuring the 3D model accurately reflects the primary thermal structure. This approach provides robust resistivity constraints for precise identification of deep heat source pathways.

4.2. 3D Electrical Structure Characteristics Analysis

Based on the high-precision resistivity model obtained from 3D inversion, this study systematically reveals the electrical structure controlling thermal anomalies in the southwestern A'nan Sag by extracting horizontal slices at depths of 3 km, 5 km, 10 km, 15 km, and 20 km, combined with eight intersecting profiles (four NNE-striking lines L1–L4 and four NW-striking lines L5–L8) covering the core thermal anomaly zone (Figure 4). As illustrated in Figure 5, the horizontal slice analysis demonstrates significant vertical heterogeneity: (1) Shallow depths (3–5 km): Three low-resistivity zones (D_1, D_2, D_3) exhibit spatial segregation from high-resistivity areas, with D_1 and D_2 showing progressively decreasing resistivity values and laterally expanding coverage with in-creasing depth, ultimately forming a vertical interconnection with the ultra-low-resistivity Anomaly C (0–5 Ωm) at 10 km depth. (2) Mid-depth (10 km): Anomaly C attains its maximum spatial extent, while a regionally extensive moderate-low resistivity background (10–40 Ωm) dominates the area. (3) Deep levels (15–20 km): Anomaly C diminishes in scale, and the regional moderate-low resistivity back-ground expands laterally at 15 km depth but contracts vertically from 15–20 km, ac-companied by a significant reduction or attenuation of the shallow high-resistivity background (>200 Ωm), likely associated with enhanced deep-seated partial melting processes.

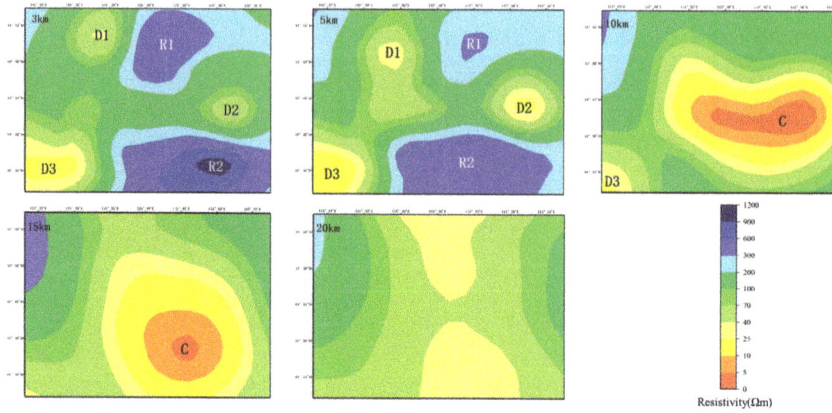

Figure 5. Horizontal slice maps at multiple depths of the 3D inversion model in the study area.

Vertical profile analysis further delineates the geometry of deep-seated heat sources (Anomalous Body C) and their structural connections. The NNE-striking profiles (L1–L4) reveal that: (1) Anomalous Body C (0–5 Ωm) is clearly defined at 10–15 km depths in profiles L1–L3; (2) It extends upward along faults (e.g., F_1 in L1, F_2 in L3), forming upward heat-conducting pathways while extending downward; (3) Its absence in the deep section of L4 indicates eastward pinch-out (Figure 6).

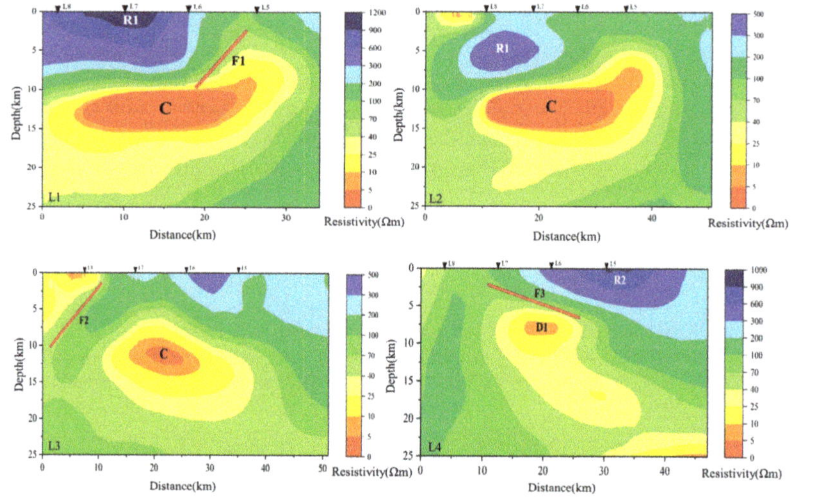

Figure 6. Cross sections along survey lines L1–L4 of the 3D inversion model in the study area.

The NW-striking profiles (L5–L8) demonstrate NNE-striking continuity of Anomaly C, with its spatial extent and geometry consistent with NNE-oriented profiles (L6 aligns with L1/L2), and reveal a shallow low-resistivity body D_1 with L6 correlating with L4 in delineating this feature. Profile L8 displays a moderate-low resistivity zone (10–50 Ωm; Figure 7) but lacks a distinct deep Anomaly C (0–5 Ωm). All profiles resolve shallow high-resistivity domains (R1/R2), with partial sections revealing thermally conductive faults (F1, F2, F3). This electrical structure provides critical constraints for delineating shallow geothermal reservoir boundaries, characterizing deep-seated heat-conducting pathways, and clarifying the thermo-tectonic framework.

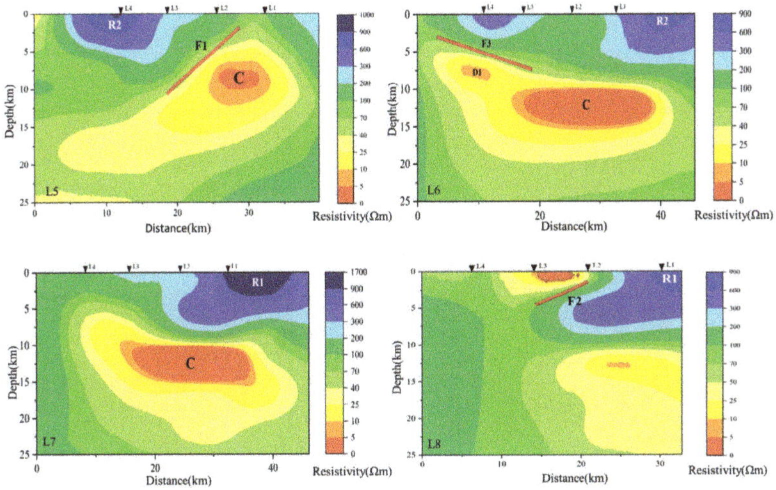

Figure 7. Cross sections along survey lines L5–L8 of the 3D inversion model in the study area.

5. Genetic Mechanism Analysis of Thermal Anomalies in the Southwestern A'nan Sag Structure Characteristics of the Southwestern A'nan Sag

5.1. Analysis of the Geothermal System in the Study Area

5.1.1. Geoelectrical Model Analysis of the Study Area

The high-precision resistivity model constructed from 3D MT inversion reveals that the southwestern A'nan Sag exhibits electrical signatures typical of a convective geothermal system dominated by hydrothermal convection as the primary heat transfer mechanism [50,51]. Fracture systems act as conduits for hydrothermal fluid migration, facilitating vertical transport of deep geothermal energy. Deep-seated thermal fluids ascend along fractures to heat host rock pore fluids, establishing convective circulation through mixing with shallow cold groundwater [51]. The spatial heterogeneity of electromagnetic responses is primarily controlled by the fracture network: high-salinity fluids saturating fracture networks generate pronounced low-resistivity anomalies (typically <10 Ωm), while hydrothermal alteration under elevated temperatures induces resistivity reduction of 1–2 orders of magnitude in altered zones relative to the background (>100 Ωm) [52].

The 3D resistivity model delineates key electrical units: A significant low-resistivity Anomaly C (0–5 Ωm) is developed at 10–15 km depth, which is connected to shallow levels via deep-rooted fault zones. Under hydrothermal fluid upwelling influence, a relatively low-resistivity layer (<10 Ωm) forms in shallow strata above these fault zones. Notably, this shallow conductive layer is capped by a rock unit with resistivity values of 40–100 Ωm—higher than underlying strata but lower than the regional background (>200 Ωm) at equivalent depths. The model's core characteristic lies in the effective spatial connectivity between deep Anomaly C and the shallow conductive layer through a moderate-low resistivity zone (10–40 Ωm). Vertical transects across fault zones reveal a sequential resistivity transition from deep to shallow: "Anomaly C—Moderate-low resistivity connectivity zone—Relatively low-resistivity layer—Relatively high-resistivity layer". This complete vertical resistivity stratification and spatial connectivity further confirm that the study area conforms to the geoelectrical signature of a typical convective geothermal system.

5.1.2. Analysis of Heat Source in the Geothermal System

The 3D resistivity model derived from inversion identifies a significant low-resistivity Anomaly C (0–5 Ωm) at 10–15 km depth. This anomaly exhibits effective electrical connectivity with the shallow low-resistivity layer (5–10 Ωm) via a moderate-low resistivity channel (10–40 Ωm), indicating that it likely constitutes the deep-seated heat source of the geothermal system. Potential mechanisms for crustal low-resistivity anomalies include partial melting, saline fluids, or conductive mineral phases [53,54]. Given the regional scale of Anomaly C, while metal oxides and graphite layers typically manifest as localized small-scale features, they can be discounted as primary contributors. Although the study area is adjacent to the Abaga Volcanic Field and partial melting is commonly invoked to explain crustal low-resistivity in volcanic settings [55–57], the temperature at the depth of Anomaly C (10–15 km) is ~400 °C according to the 3D thermal structure model of East Asia by Sun et al. (2022) [58]. This temperature is substantially lower than the solidus of mafic rocks, precluding partial melting as a viable mechanism. Integrating these eliminative analyses, the low-resistivity characteristics of Anomaly C are primarily attributed to saline fluids [59,60].

NaCl is the dominant solute in crustal fluids, with its mass fraction (wt%) defining the quantitative salinity metric [61]. The resistivity of saline fluids is primarily controlled by NaCl concentration while being modulated by temperature and pressure [61,62]. High-conductivity zones in the continental lower crust are commonly attributed to saline fluid accumulation, exhibiting NaCl concentrations of 0.2–7.0 wt% [63,64]. Anomaly C exhibits resistivity values of 0.1–5 Ωm (mean ~3 Ωm) at an in-situ temperature of ~400 °C (constrained by the East Asia thermal model) [58]. To quantify fluid salinity and content within Anomaly C, we employ the saline fluid resistivity relationship established by Dong et al. (2023) based on the Hashin–Shtrikman (H-S) model [14]. This function defines the quantitative correlation between rock resistivity, fluid volume fraction, NaCl salinity (1–5 wt%) at 400 °C and 400 MPa. As illustrated in Figure 8, for the mean resistivity of Anomaly C (3 Ωm), the required fluid volume fractions are approximately 1.44%, 2.11%, and 4.63% at NaCl salinities of 1 wt%, 3 wt%, and 5 wt%, respectively.

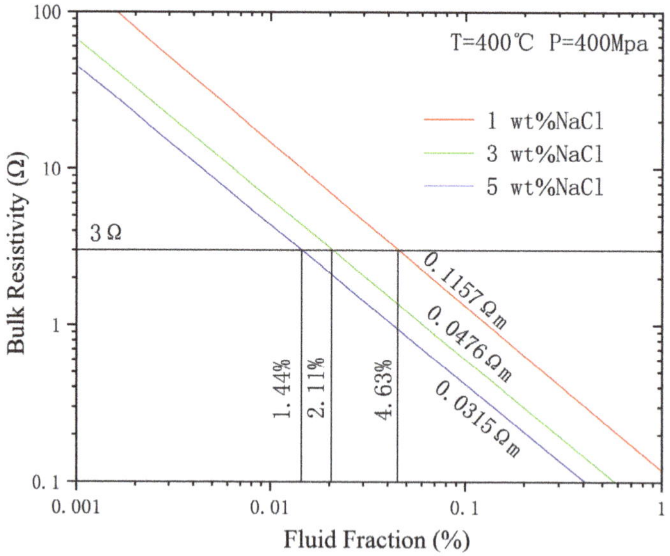

Figure 8. Relationship between electrical resistivity and fluid fraction for 1 wt%, 3 wt%, and 5 wt% nacl saline fluids, modified from Dong et al. (2023) [14].

To validate the H-S model estimates, this study further applies Archie's law to derive the porosity of the low-resistivity zone [65], assuming full saturation of rock voids, expressed as:

$$\gamma = \left(\frac{\rho}{\rho_w}\right)^{\frac{1}{m}} \quad (1)$$

where: γ: porosity; ρ: resistivity of the rock; ρ_w: resistivity of the pore fluid; m: cementation exponent, ranging between 1 and 2. Given that Anomaly C is situated at depths > 10 km under high-temperature–high-pressure conditions, the highly cemented rocks justify assigning $m = 1$.

Based on the 3D resistivity inversion model, the country rock resistivity is constrained to ~200 Ωm, while Anomaly C exhibits resistivity values of 0–5 Ωm (mean: 3 Ωm). Substituting these parameters into Archie's formula yields a porosity of ~1.5% for the low-resistivity zone. This value aligns with the range of 1.44% to 2.11% derived from the resistivity-salinity-fluid fraction relationship (Figure 8).

In summary, the electrical anomaly characteristics of low-resistivity Anomaly C can be reasonably explained by saline fluids with a fluid volume fraction of 1.5% and NaCl salinity of 3–5 wt%. T The in-situ temperature at anomaly depths is ~400 °C, and its sustained thermal energy contribution provides a stable deep heat source for regional geothermal resource accumulation. The porosity derived from Archie's law (1.50%) exhibits close agreement with the fluid fraction range (1.44–2.11 vol%) calculated by the H-S model at 3–5 wt% salinity, corroborating the reliability of this quantitative interpretation.

5.1.3. Thermal Conduits of the Geothermal System

The fault system functions as dual conduits for thermal conduction and hydro-thermal transport in the formation of thermal anomalies within the study area, acting as a key control via spatiotemporal coupling with heat sources and thermal reservoirs [51]. Thermal energy from deep heat sources is primarily vertically advected along deep-seated fault zones, directly heating fluids and country rocks to form localized thermal anomalies [66–68]. The heated fluids subsequently migrate upward along secondary fractures or damage zones, undergo thermo-chemical exchange with shallow groundwater, potentially triggering surface thermal manifestations such as hot springs [69]. Within the tectonic context of a rift basin, the deep-seated fault system also forms the framework for conductive geothermal systems, significantly enhancing the vertical transfer efficiency of deep thermal energy and shaping efficient thermal reservoirs at shallow depths [66,70]. Based on MT inversion results, the F1 and F3 faults in the southwestern A'nan Sag are identified as regionally dominant thermal conductive faults. Their deep-seated characteristics effectively connect the deep low-resistivity Anomaly C with shallow low-resistivity zones, constituting the primary channels for hydrothermal fluid migration. In contrast, the F2 fault, constrained by its boundary conditions and resistivity structure evidence, exhibits no direct connectivity with the deep low-resistivity Anomaly C.

5.1.4. Reservoir and Caprock of the Geothermal System

The thermal reservoir is characterized by shallow low-resistivity anomalies (5–10 Ωm), indicating high-permeability volcaniclastic conglomerates and fractured sandstones enriched with geothermal fluids, with secondary porosity dominated by tectonic fractures. In contrast, the overlying cap rock exhibits relatively high resistivity (40–100 Ωm); although substantially exceeding the underlying reservoir's resistivity, it remains markedly lower than the regional background resistivity at equivalent depths (>200 Ωm). Composed of low-permeability tight sandstones interbedded with argillaceous rocks, it functions as a thermal insulator through capillary pressure differentials and mineral thermal resistance effects, thereby inhibiting vertical heat dissipation and facilitating thermal energy accumulation within the reservoir.

In summary, thermal energy from the deep heat source (low-resistivity Anomaly C) is primarily transported upward via thermal convection along fault F3, driving the convergence of geothermal fluids at ~7 km depth to form a low-resistivity thermal reservoir. Notably, from ~5 km depth to the surface, resistivity systematically increases with decreasing depth (Figure 9a). This electrical structure reveals a distinct transition in heat transfer mechanisms: fluid-dominated convection prevails at depths >5 km, whereas rock-conduction becomes predominant at shallower depths (<5 km). Furthermore, the resistivity profile across fault F1 (Figure 9b,c) displays a consistent vertical sequence: "deep low-resistivity heat source—moderate-low resistivity fault conduit (F1)—shallow low-resistivity reservoir—relatively high-resistivity cap rock". The integrated electrical structure model formed by the F1 and F3 fault systems further confirms that the geothermal system in the study area aligns with the essential characteristics of a convective-type system.

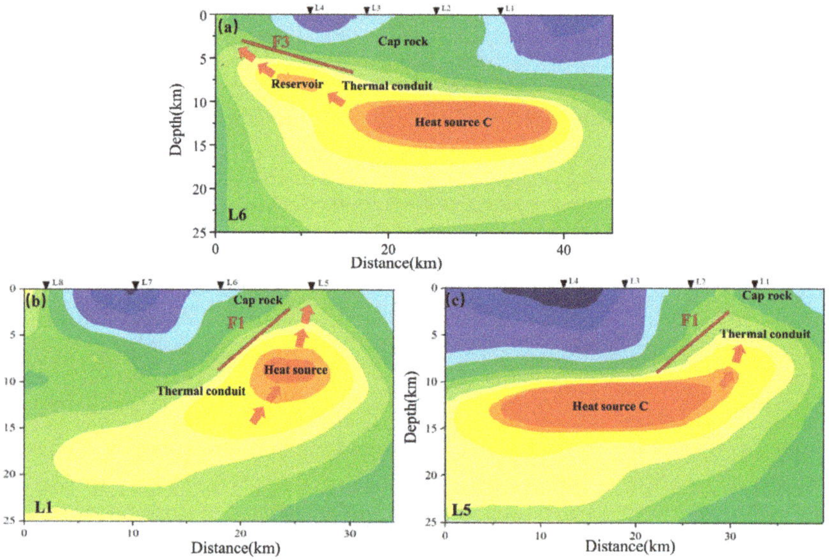

Figure 9. Geothermal system conceptual model of the study area. (**a**) profile l6-based structural-thermal model; (**b**) profile l1-based structural-thermal model; (**c**) profile l5-based structural-thermal model.

5.2. Analysis of Geothermal Anomaly Formation Mechanism

The deep heat source of the geothermal system corresponds to the prominent low-resistivity Anomaly C, with its thermal origin being pivotal to deciphering the genetic mechanism of regional thermal anomalies [15]. As a key parameter characterizing the electrical properties of the lithosphere, resistivity exhibits multi-parameter sensitivity to thermo-pressure regimes, mineral phase transitions, and structural architectures, rendering it an effective geophysical proxy for inverting deep thermal configurations [71–73]. MT inversion results from adjacent areas (Figure 10) reveal significant low-resistivity anomalies: Anomaly C_2 in southwestern Profile A spans 15–40 km depth, while Anomaly C_1 in northeastern Profile B develops at 10–40 km depth.

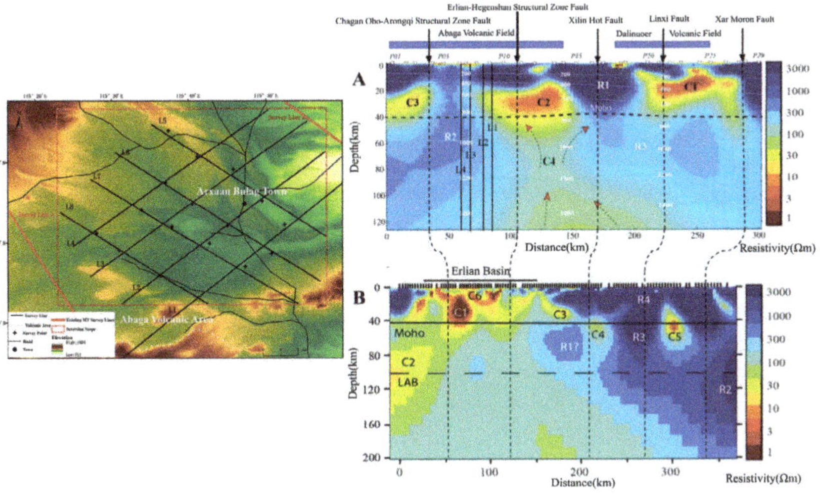

Figure 10. (**A**) Distribution of magnetotelluric survey lines in the study and adjacent areas; (**B**) resistivity inversion profiles of lines A and B, modified from Ye et al. (2022) [74].

The resistivity section of Line L8 (Figure 11a), striking subparallel to Profile A, reveals an electrical structure divisible into low-resistivity zones, relatively low-resistivity zones, high-resistivity zones, and relatively high-resistivity zones. Spatial correlation analysis confirms high consistency between Profile A and Line L8, jointly exhibiting: (1) Shallow low-resistivity zone, (2) Eastern high-resistivity zone, (3) Subjacent moderate-low resistivity zone, (4) Western moderate-high resistivity zone, and (5) Central transitional zone. Integrated with Line L1 resistivity distribution (Figure 11b), this study confirms spatial continuity between the study area's Anomaly C and the Abaga Volcanic Field Anomaly C_2, connected through a moderate-low resistivity corridor (10–40 Ωm).

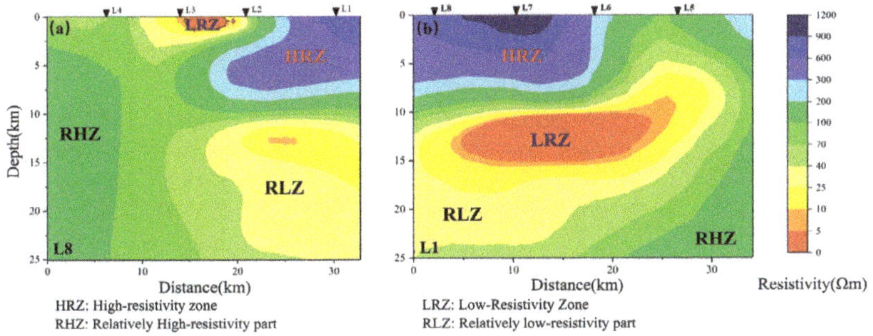

Figure 11. (**a**) Resistivity layering profile along survey line l8; (**b**) resistivity layering Profile along Survey Line L1.

Regarding the genesis of low-resistivity Anomaly C_2 in Profile A, saline fluids and partial melting constitute two primary explanatory mechanisms. However, Anomaly C is definitively attributed to hypersaline fluids. Critical differences exist in deep structural evidence, as S-wave velocity structure reveals extensive low-velocity anomalies at 20–40 km depth beneath the Abaga volcanic zone, whereas no significant velocity anomaly is detected at equivalent depths in the study area [75–78]. Liu et al.'s (2017) ambient noise tomography study further confirms that shear wave velocities within 15–70 km depth beneath this volcanic zone are significantly lower than the minimum value predicted by global orogenic belt models [79]. Additionally, Qiang et al. (2019) shear wave splitting observations indicate an isotropic medium response without detectable splitting, thus suggesting localized mantle upwelling [75]. Integrating its prominent low-resistivity characteristics and ambient thermobaric conditions (exceeding the wet basalt solidus >800 °C), Anomaly C_2 in the Abaga volcanic zone is conclusively attributed to partial melting, potentially underlain by a magma chamber. This geophysical evidence demonstrates that the partial melting layer and potential magma reservoirs at depth in the Abaga volcanic zone transmit mantle-derived melt heat and deep conductive heat to Anomaly C through a medium-low resistivity conduit (10–40 Ωm), forming a composite heat source mechanism. This hybrid mechanism provides sustained thermal energy flux to Anomaly C, driving the formation and evolution of regional thermal anomalies.

6. Conclusions

Based on the three-dimensional resistivity structure model of the southwestern A'nan Sag (Erlian Basin), constructed through (MT) data inversion and integrated analysis of key geothermal system elements and thermal anomaly genesis, the following conclusions are drawn:

(1) MT 3D inversion reveals an elliptical low-resistivity Anomaly C at depths of 10–15 km depth, with dimensions of ~20 km (long axis) × 16 km (short axis) × 5 km (thickness). Among the major faults (F1, F2, F3), Anomaly C extends upward along F1 and F3 faults, forming medium-low resistivity conduits (10–40 Ωm) that culminate in shallow relatively low-resistivity layers D1/D2 at ~5 km depth (5–10 Ωm). These interconnected features establish a vertically continuous hydrothermal migration pathway extending to 0–3 km depth.

(2) The deep-seated low-resistivity Anomaly C is primarily attributed to hypersaline fluids (1.50 vol%, 3–5 wt% NaCl) at ~400 °C, with an estimated temperature of ~400 °C, constituting the system's core heat source. The mid-depth low-resistivity layers D_1 and D_2 function as the thermal reservoirs, characterized by fracture networks saturated with geothermal fluids. Overlying these, a relatively high-resistivity layer (40–100 Ωm) forms an effective caprock through low-permeability tight sandstones that suppress vertical heat dissipation. Faults F1/F3 dominate thermal energy transport, establishing a coherent "heat source–conduit–reservoir–cap" architecture that aligns with the defining characteristics of a typical convective-type geothermal system.

(3) Integrated seismic S-wave velocity analysis reveals that the thermal energy derived from partial melt beneath the Abaga volcanic zone is supplied to Anomaly C via medium-low resistivity conduits (10–40 Ωm) primarily through thermal conduction. Superimposed with deep mantle conductive heat, forms a stable composite heat source. Driven by the fault systems, thermal energy is transported to shallow crustal levels through a coupled conduction-convection mode, generating and sustaining regional thermal anomalies.

Building on this initial analysis of the 3D resistivity structure and heat source mechanism, future research will focus on: (1) enhancing geological constraints through systematic exploration (core sampling, precision logging) to refine thermal reservoir modeling and quantify fault-controlled heat transfer; (2) advancing multi-physics coupling by integrating seismic velocity, gravity-magnetic data for joint inversion, combined with geochemical tracing of geothermal fluids, to resolve the spatiotemporal linkage between the deep heat source (Anomaly C) and fault systems (F1–F3); (3) establishing a volcano-geothermal coupling model centered on the Abaga volcanic zone, deploying integrated surveys to characterize the hierarchical development of shallow convective and mid-deep conductive-convective reservoirs, ultimately constructing a sustainable exploitation framework for regional geothermal resources.

Author Contributions: For research conceptualization, S.W. and W.X.; methodology, S.W. and T.G.; software, T.G. and S.W.; validation, S.W., W.D. and Z.W.; formal analysis, S.W.; investigation, T.G. and S.W.; data curation, T.G.; writing—original draft preparation, T.G. and S.W.; writing—review and editing, S.W.; visualization, S.W.; supervision, W.X., W.D. and S.W.; project administration, W.X.; funding acquisition, W.X. All authors have read and agreed to the published version of the manuscript.

Funding: This research was funded by the National Natural Science Foundation of China, Grant No. 42172335.

Informed Consent Statement: Not applicable.

Data Availability Statement: The data that support the findings of this study are available from the corresponding author (Wei Xu) upon reasonable request.

Acknowledgments: Our heartfelt gratitude is given to the editor and the reviewers for their scientific and linguistic revisions of the manuscript. We also sincerely thank Jiong Zhang (National Institute of Natural Hazards, Ministry of Emergency Management of China) and Ruoli Wang for his professional guidance on magnetotelluric data inversion in the A'nan Sag.

Conflicts of Interest: The authors declare no conflicts of interest.

References

1. Krieger, M.; Kurek, K.A.; Brommer, M. Global geothermal industry data collection: A systematic review. *Geothermics* **2022**, *104*, 102457. [CrossRef]
2. Van der Zwaan, B.; Dalla Longa, F. Integrated assessment projections for global geothermal energy use. *Geothermics* **2019**, *82*, 203–211. [CrossRef]
3. Bharathi N, K.; Abirami, M.; Vighneshwari, D.; Hariprasath, M. Techno-economic optimization of hybrid renewable systems for sustainable energy solutions. *Sci. Rep.* **2025**, *15*, 21114. [CrossRef]
4. Kaczmarczyk, M.; Sowiżdżał, A. Environmental and Social Dimensions of Energy Transformation Using Geothermal Energy. *Energies* **2025**, *18*, 3565. [CrossRef]
5. Ellabban, O.; Abu-Rub, H.; Blaabjerg, F. Renewable energy resources: Current status, future prospects and their enabling technology. *Renew. Sustain. Energy Rev.* **2014**, *39*, 748–764. [CrossRef]
6. Zhang, J.; Chen, L.; Sun, Y.; Xu, L.; Zhao, X.; Li, Q.; Zhang, D. Geothermal resource distribution and prospects for development and utilization in China. *Nat. Gas Ind. B* **2024**, *11*, 6–18. [CrossRef]
7. Poh, J.; Tjiawi, H.; Chidire, A.; Veerasamy, B.B.; Massier, T.; Romagnoli, A.; Wu, W.; Lu, D.; Lim, J.W.M.; Yang, L.; et al. Geothermal development in South, Southeast and East Asia: A review. *Renew. Sustain. Energy Rev.* **2025**, *209*, 115043. [CrossRef]
8. Wang, Y.; Liu, Y.; Dou, J.; Li, M.; Zeng, M. Geothermal energy in China: Status, challenges, and policy recommendations. *Util. Policy* **2020**, *64*, 101020. [CrossRef]
9. Zhu, J.L.; Hu, K.Y.; Lu, X.L.; Huang, X.X.; Liu, K.T.; Wu, X.J. A review of geothermal energy resources, development, and applications in China: Current status and prospects. *Energy* **2015**, *93*, 466–483. [CrossRef]
10. Xu, W.; Huang, S.; Zhang, J.; Zuo, Y.; Zhou, Y.; Ke, T.; Yu, R.; Li, Y. Geothermal gradient and heat flow of the Erlian Basin and adjacent areas, Northern China: Geodynamic implication. *Geothermics* **2021**, *92*, 102049. [CrossRef]
11. Xu, W.; Zhou, Y.; Zhang, J.; Li, Y. Thermo-tectonic evolution of the northern Erlian Basin (NE China): Evidence from fission track and (U–Th)/He thermochronology. *J. Asian Earth Sci.* **2023**, *248*, 105620. [CrossRef]
12. Zuo, Y.; Wang, C.; Tang, S.; Hao, Q. Mesozoic and Cenozoic thermal history and source rock thermal evolution of the Baiyinchagan sag, Erlian Basin, Northern China. *J. Pet. Sci. Eng.* **2016**, *139*, 171–184. [CrossRef]
13. Fu, S.; Liu, Z.; Song, L.; Ge, J.; Tian, N.; Yin, K.; Wang, X. A new division of late Cretaceous Erlian Basin tectonic units based on differential basement characteristics. *Geoenergy Sci. Eng.* **2025**, *245*, 213536. [CrossRef]
14. Dong, Z.; Zhan, Y.; Xiao, Q.; Li, N.; Han, B.; Sun, X.; Liu, X.; Tang, J. Electrical resistivity structure across the Late Cenozoic Abaga-Dalinor volcanic field, eastern China, from 3-D magnetotelluric imaging and its tectonic implications. *J. Asian Earth Sci.* **2023**, *258*, 105574. [CrossRef]
15. Ye, G.; Fu, H.; Jin, S.; Wei, W. Magnetotelluric study of the mechanism of the Abaga and Dalinor volcanic groups in Central Inner Mongolia, China. *Phys. Earth Planet. Inter.* **2020**, *308*, 106570. [CrossRef]

16. Zhang, J.; Huang, S.; Zuo, Y.; Zhou, Y.; Liu, Z.; Duan, W.; Wei, X. Terrestrial heat flow in the baiyinchagan sag, erlian Basin, northern China. *Geothermics* **2020**, *86*, 101799. [CrossRef]
17. Muñoz, G. Exploring for Geothermal Resources with Electromagnetic Methods. *Surv. Geophys.* **2014**, *35*, 101–122. [CrossRef]
18. Liu, Y.H.; Yin, C.C.; Cai, J.; Huang, W.; Ben, F.; Zhang, B.; Qi, Y.F.; Qiu, C.K.; Ren, X.Y.; Huang, X.; et al. Review on research of electrical anisotropy in electromagnetic prospecting. *Chin. J. Geophys.-Chin. Ed.* **2018**, *61*, 3468–3487. [CrossRef]
19. Tang, J.T.; Ren, Z.Y.; Zhou, C.; Zhang, L.C.; Yuan, Y.; Xiao, X. Frequency-domain electromagnetic methods for exploration of the shallow subsurface: A review. *Chin. J. Geophys.-Chin. Ed.* **2015**, *58*, 2681–2705. [CrossRef]
20. Thanassoulas, C. Geothermal-Exploration Using Electrical Methods. *Geoexploration* **1991**, *27*, 321–350. [CrossRef]
21. Árnason, K.; Eysteinsson, H.; Hersir, G.P. Joint 1D inversion of TEM and MT data and 3D inversion of MT data in the Hengill area, SW Iceland. *Geothermics* **2010**, *39*, 13–34. [CrossRef]
22. Araya Vargas, J.; Meqbel, N.M.; Ritter, O.; Brasse, H.; Weckmann, U.; Yáñez, G.; Godoy, B. Fluid Distribution in the Central Andes Subduction Zone Imaged with Magnetotellurics. *J. Geophys. Res. Solid Earth* **2019**, *124*, 4017–4034. [CrossRef]
23. Gao, J.; Zhang, H.; Zhang, S.; Chen, X.; Cheng, Z.; Jia, X.; Li, S.; Fu, L.; Gao, L.; Xin, H. Three-dimensional magnetotelluric imaging of the geothermal system beneath the Gonghe Basin, Northeast Tibetan Plateau. *Geothermics* **2018**, *76*, 15–25. [CrossRef]
24. He, L.; Chen, L.; Dorji; Xi, X.; Zhao, X.; Chen, R.; Yao, H. Mapping the Geothermal System Using AMT and MT in the Mapamyum (QP) Field, Lake Manasarovar, Southwestern Tibet. *Energies* **2016**, *9*, 855. [CrossRef]
25. Wu, C.; Hu, X.; Wang, G.; Xi, Y.; Lin, W.; Liu, S.; Yang, B.; Cai, J. Magnetotelluric Imaging of the Zhangzhou Basin Geothermal Zone, Southeastern China. *Energies* **2018**, *11*, 2170. [CrossRef]
26. Hata, M.; Uyeshima, M. Temperature and melt fraction distributions in a mantle wedge determined from the electrical conductivity structure: Application to one nonvolcanic and two volcanic regions in the Kyushu subduction zone, Japan. *Geophys. Res. Lett.* **2015**, *42*, 2709–2717. [CrossRef]
27. Sheng, Y.; Jin, S.; Lei, L.; Dong, H.; Zhang, L.; Wei, W.; Ye, G.; Li, B.; Lu, Z. Deep thermal state on the eastern margin of the Lhasa-Gangdese belt and its constraints on tectonic dynamics based on the 3-D electrical model. *Tectonophysics* **2020**, *793*, 228606. [CrossRef]
28. Bonnetti, C.; Malartre, F.; Huault, V.; Cuney, M.; Bourlange, S.; Liu, X.D.; Peng, Y.B. Sedimentology, stratigraphy and palynological occurrences of the late Cretaceous Erlian Formation, Erlian Basin, Inner Mongolia, People's Republic of China. *Cretac. Res.* **2014**, *48*, 177–192. [CrossRef]
29. Zhang, Y.; Liu, Z.; Fu, S.; Jiang, S.; Yao, N.; Wang, X. New understandings of the basement characteristics and evolution process of Erlian Basin. *Oil Geophys. Prospect.* **2019**, *54*, 404–416.
30. Yongqian, C.U.I.; Hengmao, T.; Xianping, L.I.; Zhen, L.I.U.; Hongbo, Z.; Changjiang, D.U. Tectonic Mechanism of the Erlian Basin during Early Yanshanian and Speculation on the Basin Prototype. *Acta Geol. Sin.* **2011**, *85*, 1265–1273.
31. Wang, Y. Coupling relationship of Mesozoic-Cenozoic Basin and Mountains in the Erlian Basin. *Spec. Oil Gas Reserv.* **2011**, *18*, 60–63,138.
32. Lu, W.; Song, T.; Li, X.; Yan, W.; Wang, J.; Zhong, X. Accumulation features and formation conditions of tight oil reservoir in A'nan Sag, Erlian Basin, China. *Nat. Gas Geosci.* **2016**, *27*, 1699–1708.
33. Dong, Y.; Ma, L.; Sun, F. Exploration prospects of low-mature oil/gas in the shallow layers of the A'nan Sag, the Erlian Basin. *Oil Gas Geol.* **2007**, *28*, 610–614,633.
34. Fu, X.; Zhang, T.; Wu, J.; Wang, X.; Zhou, J.; Jiang, T.; Li, B.; Li, C. Characteristics and main controlling factors of tight oil reservoirs in Cretaceous Tengger Formation, A'nan Sag, Erlian Basin. *Pet. Geol. Exp.* **2021**, *43*, 64–76.
35. Weiliang, D.U.; Xianping, L.I.; Yang, X.; Jiuqiang, Z.; Yahui, L.U.; Shunping, Y.U. Reverse Structures in Erlian Basin and Their Relations with Hydrocarbon. *Sci. Technol. Rev.* **2007**, *25*, 45–47.
36. Zhang, Y.; Guo, Y.; Guo, Y.; Shou, J.; Zhou, J.; Li, B.; Zhang, T.; Lan, B. Petrophysical property limits of effective tight oil reservoirs in the lower part of the first member of Cretaceous Tengger Formation, A'nan Sag, Erlian Basin, North China. *Pet. Geol. Exp.* **2016**, *38*, 551–558.
37. Xiang, X.; Dan, W.N.; Lu, S.F.; Zhang, P.F.; Li, B.; Cao, L.Z.; Wei, Y.; Li, X.W.; Fan, J.; Wang, J.J.; et al. Insight into Fluid Occurrence and Pore Structure of Lacustrine Shale from the Cretaceous Tengger Formation, A'nan Sag, Erlian Basin, China. *Energy Fuels* **2024**, *38*, 12960–12977. [CrossRef]
38. Dong, Y.; Chang, F.; Jiang, B.Y.; Li, C.; Dai, X.D.; Li, C.H. Study on Hydrocarbon Accumulation Law in Erlian Basin China. *Fresenius Environ. Bull.* **2021**, *30*, 6277–6284.
39. Wang, S.; Shao, L.Y.; Wang, D.D.; Sun, Q.P.; Sun, B.; Lu, J. Sequence stratigraphy and coal accumulation of Lower Cretaceous coal-bearing series in Erlian Basin, northeastern China. *AAPG Bull.* **2019**, *103*, 1653–1690. [CrossRef]
40. Guo, Z.X.; Zhao, X.Z.; Yang, Y.T.; Zhang, R.F.; Han, C.Y.; Shi, Y.P.; Jiang, S.Q.; Tao, M.H. Jurassic-earliest Cretaceous tectonostratigraphic evolution of the Erlian Basin, Northeast China: Records of polyphase intracontinental deformation in Northeast Asia. *Mar. Pet. Geol.* **2018**, *96*, 405–428. [CrossRef]

41. Sun, M.; Guo, Z.; Yang, Y.; Jiang, S.; Wang, H.; Wang, M. Upper Jurassic-Lower Cretaceous Stratigraphy in the Middle Wuliyasitai Depression, Erlian Basin: Evidence from Zircon U-Pb Dating. *J. Stratigr.* **2017**, *41*, 48–64.
42. Zhu, S.F.; Yue, H.; Zhu, X.M.; Sun, S.Y.; Wei, W.; Liu, X.; Jia, Y. Dolomitization of felsic volcaniclastic rocks in continental strata: A study from the Lower Cretaceous of the A'nan Sag in Er'lian Basin, China. *Sediment. Geol.* **2017**, *353*, 13–27. [CrossRef]
43. Shi, Z.W.; Cui, F.R.; Zhao, Z.D.; Bai, Z.D.; Chang, Z.G.; Zhao, J.X. First Optically Stimulated Luminescence and Radiocarbon Dating of the Late Quaternary Eruptions in the Xilinhot Volcanic Field, China. *Minerals* **2024**, *14*, 1181. [CrossRef]
44. Zhang, P.; Wang, J.; Lu, S.; Wu, H.; Liu, C.; Gao, W.; Li, W.; Zhou, N.; Chen, G.; Wu, C.; et al. Geochemical characteristics and paleo-environment of Lower Cretaceous lacustrine shale in the A'nan Sag, Erlian Basin, Northern China. *Mar. Pet. Geol.* **2025**, *178*, 107420. [CrossRef]
45. Sun, Z.; Qian, Z.; Lu, X.; Guo, Q.; Xu, J.; Shi, Y.; Hu, W. Reservoir property of the special lithologic section in the lower Tengger Formation of Anan depression, Erlian basin. *Geol. Bull. China* **2017**, *36*, 644–653.
46. Sun, Z.; Hong, Y.; Wu, Q.; Gan, C. Seismic characteristics of the gas reservoir in A'nan Sag of Erlian Basin. *Tianranqi Gongye Nat. Gas Ind.* **2001**, *21*, 26–29.
47. Xiao, B. Characteristics and favorable zone distribution of tuff reservoirt of Cretaceous in A'nan sag, Erlian Basin. *Lithol. Reserv.* **2024**, *36*, 135–148.
48. Xiong, H.Z.; Chen, K.Y.; Qi, Z.Y.; Chen, D.; Chen, W.L. Geochemical characteristics and genetic mechanism of dolomitic rocks in tengger formation in A'nan sag, China. *Chim. Oggi-Chem. TODAY* **2016**, *34*, 39–45.
49. Liu, Z.; Chen, X.; Cai, J.; Cui, T.; Zhao, G.; Tang, J.; Ouyang, B. The Design and Application of Topeak: A Threedimensional Magnetotelluric Inversion Cloud Computing System. *Seismol. Geol.* **2022**, *44*, 802–820.
50. Zhang, Y.; Feng, J.; He, Z.; Li, P. Classification of geothermal systems and their formation key factors. *Earth Sci. Front.* **2017**, *24*, 190–198.
51. Wang, G.; Lin, W. Main hydro-geothermal systems and their genetic models in China. *Acta Geol. Sin.* **2020**, *94*, 1923–1937.
52. Zhang, L.; Huang, S.P.; Fu, R.; Tang, X.Y. Application of magnetotellurics in geothermal exploration and research in volcano areas. *Acta Petrol. Sin.* **2017**, *33*, 279–290. [CrossRef]
53. Niu, Y.L. Lithosphere thickness controls the extent of mantle melting, depth of melt extraction and basalt compositions in all tectonic settings on Earth—A review and new perspectives. *Earth-Sci. Rev.* **2021**, *217*, 103614. [CrossRef]
54. Liu, S.Y.; Xu, Y.X.; Yang, B.; Guo, Z.Q.; Shi, Y.; Liu, Y. Deciphering Fine Electrical Conductivity Structures in the Crust from MT Data Using the Equivalent Conductivity Formula. *J. Geophys. Res.-Solid Earth* **2021**, *126*, e2021JB022519. [CrossRef]
55. Comeau, M.J.; Unsworth, M.J.; Cordell, D. New constraints on the magma distribution and composition beneath Volcan Uturuncu and the southern Bolivian Altiplano from magnetotelluric data. *Geosphere* **2016**, *12*, 1391–1421. [CrossRef]
56. Gao, J.; Zhang, H.J.; Zhang, S.Q.; Xin, H.L.; Li, Z.W.; Tian, W.; Bao, F.; Cheng, Z.P.; Jia, X.F.; Fu, L. Magma recharging beneath the Weishan volcano of the intraplate Wudalianchi volcanic field, northeast China, implied from 3-D magnetotelluric imaging. *Geology* **2020**, *48*, 913–918. [CrossRef]
57. Wen, L.M.; Kang, G.F.; Bai, C.H.; Gao, G.M. Studies on the relationships of the Curie surface with heat flow and crustal structures in Yunnan Province, China, and its adjacent areas. *Earth Planets Space* **2019**, *71*, 85. [CrossRef]
58. Sun, Y.J.; Dong, S.W.; Wang, X.Q.; Liu, M.A.; Zhang, H.; Shi, Y.L. Three-Dimensional Thermal Structure of East Asian Continental Lithosphere. *J. Geophys. Res.-Solid Earth* **2022**, *127*, e2021JB023432. [CrossRef]
59. Feucht, D.W.; Sheehan, A.F.; Bedrosian, P.A. Magnetotelluric Imaging of Lower Crustal Melt and Lithospheric Hydration in the Rocky Mountain Front Transition Zone, Colorado, USA. *J. Geophys. Res.-Solid Earth* **2017**, *122*, 9489–9510. [CrossRef]
60. Sun, Y.J.; Dong, S.W.; Liu, M.; Zhang, H.; Shi, Y.L. The rheological structure of East Asian continental lithosphere. *Tectonophysics* **2025**, *895*, 230575. [CrossRef]
61. Sinmyo, R.; Keppler, H. Electrical conductivity of NaCl-bearing aqueous fluids to 600 °C and 1 GPa. *Contrib. Mineral. Petrol.* **2017**, *172*, 4. [CrossRef]
62. Guo, H.H.; Keppler, H. Electrical Conductivity of NaCl-Bearing Aqueous Fluids to 900°C and 5GPa. *J. Geophys. Res.-Solid Earth* **2019**, *124*, 1397–1411. [CrossRef]
63. Heinrich, C. The physical and chemical evolution of low-salinity magmatic fluids at the porphyry to epithermal transition: A thermodynamic study. *Miner. Depos.* **2005**, *39*, 864–889. [CrossRef]
64. Sakuma, H.; Ichiki, M. Electrical conductivity of NaCl-H_2O fluid in the crust. *J. Geophys. Res.-Solid Earth* **2016**, *121*, 577–594. [CrossRef]
65. Ucok, H.; Ershaghi, I.; Olhoeft, G.R. Electrical-Resistivity of Geothermal Brines. *J. Pet. Technol.* **1980**, *32*, 717–727. [CrossRef]
66. Wang, G.; Lin, W.; Liu, F.; Gan, H.; Wang, S.; Yue, G.; Long, X.; Liu, Y. Theory and survey practice of deep heat accumulation in geothermal system and exploration practice. *Acta Geol. Sin.* **2023**, *97*, 639–660.
67. Jiang, G.; Wang, Y.; Hu, J.; Zhang, C.; Wang, Y.; Zuo, Y.; Tang, X.; Ma, F.; Hu, S. Medium-high temperature geothermal resources in China: Exploration directions and optimizing prospecting targets. *Sci. Technol. Rev.* **2022**, *40*, 76–82.
68. Wang, G.; Lin, W. The thermal status of Chinas land areas and heat-control factors. *Earth Sci. Front.* **2024**, *31*, 1–18.

69. Qiu, N.; Tang, B.; Zhu, C. Deep thermal background of hot spring distribution in the Chinese continent. *Acta Geol. Sin.* **2022**, *96*, 195–207.
70. Liu, X.H.; Liu, J.X.; Ma, Z.Q.; Wang, Y.X.; Zhang, H. Development simulation in Gucheng County, Hebei Province: Comparative study of the thermal storage development of clastic and carbonate rocks. *Nat. Gas Ind. B* **2024**, *11*, 95–105. [CrossRef]
71. Zhou, Z.K.; Wang, D.L.; Wang, T.Q.; Weng, A.H.; Li, Y.B.; Guo, J.H.; Wang, X.Q. Undulating electrical lithosphere-asthenosphere boundary beneath Northeast China; as revealed by long-period magnetotelluric data. *Tectonophysics* **2023**, *851*, 229770. [CrossRef]
72. Mu, Q.; Han, J.T.; Liu, L.J.; Hou, H.S.; Liu, W.Y.; Liu, G.X.; Liang, H.D.; Wu, Y.H.; Xin, Z.H. The deep section cutting through the Xing'an Mongolia Orogenic Belt: Lithospheric resistivity structure and modification model. *Chin. J. Geophys.-Chin. Ed.* **2023**, *66*, 1603–1617. [CrossRef]
73. Hanneson, C.; Unsworth, M.J.; Currie, C.A. Lithospheric thickness and the backarc-craton lithosphere step in southwestern Canada imaged by a 3D magnetotelluric study. *Can. J. Earth Sci.* **2025**, *62*, 843–864. [CrossRef]
74. Ye, G.F.; Xia, Z.H.; Unsworth, M.; Wei, W.B.; Jin, S.; Liu, Z.L. Ongoing Asthenospheric Upwelling and Delamination-Style Down-Welling Beneath Northeast China: Evidence from High-Resolution Magnetotelluric Profiles. *J. Geophys. Res.-Solid Earth* **2022**, *127*, e2021JB022100. [CrossRef]
75. Qiang, Z.Y.; Wu, Q.J. Upper mantle anisotropy beneath Abag area in Inner Mongolia from shear wave splitting. *Chin. J. Geophys.-Chin. Ed.* **2019**, *62*, 2510–2526. [CrossRef]
76. He, J.; Wu, Q.J.; Zhang, R.Q.; Lei, J.S. Crustal structure beneath the Abaga area of Xing'an-Mongolia Orogenic Belt using teleseismic receiver functions. *Chin. J. Geophys.-Chin. Ed.* **2018**, *61*, 3676–3688. [CrossRef]
77. Xu, H.; Wu, Q. Crustal S-wave velocity structure of the Abaga area of Xing'an-Mongolia Orogenic Belt from ambient noise. *Acta Seismol. Sin.* **2025**, *47*, 54–72.
78. Hou, J.; Wu, Q.J.; Yu, D.X.; Ye, Q.D.; Zhang, R.Q. Study on surface-wave tomography in Abaga volcanic area, Inner Mongolia. *Front. Earth Sci.* **2023**, *11*, 1131393. [CrossRef]
79. Liu, Y.N.; Niu, F.L.; Chen, M.; Yang, W.C. 3-D crustal and uppermost mantle structure beneath NE China revealed by ambient noise adjoint tomography. *Earth Planet. Sci. Lett.* **2017**, *461*, 20–29. [CrossRef]

Disclaimer/Publisher's Note: The statements, opinions and data contained in all publications are solely those of the individual author(s) and contributor(s) and not of MDPI and/or the editor(s). MDPI and/or the editor(s) disclaim responsibility for any injury to people or property resulting from any ideas, methods, instructions or products referred to in the content.

applied sciences

MDPI

Article

A Scalable and Consistent Method for Multi-Component Gravity-Gradient Data Processing

Larissa Silva Piauilino, Vanderlei Coelho Oliveira Junior and Valeria Cristina Ferreira Barbosa *

Observatório Nacional, Gal. José Cristino, 77, Rio de Janeiro 20921-400, RJ, Brazil; larissasilva@on.br (L.S.P.); vanderlei@on.br (V.C.O.J.)

* Correspondence: valcris@on.br

Abstract: We demonstrate the potential of using the convolutional equivalent layer to jointly process large gravity-gradient datasets. Based on the equivalent-layer principle, we assume a single fictitious physical property distribution on a planar layer can approximate all components of the gravity-gradient tensor. Estimating this distribution using the classical technique ensures physical consistency among components. However, the classical approach becomes computationally prohibitive for large datasets due to the need to solve a large-scale inversion with a massive sensitivity matrix. To overcome this limitation, we exploit the block-Toeplitz Toeplitz-block structure of the sensitivity matrix for data on a regular horizontal grid. This structure significantly reduces computational cost—by orders of magnitude—compared to the classical method. Applications to synthetic and real datasets show that our method offers a computationally efficient alternative for processing large gravity-gradient data from various acquisition systems (AGG and FTG), even when data are irregularly spaced or flight lines are misaligned. On a standard laptop configuration, our method processed over 290,000 AGG data points in a few tens of seconds. It also handled between 726,000 FTG and 1,250,000 AGG data points within seconds to a couple of minutes, demonstrating practical computational efficiency for large-scale datasets.

Keywords: large-scale gravity-gradient data; jointly processing; FTG andFalcon AGG acquisition systems; equivalent-layer technique; reduced floating-point operations; Kauring gravity gradiometry test site; Vinton salt dome; Canobie airborne gravity survey

1. Introduction

The advancement of airborne gradiometry allows the use of gravity-gradient data to facilitate the interpretation of gravity sources in mining and hydrocarbon explorations (e.g., [1–8]). The equivalent-layer and Fast Fourier Transform (FFT) techniques are feasible approaches for de-noising and/or transforming between conventional gravity and gravity-gradient components. For example, refs. [9,10] developed techniques to calculate the full gravity-gradient tensor from pre-existing vertical components of the gravitational attraction (g_z-component) by using FFT and a fast equivalent-layer technique, respectively. Ref. [11] used a data-adaptive 1D wavelet filtering technique to de-noise gravity-gradient data. Ref. [12] also computed the tensor components from the g_z-component; however, using the cosine transform technique shows it is less affected by noise than the FFT approach. Ref. [13] presented the first method to assess and highlight the internal consistency of airborne gravity-gradient components by representing all measured data through a common potential function that accounts for both lateral and vertical variations. Ref. [14] introduced a de-noising method for full-tensor gravity-gradiometer data using translation-invariant wavelets with mixed and adaptive Bayesian thresholding to effectively suppress noise while preserving high-frequency details and minimizing artifacts. Ref. [15] presented a joint inversion algorithm for gravity and magnetic data based on a sequential strategy and cross-gradient structural coupling, deriving a novel formula for minimizing the structural similarity term by combining an alternating strategy with the randomized singular value decomposition algorithm to improve computational performance. Ref. [16] proposed a real-time joint filtering method for gravity and gravity-gradient data that uses the derivative relationship between the two to construct the system and measurement equations of a Kalman filter. An improved Kalman filter is then applied, enabling simultaneous noise reduction in both datasets while ensuring that the filtered outputs satisfy the Laplace equation constraint. By using variational mode decomposition and the Hilbert transform to extract time-frequency features of airborne gravity-gradient data, ref. [17] applied a time-frequency feature threshold to suppress noise while preserving high-frequency features of the data.

The equivalent layer is a well-known method based on the potential theory. Following this theory, a set of potential-field data produced by a 3D source can be reproduced by a 2D infinite layer. In practice, ref. [18] showed that this layer can be represented by a finite and discrete physical-property distribution (e.g., point masses). The classical approach consists of estimating the physical-property distribution by solving a linear system subject to fitting the observed data. Nevertheless, the processing of a large potential-field dataset by using the classical equivalent-layer method may lead to costly computational schemes. As a result, some authors search for ways to increase the computational efficiency of the equivalent-layer technique, (e.g., [5,19–32]).

From the above-mentioned authors, there are just a few works for processing measured gravity-gradient measurements. Ref. [24] were the first to use the equivalent-layer technique for processing gravity-gradient data. Ref. [27] compared the equivalent-layer technique and FFT for processing airborne full-tensor gravity-gradient data. Ref. [5] used the equivalent-layer technique for de-noising gravity-gradient data and estimating the errors associated with the de-noised data. Ref. [29] combined the Gauss–FFT with Landweber's iteration to propose a fast equivalent layer technique for processing two-component of the gravity-gradient data simultaneously.

Recently, refs. [33,34] developed an efficient and rapid equivalent layer technique for processing gravity and magnetic data by using an equivalent source to carry the calculations of the forward problem of iterative methods to estimate the physical property distribution through a 2D discrete convolution. This method reduces the memory RAM usage taking advantage of the symmetry in Block-Toeplitz Toeplitz-block (BTTB) contained in the sensitivity matrices. However, it is important to note that neither [33] nor [34] explicitly mentioned that this specific structure, the BTTB form, emerges only when the sensitivity matrix is defined in the coordinate system of the observed data. Consequently, all computational operations in this study were performed in the data reference frame, which is aligned with the acquisition geometry axes.

We demonstrate an efficient equivalent layer technique for jointly processing gravity-gradient data by exploiting the block Toeplitz with Toeplitz blocks (BTTB) structure of the sensitivity matrices. Our approach combines the conjugate gradient least squares (CGLSs) algorithm with 2D FFT, following a structure similar to the method proposed by [34]. By estimating a single equivalent layer, our method is capable of handling all available gravity-gradient components simultaneously. Although inspired by the same general objective as [34], our formulation differs in key aspects. In [34], the sensitivity matrix is constructed as a weighted sum of directional components derived from magnetic data. In contrast, our method uses gravity-gradient observations and constructs the sensitivity matrix by stacking contributions from each measured component. Additionally, unlike the method proposed in [34], which assumes acquisition geometries aligned with north–south and east–west directions, our approach is designed to remain effective even when the flight lines follow rotated geometries. Furthermore, it can handle data distributed over irregularly spaced points or uneven observation surfaces. Tests on both synthetic and real datasets confirm the robustness and flexibility of the proposed method under a variety of acquisition configurations.

2. Methodology

2.1. Gravity-Gradient Data

Consider a right-handed Cartesian coordinate system with the x and y oriented horizontally, pointing north and east, respectively, and the z-axis pointing vertically downward. Gravity-gradient data are commonly defined in terms of second derivatives $g_i^{\alpha\beta}$, $\alpha, \beta \in \{x, y, z\}$, of the gravitational potential at the ith observation point (x_i, y_i, z_i), $i \in \{1, \ldots, N\}$. In the source-free region, these second derivatives form a symmetric 3×3 gravity-gradient tensor given by

$$\mathbf{G}_i = \begin{bmatrix} g_i^{xx} & g_i^{xy} & g_i^{xz} \\ g_i^{xy} & g_i^{yy} & g_i^{yz} \\ g_i^{xz} & g_i^{yz} & g_i^{zz} \end{bmatrix}, \tag{1}$$

with trace equal to zero, i.e., $g_i^{xx} + g_i^{yy} + g_i^{zz} = 0$, due to Laplace's equation. This means that there are five linearly independent components g_i^{xx}, g_i^{xy}, g_i^{xz}, g_i^{yy}, and g_i^{yz} [35].

The two most common airborne systems that can provide gravity-gradient data are the Full Tensor Gradiometry (FTG) system and the single axial Falcon airborne gravity gradiometer (AGG) system. The FTG system measures five linearly independent components of the gravity-gradient tensor (i.e., g_i^{xx}, g_i^{xy}, g_i^{xz}, g_i^{yy} and g_i^{yz}). The Falcon AGG system measures the two horizontal differential curvature gradients g_i^{xy} and $g_i^{uv} = (g_i^{yy} - g_i^{xx})/2$ [36].

2.2. Classical Equivalent Layer

Let $\mathbf{g}^{\alpha\beta}$ be an $N \times 1$ vector whose ith element is a gravity-gradient measurement of the $\alpha\beta$-component $g_i^{\alpha\beta}$, $\alpha, \beta \in \{x, y, z\}$, at the observation point (x_i, y_i, z_i), $i \in \{1, \ldots, N\}$. According to the equivalent-layer principle, the observed data vector $\mathbf{g}^{\alpha\beta}$ can be approximated by

$$\mathbf{d}^{\alpha\beta}(\mathbf{m}) = \mathbf{H}^{\alpha\beta}\mathbf{m}\,, \quad \alpha, \beta \in \{x, y, z\}\,, \tag{2}$$

where $\mathbf{d}^{\alpha\beta}$ is the predicted gravity-gradient data vector of size $N \times 1$ due to a discrete set of M point masses located below the observation points, at the coordinates (x_j, y_j, z_c), $j \in \{1, \ldots, M\}$, on a horizontal plane with constant vertical coordinate $z_c > z_i$, $\forall i \in \{1, \ldots, N\}$. In Equation (2), $\mathbf{m}$ is the $M \times 1$ parameter vector whose jth element is the mass of the jth point mass and $\mathbf{H}^{\alpha\beta}$ is an $N \times M$ sensitivity matrix whose element ij is defined as

$$h_{ij}^{\alpha\beta} = \begin{cases} c_g \gamma \dfrac{3(\alpha_i - \alpha_j)^2 - r_{ij}^2}{r_{ij}^5} & \text{if} \quad \alpha = \beta \\ c_g \gamma \dfrac{3(\alpha_i - \alpha_j)(\beta_i - \beta_j)}{r_{ij}^5} & \text{if} \quad \alpha \neq \beta \end{cases}, \tag{3}$$

where $\alpha, \beta \in \{x, y, z\}$, $c_g = 10^9$ is a transformation constant from $1/\text{s}^2$ to eotvos, γ is the gravitational constant in $\text{m}^3/(\text{kg s}^2)$, and r_{ij} is the Euclidean distance between the ith observation point and the jth point mass given by

$$r_{ij} = \left[(x_i - x_j)^2 + (y_i - y_j)^2 + (z_i - z_c)^2\right]^{\frac{1}{2}}. \tag{4}$$

The classical equivalent-layer technique applied to gravity-gradient data involves estimating a single mass distribution $\hat{\mathbf{m}}$ that simultaneously fits all the $\alpha\beta$-components of the gravity-gradient data. It can be accomplished by minimizing the data-misfit function

$$\Phi(\mathbf{m}) = \frac{1}{N_t}\|\mathbf{g}^o - \mathbf{H}\,\mathbf{m}\|_2^2, \tag{5}$$

which is subjected to some constraint, where N_t is the total number of observations, $\|\cdot\|_2$ denotes the Euclidean norm, $\mathbf{g}^o$ is an observed data vector containing all measured $\alpha\beta$-components of the gravity-gradient data, and $\mathbf{H}$ is a sensitivity matrix depending on the specific airborne system. For example, if we consider gravity-gradient data measured by the FTG system, $\mathbf{g}^o$ is a $N_t \times 1$ vector given by

$$\mathbf{g}^o = \left[\mathbf{g}^{xx\top}, \mathbf{g}^{xy\top}, \mathbf{g}^{xz\top}, \mathbf{g}^{yy\top}, \mathbf{g}^{yz\top}\right]^\top, \tag{6}$$

and $\mathbf{H}$ is the $N_t \times M$ matrix composed by the sub-matrices $\mathbf{H}^{\alpha\beta}$, $\alpha, \beta \in \{x, y, z\}$ (Equation (3))

$$\mathbf{H} = \left[\mathbf{H}^{xx\top}, \mathbf{H}^{xy\top}, \mathbf{H}^{xz\top}, \mathbf{H}^{yy\top}, \mathbf{H}^{yz\top}\right]^\top, \tag{7}$$

where $N_t = 5N$. Otherwise, if we consider the two components measured by the Falcon AGG system, $\mathbf{g}^o$ is a $N_t \times 1$ vector given by

$$\mathbf{g}^o = \begin{bmatrix}\mathbf{g}^{xy} \\ \mathbf{g}^{uv}\end{bmatrix}, \tag{8}$$

and $\mathbf{H}$ is the $N_t \times M$ matrix

$$\mathbf{H} = \begin{bmatrix}\mathbf{H}^{xy} \\ \mathbf{H}^{uv}\end{bmatrix}, \tag{9}$$

where $N_t = 2N$ and

$$\mathbf{H}^{uv} = \frac{1}{2}(\mathbf{H}^{yy} - \mathbf{H}^{xx}). \tag{10}$$

The estimated mass distribution $\hat{\mathbf{m}}$ minimizing the data-misfit function $\Phi(\mathbf{m})$ (Equation (5)) without any constraint can be obtained by solving the normal equations

$$\left(\mathbf{H}^\top\mathbf{H}\right)\hat{\mathbf{m}} = \mathbf{H}^\top\mathbf{g}^o. \tag{11}$$

Notice that, for large datasets, there exists a very large computational demand to store the required matrices and vectors, solve the normal equations (Equation (11)), and perform matrix–vector operations. Our approach to overcome this computational demand is adapting the convolutional equivalent-layer methods proposed by [33] and [34] to take advantage of the particular structure of matrix $\mathbf{H}$ (Equations (7) and (9)).

We call attention to the fact that the Equations (2)–(10) highlight key differences between our method and the convolution-based formulation proposed by [34]. Unlike their approach, which constructs the sensitivity matrix as a weighted sum of directional components derived from magnetic data, our formulation is based on gravity-gradient observations and builds the sensitivity matrix by vertically stacking the individual contributions from each measured component (Equations (7) and (9)). Moreover, whereas the method proposed in [34] presumes data-acquisition geometries strictly aligned with the north–south and east–west directions, our approach is designed to maintain effectiveness when flight lines are oriented along rotated geometries that do not conform to these alignments.

2.3. Convolutional Equivalent Layer in Rotated Coordinates

Consider that the observed gravity-gradient data are distributed over a horizontal plane with constant vertical coordinate $z_c = z_i, i = 1, \ldots, N$, at a regularly spaced grid of points along the x and y axes of our Cartesian coordinate system, which we refer to as the *north-east system* in which the x-axis points north and the y-axis to the east. In practice, however, gravity-gradient data are often referenced to a different coordinate system whose horizontal axes are not aligned with geographic directions. We refer to this as the *data system*, which is aligned with the acquisition geometry, with local coordinates $\tilde{x}$-axis and $\tilde{y}$-axis (Figure 1). The data system is rotated relative to the north-east system by an angle ψ from north. To relate these two systems, we apply a coordinate rotation. The gravity-gradient tensor $\tilde{\mathbf{G}}$ defined in the data system can be obtained from the tensor $\mathbf{G}$, which in turn is defined in to north-east system by Equation (1) through the transformation:

$$\tilde{\mathbf{G}} = \mathbf{R}\mathbf{G}\mathbf{R}^{\top}, \tag{12}$$

where $\mathbf{R}$ is a rotation matrix given by

$$\mathbf{R} = \begin{bmatrix} \cos\psi & \sin\psi & 0 \\ -\sin\psi & \cos\psi & 0 \\ 0 & 0 & 1 \end{bmatrix}. \tag{13}$$

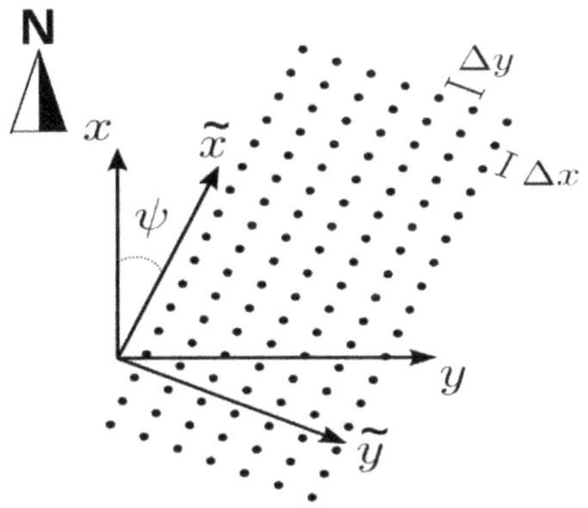

Figure 1. Schematic representation of a regular grid of gravity-gradient data (black dots) measured on a horizontal plane, with constant grid spacings Δx and Δy. These data are aligned with a local Cartesian system having horizontal axes $\tilde{x}$ and $\tilde{y}$ (data system). The data system, in turn, is rotated by an angle ψ with respect to the north-east system, with x- and y-axes pointing north and east, respectively. Both systems share the same z-axis pointing vertically downward.

Let us consider an equivalent layer located on a horizontal plane below the grid of observations, with constant vertical coordinate $z_c > z_i$, having one point mass directly below each observation point, so that the number of point masses M is equal to that of observation points N. In this case, ref. [34] described the particular structure of the sensitivity matrix $\mathbf{H}$ (Equations (7) and (9)); however, they did not mention that this specific structure, namely, the block-Toeplitz Toeplitz-block (BTTB) structure, emerges only when $\mathbf{H}$ is defined in the coordinates of the data system (Figure 1). For this reason, all computational operations in this work were performed in the data system, which is aligned with the $\tilde{x}$-

and $\tilde{y}$-axes (the acquisition geometry). These operations include the calculation of the BTTB matrix eigenvalues of **H**, parameter estimation, prediction of gravity-gradient tensor components, and gravity-gradient data transformations.

After completing all computational operations in the data system, the results should be transformed back to the north–east system, which is aligned with geographic directions. This is conducted using the following transformation:

$$\mathbf{G} = \mathbf{R}^{\top} \tilde{\mathbf{G}} \, \mathbf{R}. \tag{14}$$

We stress that the north-east system was used solely for visualizing the results, including the predicted gravity-gradient data, estimated parameters, and transformed data, since it corresponds to the spatial arrangement of the original observations.

To estimate the mass distribution $\hat{\mathbf{m}}$, we solve the normal equations (Equation (11)) iteratively by using the *standard conjugate gradient least squares (CGLSs) method* ([37], p. 166) that requires three matrix-vector products: one out of the loop and two per iteration. These products can be written generically as

$$\mathbf{H}\mathbf{v} = \mathbf{H}_1\mathbf{v}_1 + \cdots + \mathbf{H}_L\mathbf{v}_L, \tag{15}$$

and

$$\mathbf{H}^{\top}\mathbf{v} = \mathbf{H}_1^{\top}\mathbf{v}_1 + \cdots + \mathbf{H}_L^{\top}\mathbf{v}_L, \tag{16}$$

in terms of a generic partitioned matrix

$$\mathbf{H} = \left[\mathbf{H}_1^{\top}, \ldots, \mathbf{H}_L^{\top}\right]^{\top}, \tag{17}$$

and a generic partitioned vector

$$\mathbf{v} = \left[\mathbf{v}_1^{\top}, \ldots, \mathbf{v}_L^{\top}\right]^{\top}. \tag{18}$$

The matrix–vector product $\mathbf{H}_\ell\mathbf{v}_\ell$ or $\mathbf{H}_\ell^{\top}\mathbf{v}_\ell$ in Equations (15) and (16), $\ell \in \{1, \ldots, L\}$, is computed using the fast discrete convolution approach introduced by [33]. It is important to note that L (Equations (15)–(18)) represents the number of measured gravity-gradient components. In the case of gravity-gradient data obtained from the FTG system, for example, the matrix **H** (Equation (17)) consists of $L = 5$ sub-matrices, as defined by Equation (7). Otherwise, when the data is measured by the AGG system, the matrix is defined with $L = 2$ sub-matrices, as specified in Equation (9). Our convolutional equivalent-layer method can be schematically defined as follows:

The convergence criterion of the algorithm (step 3) is based on the invariance of the normalized data-misfit function (Equation (5)) between successive iterations. The algorithm stops when

$$\delta = \frac{\left|\Phi(\mathbf{m})^{(i)} - \mathbf{m})^{(i+1)}\right|}{\Phi(\mathbf{m})^{(i)}} \leq \epsilon, \tag{19}$$

where ϵ varies from 10^{-2} to 10^{-5}, $\Phi(\mathbf{m})^{(i)}$ and $\Phi(\mathbf{m})^{(i+1)}$ are the data-misfit function at the ith and $(i+1)$th iterations, respectively.

2.4. Computational Performance

To quantify the computational performance of our method, we use the total number of flops associated with the required matrix operations ([38] p. 12). We also refer the reader to ([39] Section 7) for a detailed explanation of the flops computation presented below. Previously, we established that the total number of observations and the total number of equivalent sources are given by $N_t = NL$, where L is the number of measured gravity-gradient components (see Equations (15)–(18)) and N the number of observation points of each component.

We now determine the total number of flops required to solve the normal equations (Equation (11)) using the classical Cholesky decomposition approach ([38] p. 262); see also ([39] Equation (27)):

$$f_{\text{classical}} = \frac{7}{3}(NL)^3 + 4(NL)^2. \tag{20}$$

The number of flops associated with the standard CGLS algorithm without exploring the structure of matrix **H** (see ([38] Section 11.3) and ([37] p. 165); see also ([39] Equation (28)) is:

$$f_{CGLS} = L\Big(2N^2 + it(4N^2 + 4N)\Big), \tag{21}$$

where "it" is the number of iterations.

If we now define the number of flops associated with the CGLS algorithm by considering that the matrix–vector products are efficiently computed as fast discrete convolutions, and the resulting flops for our method are given by:

$$f_{our} = (\ L(\ \kappa(16N)\ \log_2(4N) + 26N + it(\kappa(16N)\log_2(4N) + 58N)), \tag{22}$$

where $\kappa = 5$. We refer the reader to ([39], Section 7.7, Equation (67) and Algorithm 7) for a detailed explanation of the flops computation underlying the method, which has been adapted in this work to handle gradient-gravity data. It is important to note that we did not consider the flops associated with data transformations (Algorithm 1).

Algorithm 1 Convolutional equivalent-layer for gravity-gradient data in rotated coordinates.

Input: $\mathbf{g}^o$ (defined in the data system), angle ψ (Figure 1)
1: Transform the observed data from north-east system to data system (Equation (12)).
2: Interpolate the rotated data to produce a dataset closer to a regular grid (optional).
3: Estimate the discrete equivalent layer from the interpolated data by combining the Algorithm *standard Conjugate Gradient Least Squares (CGLS) method* ([37] p. 166) and the fast discrete convolution approach.
4: Compute the gravity-gradient data produced by the equivalent layer (predicted data).
5: Transform the observed and predicted data from data system to north-east system using Equation (14).

Figure 2 shows the number of flops using the classical (black line), standard CLGS (red line), and our method (blue line). The latter two methods were computed considering $it = 50$ and 5 components. This result demonstrates the excellent efficiency of our convolutional equivalent-layer method for processing gravity-gradient data. Note that the number of flops required to estimate the mass distribution by our method decreased by approximately 8 orders of magnitude compared with the classical approach.

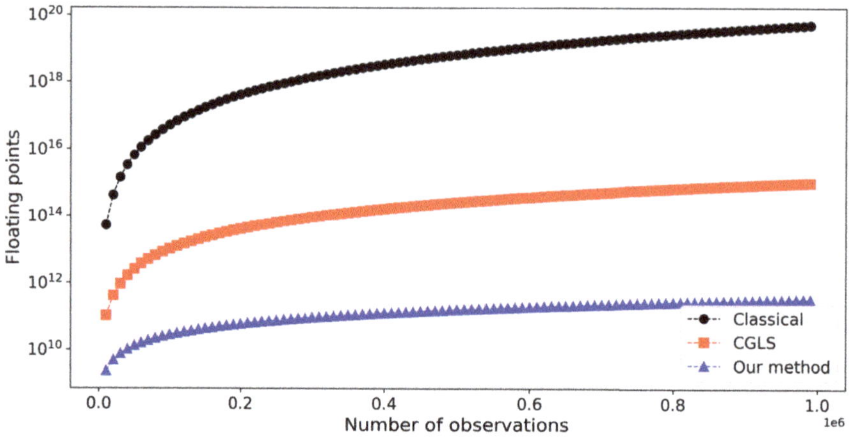

Figure 2. Graph comparing the number of *flops* versus the number of observation points for the classical equivalent layer technique (black line, Equation (20)), standard CGLS (red line, Equation (21)), and our method (blue line, Equation (22)), using $it = 50$ iterations and $L = 5$ components, both selected arbitrarily. The computational cost of our method is approximately eight orders of magnitude lower than that of the classical approach.

Figure 3 shows the runtime to solve the linear system with the maximum value of N equal to 10,000 points and $L = 5$ components, i.e., N_t 50,000 gravity-gradient data using the classical (black line), standard CLGS (red line), and our method (blue line). The quantity of data for runtime comparison is limited by using the classical equivalent layer technique on a processor Intel(R) Core(TM) i9-9880H CPU @ 2.30 GHz with 16 GB of RAM memory.

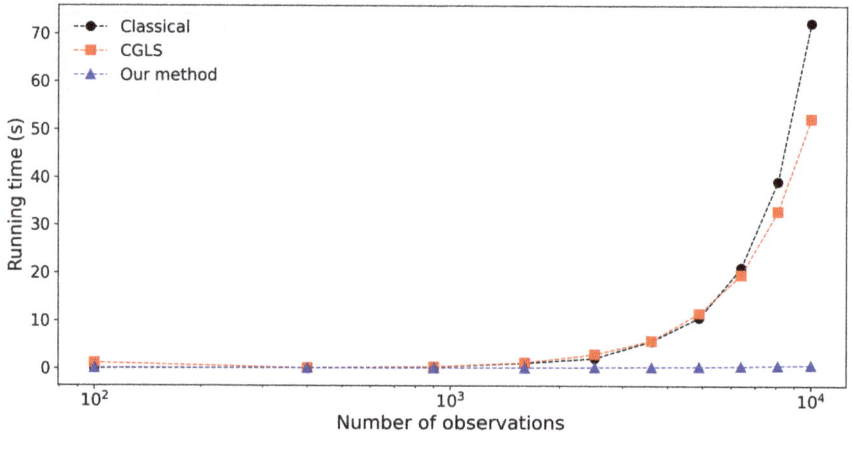

Figure 3. Graph comparing the runtime versus the number of observation points for the classical equivalent layer technique (black line), standard CGLS (red line), and our method (blue line) with $it = 50$ iterations and $L = 5$ components, both selected arbitrarily. The runtime of our method remains negligible as the number of observations increases, in contrast to the rapidly growing runtime of the classical and standard CGLS methods.

We emphasize that the values of it (in Figure 2) and the maximum number of observations N (in Figure 3) were selected arbitrarily and serve solely illustrative purposes. These choices aim to highlight the scalability behavior of the methods. Specifically, (i) the computational cost of our method (blue line in Figure 2) is approximately eight orders of magnitude lower than that of the classical approach (black line) and (ii) the runtime of both the classical (black line) and standard CGLS (red line) methods in Figure 3 increases rapidly with the number of observations, while the runtime of our method (blue line) remains negligible.

2.5. Criterion for Selecting Layer Depth

There is no analytical criterion in the literature for choosing the depth of the equivalent layer. In our work, the criterion for selecting the layer depth was based on [18]. This criterion asserts that the depth of the layer in relation to the observed data should fall within a range of 2.5 to 6 times the data spacing. Specifically, we first compute the average elevation of the observation points using the z values. The equivalent layer is then positioned at a depth offset determined by the maximum data spacing, scaled by a factor within the recommended range.

2.6. Interpolating Data

To reduce the effect of over sampling along flight and tight lines, we applied a block-averaging approach similar to that proposed by [31] in the data system (Figure 1). We have first divided the survey area into a regular grid of adjacent horizontal blocks along the x and y directions, respectively. To the horizontal center of each block, we assign the averaged gravity-gradient data and vertical coordinate (z-coordinates) of the observation points within the corresponding block. The result is a new dataset on a regular grid of points, with block-averaged gravity-gradient data and vertical coordinates.

3. Synthetic Data Application

We simulated Falcon AGG and FTG data to demonstrate the efficiency of our method for jointly processing gravity-gradient tensor components. These datasets were produced by multiple simulated geologic sources generating strong-interfering anomalies more closely to a real-world scenario. Specifically, the model consists of 302 rectangular prisms, divided into two groups. The first group represents the main geologic bodies and consists of two prisms, each with a density contrast of 1.0 g/cm^3, a top depth of 100 m, and dimensions of $2000 \times 1000 \times 300$ m (the smaller body) and $5000 \times 2000 \times 300$ m (the larger body) along the x-, y-, and z-axes, respectively. The second group, representing geologic noise, consists of 300 small, shallow cubes with side lengths and top depths uniformly distributed between 100 and 200 m, and between 0 and 10 m, respectively. These sources produce anomalies with predominantly short-wavelength spectral content.

Both data simulations followed the flight pattern of a real airborne gravity-gradient survey over the Kauring test site, Australia [40]. To each gravity-gradient component, we introduced additive Gaussian-distributed noise, with zero mean and a standard deviation equal to 7% of the maximum absolute value of the corresponding noise-free gravity-gradient data. Specifically, the standard deviations of the noise (in eotvos) for each component were the following: $xx = 2.19$, $xy = 1.94$, $xz = 4.47$, $yy = 3.83$, $yz = 4.59$, $zz = 5.46$, and $uv = 1.48$.

The simulated synthetic data consisted of 114 flight lines each with 1275 points, totaling 145,350 observations, at vertical coordinates z (Figure 4) ranging from -296 to -440 m and a mean and standard deviation of -353.34 m and 23.56 m, respectively.

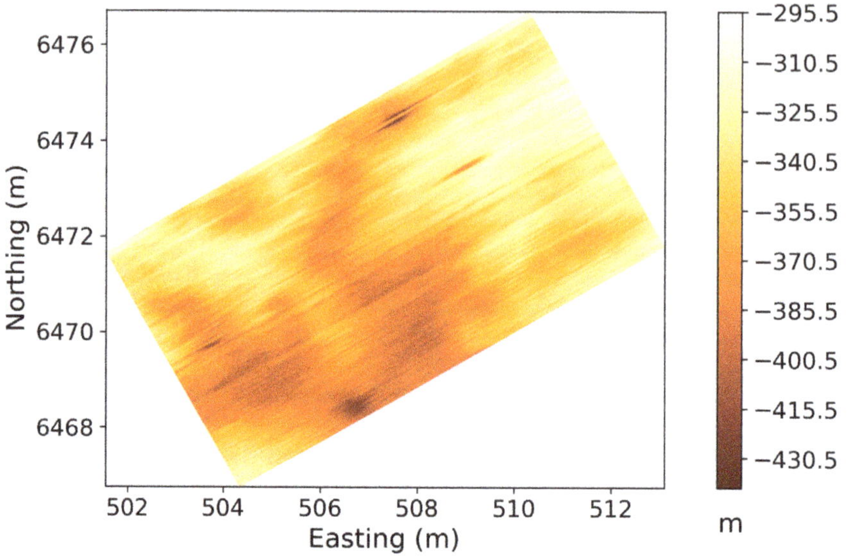

Figure 4. GPS altitude (WGS84) was converted to downward-positive z-coordinates over the Kauring test site, Australia.

We applied our method to simulated Falcon and FTG data using a single equivalent layer with one source beneath each observation point at a constant depth of z_c = 23.54 m. Transformations between the north-east and data systems (Figure 1 and steps 1 and 5 of Algorithm 1) were computed with $\psi = -30°$.

Here, we illustrate the transformation by predicting the $\mathbf{g}_{zz}$ component and a quantity we refer to as the gradient magnitude of $\mathbf{g}_z$ (i.e., $\|\nabla \mathbf{g_z}\|$), defined as the square root of the sum of the squares of the predicted $\mathbf{g}_{xz}$-, $\mathbf{g}_{yz}$-, and $\mathbf{g}_{zz}$-components.

3.1. Falcon AGG Data

The simulated Falcon AGG components $\mathbf{g}_{xy}$ and $\mathbf{g}_{uv}$ are shown in Figure 5. Each component has N = 145,350 observation points, totaling N_t = 290,700 simulated data. After transforming the simulated data from the north-east system to the data system (step 1 of Algorithm 1), we estimated the mass distribution at 39.02 s and computed the predicted data in the data system (steps 3 and 4 of Algorithm 1). Refer to the Supplementary Material for the Falcon AGG data transformed into the data system (see Figure S1).

Figure 6a,b show, in the north–east system, the residuals between the simulated data (Figure 5) and the predicted data for the two components of the Falcon AGG system. The insets display histograms of the residuals, with means of 0.16 and 0.18 eotvos, and standard deviations of 2.04 and 1.70 eotvos, for the $\mathbf{g}_{xy}$ and $\mathbf{g}_{uv}$ components, respectively. The bell-shaped histograms and the estimated standard deviations are close to those of the added Gaussian noise (1.94 and 1.48 eotvos), which suggests an excellent fit to the data. Figure 6c shows the root mean square error (RMSE) as a function of iterations. The RMSE curve flattens just below 2 eotvos, consistent with the noise levels, further supporting that our method has converged properly. Taken together, the residual distribution and the RMSE convergence indicate that the solution fits the data within the expected noise level, without evidence of overfitting.

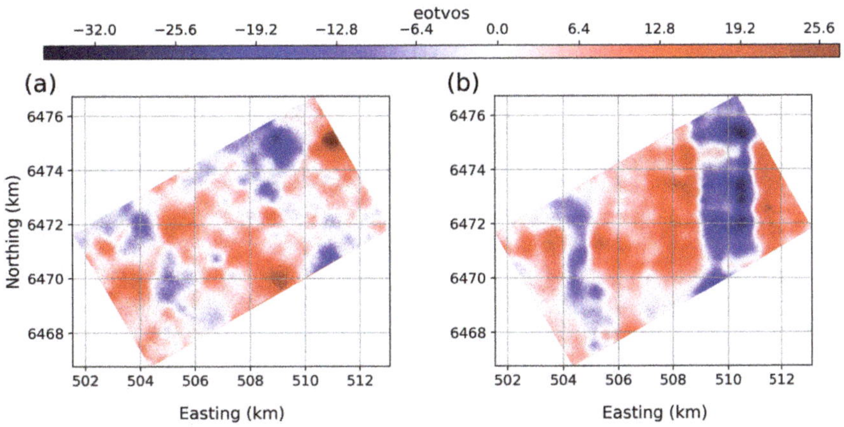

Figure 5. Synthetic Falcon AGG data in the north-east system. Simulated noise-corrupted components (**a**) $\mathbf{g}_{xy}$ and (**b**) $\mathbf{g}_{uv}$ at 145,350 observation points.

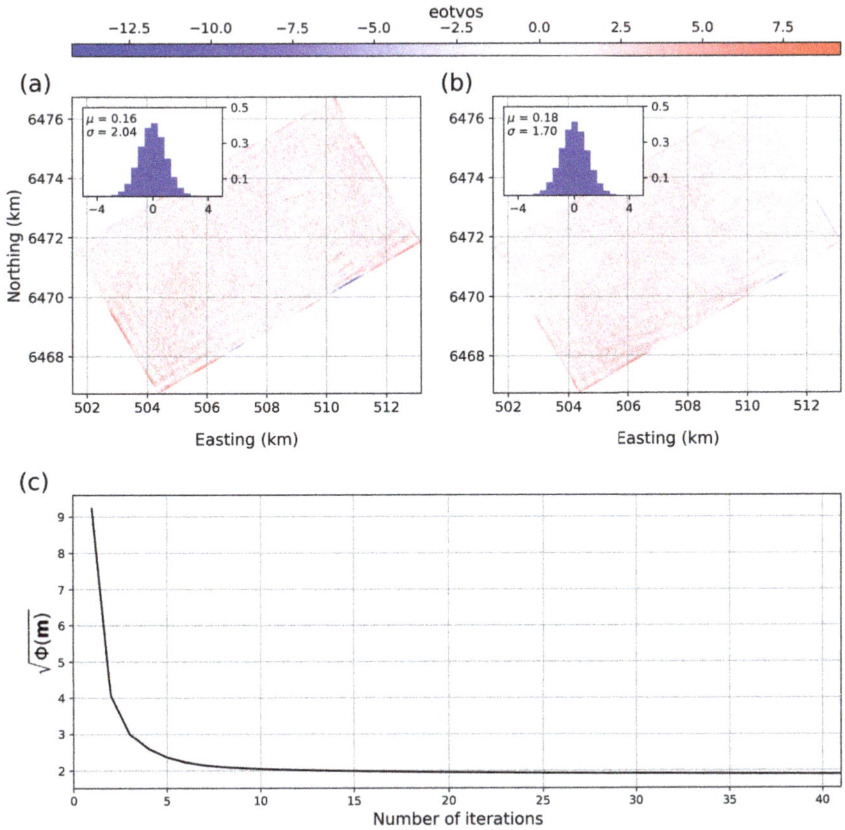

Figure 6. Synthetic Falcon AGG data. The residuals and the histograms of the residuals are shown as insets (**a**,**b**) between the simulated (Figure 5a,b) and the corresponding predicted (**a**) $\mathbf{g}_{xy}$ and (**b**) $\mathbf{g}_{uv}$ components (in eotvos). The mean μ and standard deviations σ (in eotvos) are shown in each histogram. (**c**) Convergence curve, showing the root mean square error (RMSE) along successive iterations.

Figure 7 shows two transformation results predicted by the estimated equivalent layer using our method. Despite the strong presence of geologic noise, simulated by 300 small sources, both the predicted $\mathbf{g}_{zz}$ (Figure 7a) and the predicted $\|\nabla \mathbf{g_z}\|$ (Figure 7b) provide a clear definition of the boundaries of the two main geologic bodies. The geologic noise is visible as short-wavelength components in these transformation results.

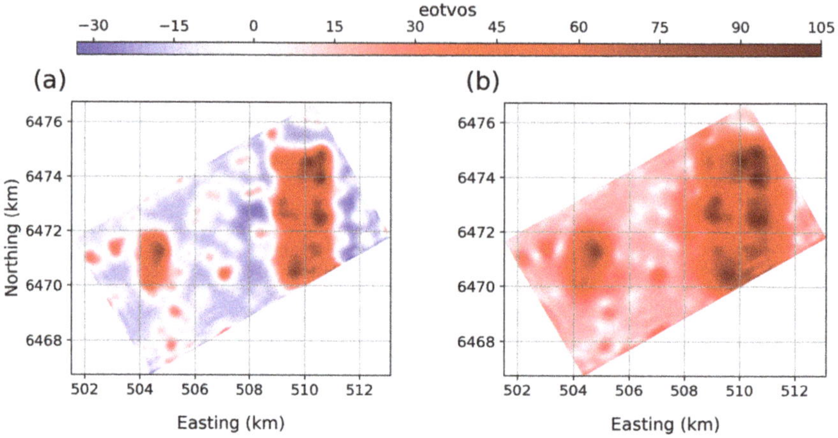

Figure 7. Synthetic Falcon AGG data. The transformation results: (**a**) the predicted $\mathbf{g_{zz}}$ component and (**b**) the predicted $\|\nabla \mathbf{g_z}\|$.

3.2. FTG Data

Figure 8 shows the simulated FTG components $\mathbf{g}_{xx}$, $\mathbf{g}_{xy}$, $\mathbf{g}_{xz}$, $\mathbf{g}_{yy}$, and $\mathbf{g}_{yz}$. Each component has N = 145,350 observation points. After transforming the simulated data from the north-east system to the data system (step 1 of Algorithm 1), we estimated the mass distribution at 98.20 s using a total of N_t = 726,750 simulated data and computed the predicted data by applying steps 3 and 4 of Algorithm 1. The simulated FTG data, transformed into the data system, is provided in the Supplementary Material (see Figure S2).

The estimated mass distribution produces the predicted data that acceptably fit the simulated FTG components (Figure 8a–e). This is corroborated by Figure 9a–e, which shows the residuals between the simulated (Figure 8a–e) and predicted FTG data. The bell-shaped histograms of the residuals (shown in Figure 9 as insets) also confirm the acceptability of the data fit with means varying between −0.77 and 0.32 eotvos, as well as the standard deviations varying between 2.24 and 4.52 eotvos. The RMSE convergence curve in Figure 8f stops decreasing below 4 eotvos, which matches the level of random noise added to the data. This indicates that the predicted gravity-gradient components have minimized the random errors, and further improvement is limited by the noise floor.

Similar to the previous transformation results, Figure 10a,b also defines the boundaries of the two main geologic bodies, despite the presence of short-wavelength components.These components are not caused by random noise, but are primarily due to geologic noise simulated by 300 small, shallow cubes.

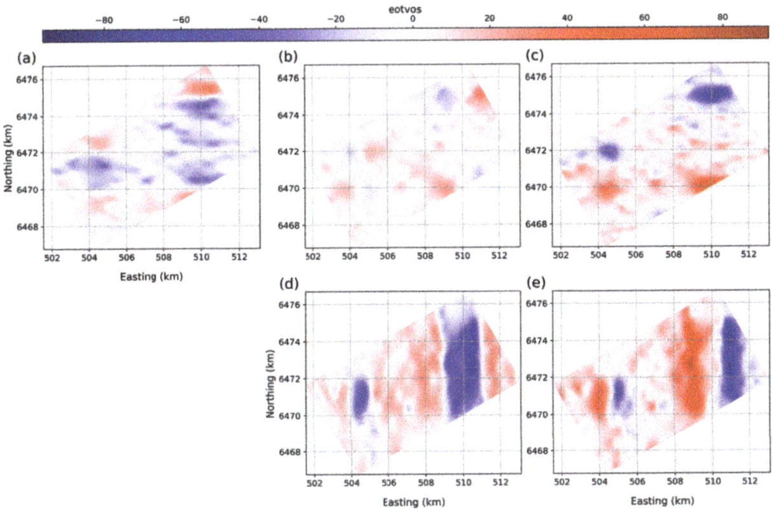

Figure 8. Synthetic FTG data in the north-east system. Simulated noise-corrupted components **(a)** g_{xx}, **(b)** g_{xy}, **(c)** g_{xz} **(d)** g_{yy}, and **(e)** g_{yz} at 145,350 observation points.

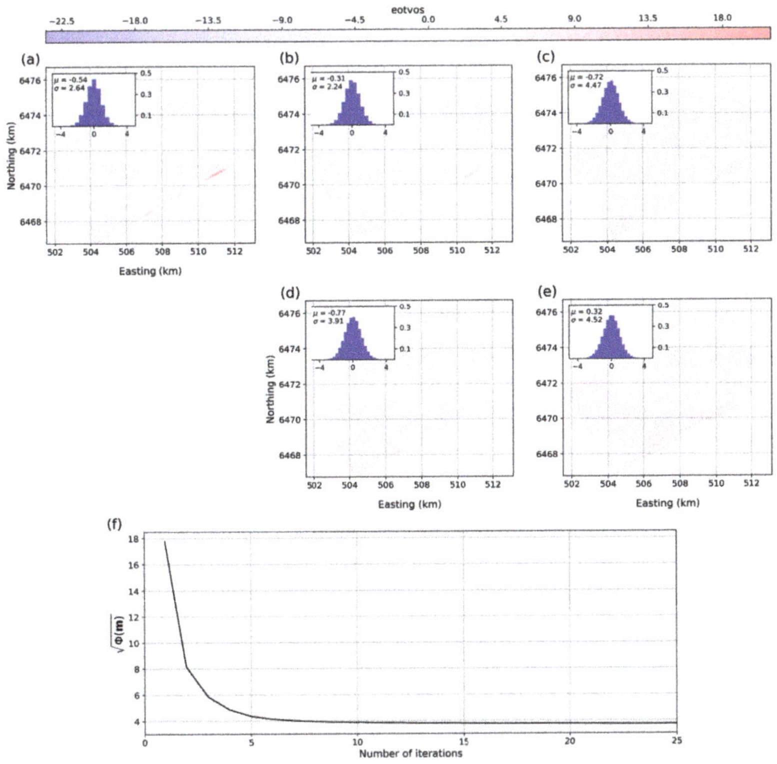

Figure 9. Synthetic FTG data. The residuals and the histograms of the residuals are shown as insets **(a–f)** between the simulated (Figure 8a–f) and the corresponding predicted **(a)** g_{xx}, **(b)** g_{xy}, **(c)** g_{xz}, **(d)** g_{yy} and **(e)** g_{yz}, components (in eotvos). The mean μ and standard deviations σ (in eotvos) are shown in each histogram. **(f)** Convergence curve, showing the root mean square error (RMSE) along successive iterations.

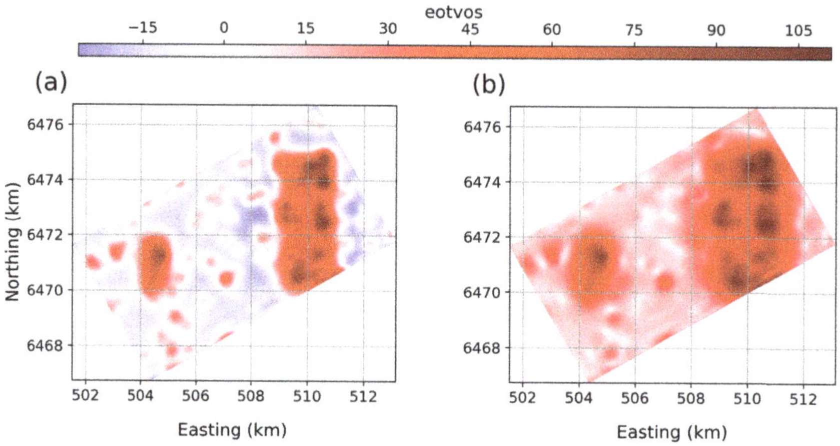

Figure 10. Synthetic FTG data. The transformation results: (**a**) the predicted $\mathbf{g}_{zz}$ component and (**b**) the predicted $\|\nabla \mathbf{g_z}\|$.

A quantitative comparison between our proposed method and the conjugate gradient for least squares (CGLSs) algorithm is provided in the Supplementary Material. Using reduced versions of the synthetic Falcon AGG and FTG datasets, we assessed the accuracy and performance of both methods through residual maps, histograms, and RMSE convergence curves. The results demonstrate that BTTB achieves comparable accuracy to CGLS while offering significantly lower computational cost (Figures S4–S11, in the Supplementary Materials).

To evaluate the robustness of our inversion method, we conducted a sensitivity analysis using synthetic Falcon AGG datasets contaminated with Gaussian noise at three levels, 5%, 7%, and 9%, of the maximum signal amplitude. For each level, six independent noise realizations were generated and inverted. Under all noise conditions examined, our method demonstrated stable convergence behavior, with residuals closely matching the noise characteristics: unbiased, Gaussian-distributed, and spatially uncorrelated. The final RMSE values converged to levels consistent with the added noise, and the transformation results consistently delineated the boundaries of the modeled geologic bodies. These findings confirm that the method is robust, repeatable, and maintains reliable performance under increasing noise conditions. Full details and figures are provided in the Supplementary Materials (see Figures S12–S20).

4. Real Data Applications

We applied our method to the Falcon AGG data from the Kauring gravity gradiometry test site, Australia [40] and FTG data from the Vinton salt dome, LA, USA [3,41]. In these real-world applications, whether using Falcon AGG data over Kauring gravity gradiometry or FTG data over the Vinton salt dome, we emphasize that the fitted data were computed at the original observation points, not at the reduced points. Aiming to process large gravity-gradient data, we also applied the method to more than 1,250,000 Falcon AGG data points from the Canobie Airborne Gravity Survey, Australia.

4.1. Kauring

With funding from the Western Australian government's Exploration Incentive Scheme, a matching industry contribution, and support from the Geological Survey of Western Australia and Geoscience Australia, the Kauring test site situated near York in

Western Australia, was established to support coordinated geophysical data collection [42]. The site is located in a farming region characterized by gently undulating granitic terrain, featuring occasional incised valleys, with a maximum topographic relief of about 75 m. According to [43], the gravity anomaly at the Kauring test site is attributed to iron-rich rocks underlying a small topographic rise near the center of the map. Ref. [44] interpreted that the amplitude of the complete Bouguer gravity anomaly field in the Kauring test site gradually decreases from west to east. Higher anomaly values are concentrated in the southwestern part of the study area, while lower values are found in the northeast. Although it remains uncertain whether the linear gravity ridge is produced by similar rocks beneath the soil cover, airborne geophysical surveys support this interpretation.

The observed Falcon AGG components $\mathbf{g}_{xy}$ and $\mathbf{g}_{uv}$ (Figure 11a,b) over the Kauring test site, Australia, consists of 114 northwest–southeast oriented lines with line spacing of 50 m, totaling 164,624 observed points at the z-coordinates shown in Figure 4. We have transformed the gravity-gradient data from the north-east system to the data system (step 1 of Algorithm 1) by using $\psi = -30°$ (Equation (12)). The observed Falcon AGG data, transformed to the data system, is available in the Supplementary Materials (see Figure S3). We then interpolated the dataset in the data system (step 2 of Algorithm 1) to obtain 1275 points in each of the 114 flight lines, which corresponds to 145,350 reduced points. We placed one equivalent source directly beneath each interpolated point, at a constant depth of z_c = 23.54 m.

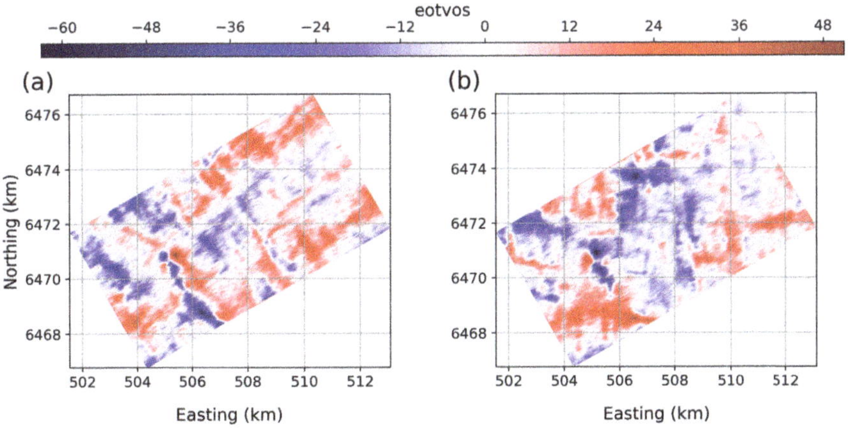

Figure 11. Real data over Kauring test site, Australia. Observed Falcon AGG components (**a**) $\mathbf{g}_{xy}$ and (**b**) $\mathbf{g}_{uv}$ obtained with the Falcon AGG system at 164,624 observation points.

After approximately 120 iterations and just a few seconds of computation (Figure 12c), we estimated a mass distribution that fits the $\mathbf{g}_{xy}$ and $\mathbf{g}_{uv}$ components. Using this model, we calculated the predicted data and the corresponding residuals (Figure 12a,b) for the two AGG components in the north–east coordinate system. The insets in these figures show the histograms of the residuals, which have means of 0.11 and −1.22 eotvos, and standard deviations of 3.88 and 3.40 eotvos, for the $\mathbf{g}_{xy}$ and $\mathbf{g}_{uv}$ components, respectively. Combining the RMSE curve (Figure 12c), which stabilizes below 4 eotvos, with the residual distributions (insets in Figure 12a,b) supports the conclusion that the data are well-fitted within the expected noise level.

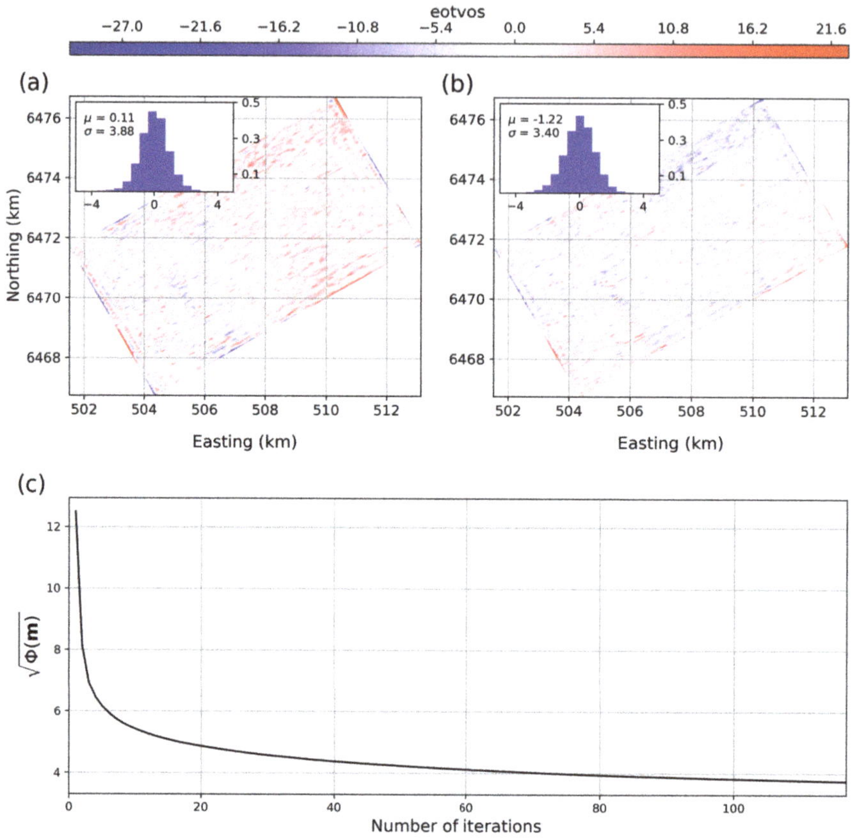

Figure 12. Real data over the Kauring test site, Australia. The residuals and histograms of the residuals are shown as insets (**a**,**b**) between the real (Figure 11) and the corresponding predicted (**a**) $\mathbf{g}_{xy}$ and (**b**) $\mathbf{g}_{uv}$ components (in eotvos) in the data system. The mean μ and standard deviations σ (in eotvos) are shown in each histogram. (**c**) Convergence curve, showing the root mean square error (RMSE) along successive iterations.

The predicted $\mathbf{g}_{zz}$ and $\|\nabla \mathbf{g_z}\|$, shown in Figures 7b and 13a, respectively, reveal strong linear anomalies in the southwestern part and a set of isolated anomalies in the northeastern part. These results are consistent with those reported in [44].

4.2. Vinton

The Vinton salt dome, located in southwestern Louisiana within the onshore Gulf of the Mexico Basin, is a geological structure situated in a region known for its long-standing oil and gas development history [45]. Geologically, the dome consists of a massive salt core overlain by a well-defined cap rock, which exhibits a vertical gradation from limestone at the top to gypsum and anhydrite at the bottom. This cap rock is embedded within a sedimentary sequence of interbedded sandstone and shale, reflecting the region's complex depositional and diagenetic history. Analysis of lithological density ranges led [3] to conclude that the gravity-gradient response is largely attributable to the cap rock. Geophysical interpretations reveal a 3D cap rock structure with a subtly elongated shape, consistent with the strike of the main fault in the study area [3,46].

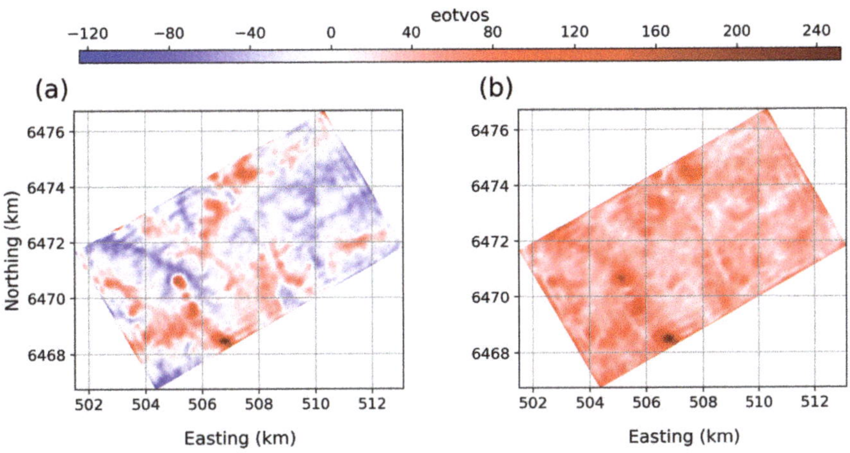

Figure 13. Real data over the Kauring test site, Australia. The transformation results of (**a**) the predicted $\mathbf{g}_{zz}$ component and (**b**) the predicted $\|\nabla \mathbf{g_z}\|$.

Figure 14 shows the observed components $\mathbf{g}_{xx}$, $\mathbf{g}_{xy}$, $\mathbf{g}_{xz}$, $\mathbf{g}_{yy}$ and $\mathbf{g}_{yz}$ over the Vinton salt dome. The selected area over the dome has flight lines, each with 9349 points, totaling 46,745 observations, at vertical coordinates z-coordinates ranging from -62.14 to -109.32 m. Because the flight lines are oriented along the north–south direction, the data and north-east systems coincide and we by-passed the step 1 of Algorithm 1. We interpolated the data to obtain 6000 points, totaling 30,000 observations. We placed one equivalent source directly beneath each interpolated point, at a constant depth of 1981.81 m.

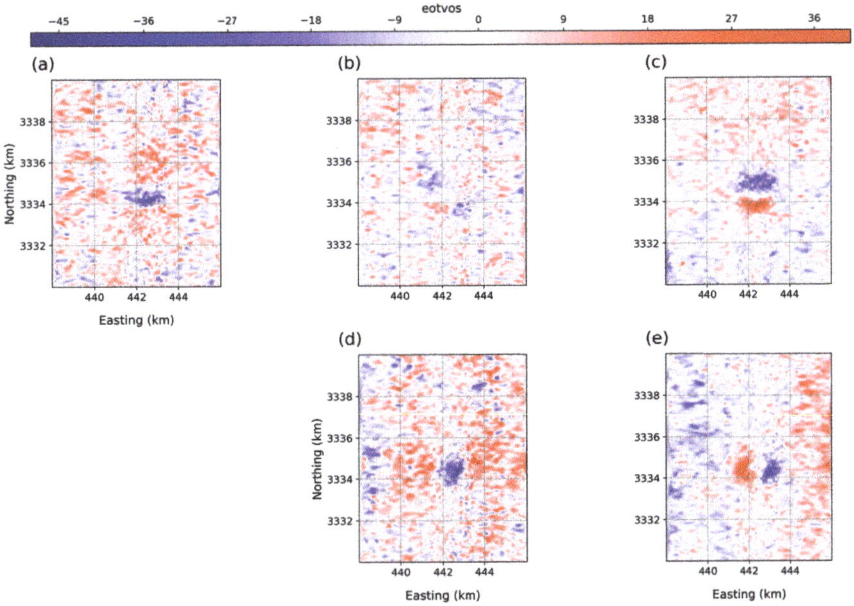

Figure 14. Real data over the Vinton salt dome, USA. (**a**) $\mathbf{g}_{xx}$, (**b**) $\mathbf{g}_{xy}$, (**c**) $\mathbf{g}_{xz}$, (**d**) $\mathbf{g}_{yy}$, and (**e**) $\mathbf{g}_{yz}$ components obtained with the FTG system.

After 500 iterations (Figure 15f) and 4.3 s, we estimated the mass distribution using the measured components, i.e, g_{xx}, g_{xy}, g_{xz}, g_{yy}, and g_{yz} components. This estimated model produces predicted data that fit the observed FTG components reasonably well (Figure 14a–e). This result is corroborated by the residuals shown in Figure 15a–e, which compare the observed data (Figure 14a–e) with the predicted values. The histograms of the residuals (insets in Figure 15a–e) exhibit approximately Gaussian distributions, with means ranging from −0.29 to 0.26 eotvos and standard deviations between 5.79 and 8.83 eotvos, further confirming the quality of the data fit. As in the synthetic tests, the RMSE curve (Figure 15f) reaches a value below 7.4 eotvos, which is consistent with the residual distributions shown in the insets of Figure 15a–e.

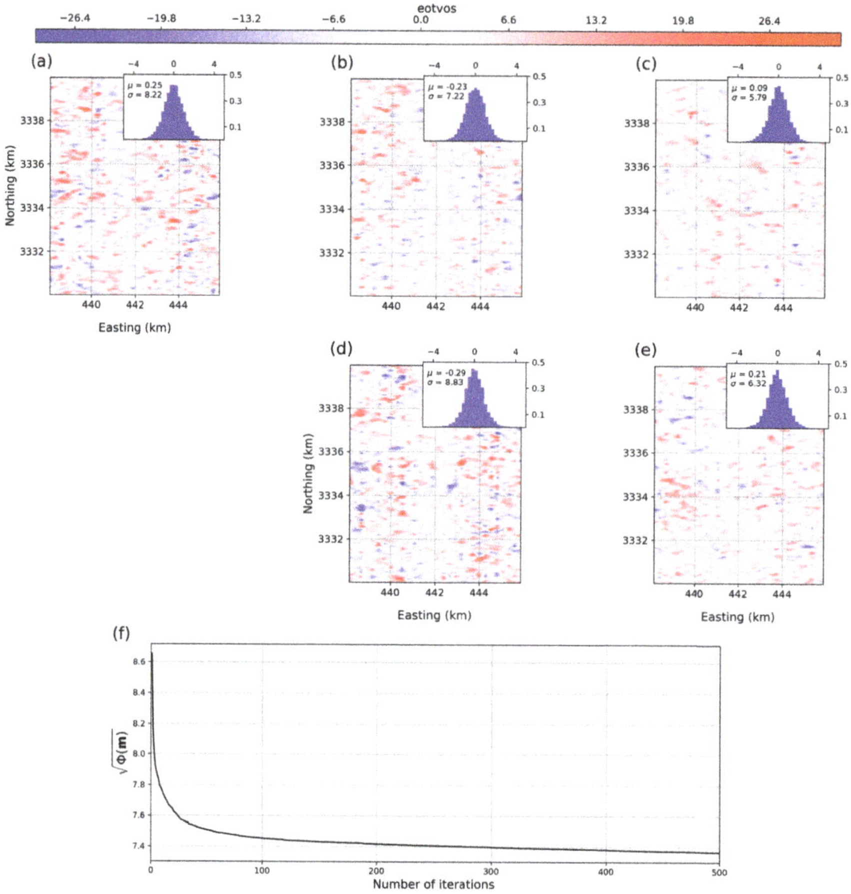

Figure 15. Real data over the Vinton salt dome, USA. The residuals and the histograms of the residuals are shown as insets (**a–e**) between the real (Figure 14a–e) and the corresponding predicted (**a**) g_{xx}, (**b**) g_{xy}, (**c**) g_{xz}, (**d**) g_{yy}, and (**e**) g_{yz} components (in eotvos). The mean μ and standard deviations σ (in eotvos) are shown in each histogram. (**f**) Convergence curve, showing the root mean square error (RMSE) along successive iterations.

The transformation results, shown in Figure 16, clearly delineate the boundary of the cap rock of the Vinton salt dome. We call attention to the presence of short-wavelength components in both the predicted g_{zz} (Figure 16a) and the predicted $\|\nabla \mathbf{g_z}\|$ (Figure 16b). Based on synthetic test results (Figures 7 and 10), where simulated geologic noise appears as short-wavelength components in the transformation outputs, we infer that the short-wavelength features in Figure 15 may be due to shallow, small-scale geologic sources with relative high density contrasts.

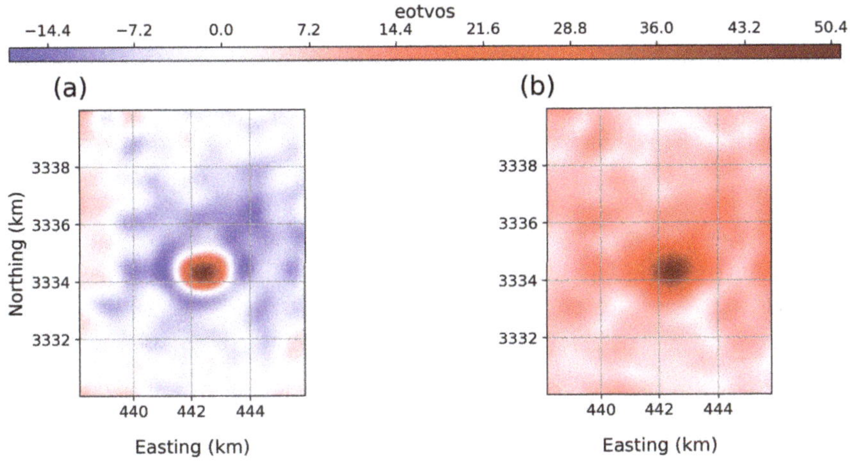

Figure 16. Real data over the Vinton salt dome, USA. The transformation results: (**a**) the predicted $\mathbf{g}_{zz}$ component and (**b**) the predicted $\|\nabla \mathbf{g_z}\|$.

4.3. Canobie

The Canobie survey area, located in northern Queensland, Australia, is being explored under contract with Geoscience Australia. It spans latitudes 19°10′ S to 19°52′ S and longitudes 140°27′ E to 141°01′ E [47]. The area lies within the Mount Isa Province, specifically within the Canobie Domain, which is part of the Eastern Fold Belt Subprovince [48]. The Mount Isa Province is a Paleoproterozoic to Mesoproterozoic terrane that forms part of the North Australian Craton. It is recognized as a major metallogenic province, hosting significant deposits of copper, gold, lead, zinc, silver, and uranium. The Canobie Domain is characterized by folded Proterozoic sedimentary and volcanic basement rocks typical of the eastern fold belt of the Mount Isa Inlier. Its definition is based primarily on geophysical characteristics, notably a distinctively high magnetic response [48,49]. The Eastern Fold Belt Subprovince, which includes the Canobie Domain, is particularly known for shear- and vein-hosted copper deposits, as well as iron oxide–copper–gold (IOCG) mineralization, often associated with mafic magmatism and structural features.

The Canobie Airborne Gravity Survey of Northern Queensland, Australia, aims to attract exploration to greenfield areas within the Mount Isa Basin. This survey measured the Falcon AGG components $\mathbf{g}_{xy}$ and $\mathbf{g}_{uv}$ using two acquisition channels, A and B. The observed $\mathbf{g}_{xy}$ component was computed as the average of the measurements from channels A and B, and the same procedure was applied to obtain the observed $\mathbf{g}_{uv}$ component. These averaged results are shown in Figures 17a and 17b, respectively.

The estimated mass distribution results in predicted AGG components (Figure 18) that fit the observed data with satisfactory accuracy. This is supported by the residual maps shown in Figure 19 and by the corresponding histograms, which show near-Gaussian distributions centered around zero eotvos. The standard deviations of the residuals are 4.06 eotvos for the $\mathbf{g}_{xy}$ component and 3.94 eotvos for the $\mathbf{g}_{uv}$ component. Additionally, the observed AGG data range from approximately −28.8 to 38.0 eotvos, giving a total signal amplitude of about 66.8 eotvos. In this context, an RMSE of around 4 eotvos corresponds to a signal-to-noise ratio of roughly 15:1, indicating a robust inversion result. Furthermore, the RMSE convergence curve (Figure 19c) flattens after approximately 300 iterations, converging to a value just below 4 eotvos. The agreement between the residual statistics and the final RMSE, together with the absence of large coherent patterns in the residuals, suggests that the inversion has successfully recovered the dominant geologic signal, with the remaining misfit primarily reflecting random measurement noise.

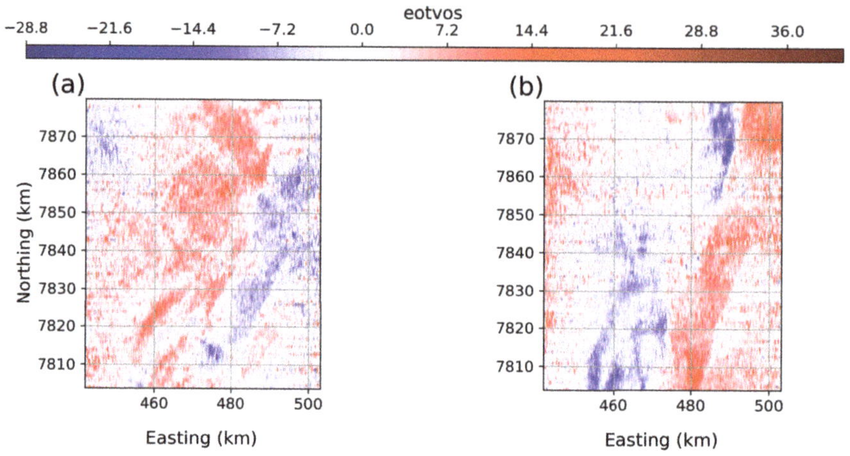

Figure 17. Canobie airborne gravity survey of Northern Queensland, Australia. The Canobie airborne gravity survey was centered at latitude 19°31′04″ S and longitudes 140°44′14″ E to the north of Cloncurry. (**a**) Observed components $\mathbf{g}_{xy}$ and (**b**) $\mathbf{g}_{uv}$, both obtained by averaging the measurements from channels A and B of the Falcon AGG system.

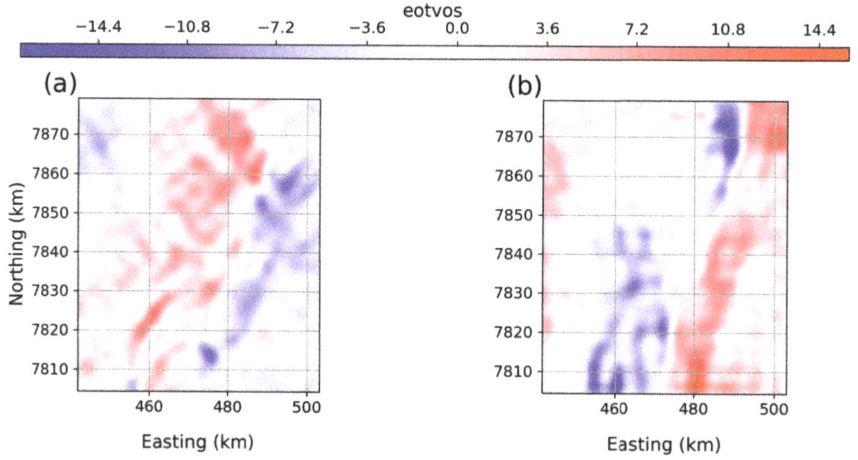

Figure 18. Canobie airborne gravity survey of Northern Queensland, Australia. The predicted (**a**) $\mathbf{g}_{xy}$ and (**b**) $\mathbf{g}_{uv}$ components.

Most of the survey area is characterized by strong, linear north–south and northeast-trending anomalies, except in the northeastern part, which displays a set of more isolated anomalies. This pattern is evident in both the predicted $\mathbf{g}_{xy}$ and $\mathbf{g}_{uv}$ components (Figure 18), as well as in the transformed results, the predicted $\mathbf{g}_{zz}$ component and $||\nabla \mathbf{g_z}||$, shown in (Figure 20).

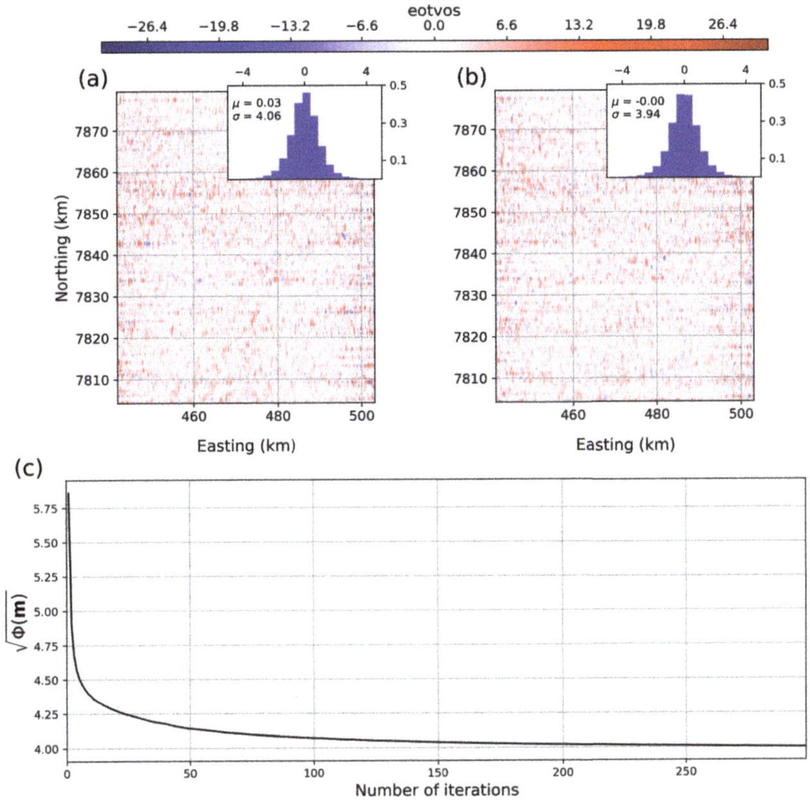

Figure 19. Canobie airborne gravity survey of Northern Queensland, Australia. The residuals and the histograms of the residuals are shown as insets between the real (Figure 15) and the corresponding predicted (**a**) $\mathbf{g}_{xy}$ and (**b**) $\mathbf{g}_{uv}$ components (in eotvos). The mean μ and standard deviations σ (in eotvos) are shown in each histogram. (**c**) Convergence curve, showing the root mean square error (RMSE) along successive iterations.

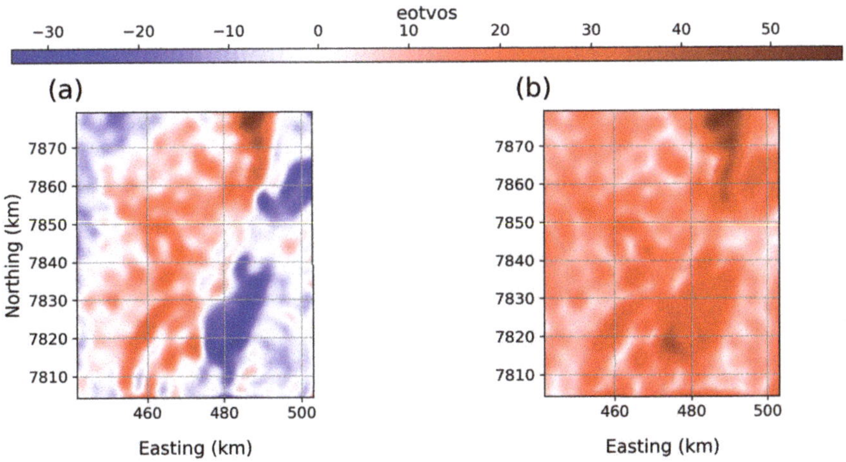

Figure 20. Canobie airborne gravity survey of Northern Queensland, Australia. The transformation results: (**a**) the predicted $\mathbf{g}_{zz}$ component and (**b**) the predicted $\|\nabla \mathbf{g_z}\|$.

5. Conclusions

We present a cost-effective equivalent-layer technique for jointly processing gravity-gradient components. The method's computational efficiency stems from the Block-Toeplitz Toeplitz-Block structure of sensitivity matrices for regular grids, enabling efficient vector–matrix product evaluations via 2D FFT. This approach estimates a mass distribution over a single equivalent layer using all gravity-gradient components, while requiring significantly fewer floating-point operations than the classical equivalent-layer technique or the CGLS method. On typical laptop hardware, the method handled over 290,000 AGG data points in just tens of seconds and is scaled to process between 726,000 FTG and 1,250,000 AGG data points in under a few minutes, demonstrating strong computational performance on large-scale data.

Tests on synthetic data confirm the method's ability to handle irregularly spaced points, non-horizontal surfaces, and flight lines with arbitrary orientations, as well as data from different acquisition systems (e.g., AGG and FTG). The synthetic tests incorporated both random Gaussian noise and complex geologic noise from multiple small sources, simulating realistic survey conditions. The estimated mass distribution produced predicted data that fit noisy observations within expected error margins. The transformation results, specifically the vertical gradient component and the magnitude of the vertical gravity-gradient vector, effectively delineated the boundaries of the main geological bodies, despite the presence of strong short-wavelength features caused by geological noise.

Applications to real datasets, including Falcon AGG data from the Kauring test site and Canobie survey in Australia, and FTG data over the Vinton salt dome in the USA, demonstrate the method's robustness across diverse geological settings, survey geometries, and acquisition systems. In all cases, the estimated mass distributions produced predicted gravity-gradient components that matched observations within expected noise levels. The residuals were near-Gaussian and uncorrelated, and convergence was achieved rapidly. The transformation results, particularly the vertical gradient and total gradient magnitude, successfully delineated key geological features such as the linear anomalies in Kauring and Canobie, and the cap rock boundary at Vinton, even in the presence of complex short-wavelength geological signals.

Sensitivity analyses across multiple noise levels demonstrated stable convergence and robustness, with residuals matching noise characteristics and consistent results across repeated runs. Moreover, comparative evaluations against the classical equivalent-layer technique and the conjugate gradient for least squares (CGLSs) method showed that our method achieves comparable accuracy while requiring significantly fewer floating-point operations and lower computational cost. These findings highlight the method's reliability, efficiency, and practicality for processing large-scale airborne gravity-gradient datasets.

Extending the method to jointly process AGG and FTG data from the same area presents no methodological obstacles, requiring only that data from both datasets be acquired over the same region.

Supplementary Materials: The following supporting information can be downloaded at: https://www.mdpi.com/article/10.3390/app15158396/s1, Figure S1: Synthetic Falcon AGG data transformed to data system, which is aligned with the simulated acquisition geometry of the data. (**a**) $\mathbf{g}_{xy}$ and (**b**) $\mathbf{g}_{uv}$ components. This figure corresponds to Figure 5 in the main article, which shows the same components in the north-east coordinate system. Figure S2: Synthetic FTG data transformed to data system, which is aligned with the simulated acquisition geometry of the data. (**a**) $\mathbf{g}_{xx}$, (**b**) $\mathbf{g}_{xy}$, (**c**) $\mathbf{g}_{xz}$ (**d**) $\mathbf{g}_{yy}$, and (**e**) $\mathbf{g}_{yz}$ components. This figure corresponds to Figure 8 in the main article, which shows the same components in the north-east coordinate system. Figure S3: Real data over Kauring Test site, Australia. Observed Falcon AGG components transformed to data system, which is aligned with the acquisition geometry of the data. (**a**) $\mathbf{g}_{xy}$ and (**b**) $\mathbf{g}_{uv}$ components. This figure corresponds to Figure 11 in the main article, which shows the same components in the north-east coordinate system. Figure S4: Synthetic Falcon AGG with reduced data size, processed using our method. The residuals and the histograms of the residuals (insets): (**a**) $\mathbf{g}_{xy}$ and (**b**) $\mathbf{g}_{uv}$ components (in eotvos). (**c**) Convergence curve, showing the root-mean-square error over iterations. Figure S5: Synthetic Falcon AGG with reduced data size, processed using our method. Predicted Falcon AGG data (**a**) $\mathbf{g}_{xy}$ and (**b**) $\mathbf{g}_{uv}$ components (in eotvos). Figure S6: Synthetic Falcon AGG with reduced data size, processed using CGLS method. The residuals and the histograms of the residuals (insets): (**a**) $\mathbf{g}_{xy}$ and (**b**) $\mathbf{g}_{uv}$ components (in eotvos). (**c**) Convergence curve, showing the root-mean-square error over iterations. Figure S7: Synthetic Falcon AGG with reduced data size, processed using CGLS method. Predicted Falcon AGG data (**a**) $\mathbf{g}_{xy}$ and (**b**) $\mathbf{g}_{uv}$ components (in eotvos). Figure S8: Synthetic FTG with reduced data size, processed using our method. The residuals and the histograms of the residuals (insets): (**a**) $\mathbf{g}_{xx}$, (**b**) $\mathbf{g}_{xy}$, (**c**) $\mathbf{g}_{xz}$ (**d**) $\mathbf{g}_{yy}$, and (**e**) $\mathbf{g}_{yz}$ components (in eotvos). (**f**) Convergence curve, showing the root-mean-square error over iterations. Figure S9: Synthetic FTG with reduced data size, processed using our method. Predicted FTG data (**a**) $\mathbf{g}_{xx}$, (**b**) $\mathbf{g}_{xy}$, (**c**) $\mathbf{g}_{xz}$ (**d**) $\mathbf{g}_{yy}$, and (**e**) $\mathbf{g}_{yz}$ components (in eotvos). Figure S10: Synthetic FTG with reduced data size, processed using CGLS method. The residuals and the histograms of the residuals (insets): (**a**) $\mathbf{g}_{xx}$, (**b**) $\mathbf{g}_{xy}$, (**c**) $\mathbf{g}_{xz}$ (**d**) $\mathbf{g}_{yy}$, and (**e**) $\mathbf{g}_{yz}$ components (in eotvos). (**f**) Convergence curve, showing the root-mean-square error over iterations. Figure S11: Synthetic FTG with reduced data size, processed using CGLS method. Predicted FTG data (**a**) $\mathbf{g}_{xx}$, (**b**) $\mathbf{g}_{xy}$, (**c**) $\mathbf{g}_{xz}$ (**d**) $\mathbf{g}_{yy}$, and (**e**) $\mathbf{g}_{yz}$ components (in eotvos). Figure S12: Sensitivity Test 1—Noise Level: 5%. Synthetic noise input data: Six noise-contaminated Falcon AGG datasets each generated using a different pseudo-random Gaussian noise realization. Each row corresponds to one noise realization. The left and right columns display the $\mathbf{g}_{xy}$ and $\mathbf{g}_{uv}$ components, respectively. Each panel includes an inset showing the histogram of the additive zero-mean Gaussian noise applied to the corresponding noise-free component. The noise standard deviation was set to 5% of the maximum absolute value of each component. Figure S13. Sensitivity Test 1—Noise Level: 5% Residual analysis and convergence behavior for the six noise-contaminated synthetic Falcon AGG datasets. The noise standard deviation was set to 5% of the maximum absolute value of each component. Each row corresponds to one noise realization. Columns show: residual maps for the $\mathbf{g}_{xy}$ and $\mathbf{g}_{uv}$ components and convergence curves displaying RMSE as a function of iteration. Insets in the residual maps show histograms of the predicted residuals obtained using the inversion method described in Algorithm 1. Figure S14. Sensitivity Test 1—Noise Level: 5% The transformation results: left column displays the predicted $\mathbf{g_{zz}}$ component and the right column shows the predicted $\|\nabla \mathbf{g_z}\|$. Both clearly delineate the boundaries of the two simulated geologic sources. The noise standard deviation was set to 5% of the maximum absolute value of each component. Figure S15: Sensitivity Test 2—Noise Level: 7%. Synthetic noise input data: Six noise-contaminated Falcon AGG datasets, each generated using a different pseudo-random Gaussian noise realization. Each row corresponds to one noise realization. The left and right columns display the the $\mathbf{g}_{xy}$ and $\mathbf{g}_{uv}$ components, respectively. Each panel includes an inset showing the histogram of the additive zero-mean Gaussian noise applied to the corresponding noise-free component. The noise standard deviation was set to 7% of the maximum absolute value of each component. Figure S16. Sensitivity Test 2—Noise Level: 7% Residual analysis and convergence behavior for the six noise-contaminated synthetic Falcon AGG datasets. The noise standard deviation was set to 7% of the maximum absolute value of each component. Each row corresponds to one noise realization. Columns show: residual maps for the $\mathbf{g}_{xy}$ and $\mathbf{g}_{uv}$ components and convergence curves displaying RMSE as a function of iteration. Insets in the residual maps show histograms of the predicted residuals obtained using the inversion method described in Algorithm 1. Figure S17. Sensitivity Test 2—Noise Level: 7% The transformation results: left column displays the predicted $\mathbf{g_{zz}}$ component and the right column shows the predicted $\|\nabla \mathbf{g_z}\|$. Both clearly delineate the boundaries of the two simulated geologic sources. The noise standard deviation was set to 7% of the maximum absolute value of each component. Figure S18: Sensitivity Test 3—Noise Level: 9%. Synthetic noise input data: Six noise-contaminated Falcon AGG datasets, each generated using a different pseudo-random Gaussian noise realization. Each row corresponds to one noise realization. The left and right columns display the $\mathbf{g}_{xy}$ and $\mathbf{g}_{uv}$ components, respectively. Each panel includes an inset showing the histogram of the additive zero-mean Gaussian noise applied to the corresponding noise-free component. The noise standard deviation was set to 9% of the maximum absolute value of each component. Figure S19. Sensitivity Test 3—Noise Level: 9% Residual analysis and convergence behavior for the six noise-contaminated synthetic Falcon AGG datasets. The noise standard deviation was set to 9% of the maximum absolute value of each component. Each row corresponds to one noise realization. Columns show: residual maps for the $\mathbf{g}_{xy}$ and $\mathbf{g}_{uv}$ components, and convergence curves displaying RMSE as a function of iteration. Insets in the residual maps show histograms of the predicted residuals obtained using the inversion method described in Algorithm 1. Figure S20. Sensitivity Test 3—Noise Level: 9% The transformation results: left column displays the predicted $\mathbf{g_{zz}}$ component and the right column shows the predicted $\|\nabla \mathbf{g_z}\|$. Both clearly delineate the boundaries of the two simulated geologic sources. The noise standard deviation was set to 9% of the maximum absolute value of each component. Table S1: Noise standard deviations (in eotvos) corresponding to 5%, 7%, and 9% of the maximum absolute value of the noise-free $\mathbf{g}_{xy}$ and $\mathbf{g}_{uv}$ components.

Author Contributions: V.C.O.J. and V.C.F.B.: conceptualization, methodology, formal analysis, supervision, validation, visualization, writing—original draft, writing—review and editing; L.S.P.: investigation, software, validation, visualization. All authors have read and agreed to the published version of the manuscript.

Funding: This study was financed by the Brazilian agencies CAPES (in the form of a scholarship), FAPERJ (grant nos. E-26 202.729/2018 and 26/202.582/2019), and CNPq (grant nos. 315768/2020-7 and 309624/2021-5).

Data Availability Statement: Data derived from public domain resources. The Falcon AGG data over Kauring test site, Australia, are available in https://www.ga.gov.au/about/projects/resources/kauring (accessed on 23 July 2025). The Falcon AGG data over Canobie, Australia, are available in https://ecat.ga.gov.au/geonetwork/srv/eng/catalog.search#/metadata/146760 (accessed on 23 July 2025).

Conflicts of Interest: The authors declare no conflicts of interest.

References

1. Droujinine, A.; Vasilevsky, A.; Evans, R. Feasibility of using full tensor gradient (FTG) data for detection of local lateral density contrasts during reservoir monitoring. *Geophys. J. Int.* **2007**, *169*, 795–820. [CrossRef]
2. Uieda, L.; Barbosa, V.C.F. Robust 3D gravity gradient inversion by planting anomalous densities. *Geophysics* **2012**, *77*, G55–G66. [CrossRef]
3. Oliveira, V.C., Jr.; Barbosa, V.C.F. 3-D radial gravity gradient inversion. *Geophys. J. Int.* **2013**, *195*, 883–902. [CrossRef]
4. Carlos, D.U.; Uieda, L.; Barbosa, V.C.F. Imaging iron ore from the Quadrilátero Ferrífero (Brazil) using geophysical inversion and drill hole data. *Ore Geol. Rev.* **2014**, *61*, 268–285. [CrossRef]
5. Martinez, C.; Li, Y. Denoising of gravity gradient data using an equivalent source technique. *Geophysics* **2016**, *81*, G67–G79. [CrossRef]
6. Vera, N.; Couder-Castañeda, C.; Hernández, J.; Trujillo-Alcántara, A.; Orozco-del Castillo, M.; Ortiz-Aleman, C. OpenMP implementation of a novel potential-field-data source-growth-based inversion approach for 3D salt imaging in deepwater gulf of Mexico. *Appl. Sci.* **2020**, *10*, 4798. [CrossRef]
7. Maag-Capriotti, E.; Li, Y. Understanding the information content in gravity gradiometry data through constrained inversions for salt bodies. *Geophysics* **2021**, *86*, G35–G53. [CrossRef]
8. Soulaimani, A.; Chakiri, S.; Soulaimani, S.; Bejjaji, Z.; Miftah, A.; Manar, A. Gradiometry Processing Techniques for Large-Scale of Aeromagnetic Data for Structural and Mining Implications: The Case Study of Bou Azzer Inlier, Central Anti-Atlas, Morocco. *Appl. Sci.* **2023**, *13*, 9962. [CrossRef]
9. Mickus, K.L.; Hinojosa, J.H. The complete gravity gradient tensor derived from the vertical component of gravity: A Fourier transform technique. *J. Appl. Geophys.* **2001**, *46*, 159–174. [CrossRef]
10. Piauilino, L.S.; Siqueira, F.C.L.; Oliveira, V.C., Jr.; Barbosa, V.C.F. Estimative of gravity-gradient tensor components via fast iterative equivalent-layer technique. In *SEG Technical Program Expanded Abstracts 2019*; Society of Exploration Geophysicists: Tulsa, OK, USA, 2019; pp. 1714–1718. [CrossRef]
11. Lyrio, J.C.S.O.; Tenorio, L.; Li, Y. Efficient automatic denoising of gravity gradiometry data. *Geophysics* **2004**, *69*, 772–782. [CrossRef]

12. Jiang, F.Y.; Huang, Y.; Yan, K. Full gravity gradient tensors from vertical gravity by cosine transform. *Appl. Geophys.* **2012**, *9*, 247–260. [CrossRef]
13. Abtahi, S.M.; Pedersen, L.B.; Kamm, J.; Kalscheuer, T. Consistency investigation, vertical gravity estimation, and inversion of airborne gravity gradient data—A case study from northern Sweden. *Geophysics* **2016**, *81*, B65–B76. [CrossRef]
14. Zhang, D.; Huang, D.; Yu, P.; Yuan, Y. Translation-invariant wavelet denoising of full-tensor gravity–gradiometer data. *Appl. Geophys.* **2017**, *14*, 606–619. [CrossRef]
15. Fang, Y.; Wang, J.; Meng, X.; Zheng, S.; Tang, H. An efficient cross-gradient joint inversion algorithm for gravity and magnetic data using a sequential strategy. *IEEE Trans. Geosci. Remote Sens.* **2022**, *60*, 1–16. [CrossRef]
16. Yuan, Y.; Qin, G.; Li, D.; Zhong, M.; Shen, Y.; Ouyang, Y. Real-time joint filtering of gravity and gravity gradient data based on improved Kalman filter. *IEEE Trans. Geosci. Remote Sens.* **2024**, *62*, 1–12. [CrossRef]
17. Jiang, T.; Ke, B.; Yu, X.; Yu, L.; Yang, M.; Fan, J.; Hu, C.; Feng, W.; Liu, H.; Zhong, M.; et al. Gravity gradient signal extraction based on time-frequency feature threshold method. *J. Appl. Geophys.* **2025**, *234*, 105643. [CrossRef]
18. Dampney, C.N.G. The equivalent source technique. *Geophysics* **1969**, *34*, 39–53. [CrossRef]
19. Leão, J.W.D.; Silva, J.B.C. Discrete linear transformations of potential field data. *Geophysics* **1989**, *54*, 497–507. [CrossRef]
20. Xia, J.; Sprowl, D.R. Correction of topographic distortion in gravity data. *Geophysics* **1991**, *56*, 537–541. [CrossRef]
21. Xia, J.; Sprowl, D.R.; Adkins-Heljeson, D. Correction of topographic distortions in potential-field data—A fast and accurate approach. *Geophysics* **1993**, *58*, 515–523. [CrossRef]
22. Cordell, L. A scattered equivalent-source method for interpolation and gridding of potential-field data in three dimensions. *Geophysics* **1992**, *57*, 629–636. [CrossRef]
23. Mendonça, C.A.; Silva, J.B.C. The equivalent data concept applied to the interpolation of potential field data. *Geophysics* **1994**, *59*, 722–732. [CrossRef]
24. Dransfield, M.H.; Lee, J.B. The FALCON airborne gravity gradiometer survey systems. In *Airborne Gravity 2004—Abstracts from the ASEG-PESA Airborne Gravity 2004 Workshop*; Lane, R., Ed.; Geoscience Australia Record 2004/18; Geoscience Australia: Canberra, Australia, 2004; pp. 15–19.
25. Guspí, F.; Novara, I. Reduction to the pole and transformations of scattered magnetic data using Newtonian equivalent sources. *Geophysics* **2009**, *74*, L67–L73. [CrossRef]
26. Li, Y.; Oldenburg, D.W. Rapid construction of equivalent sources using wavelets. *Geophysics* **2010**, *75*, L51–L59. [CrossRef]
27. Barnes, G.; Lumley, J. Processing gravity gradient data. *Geophysics* **2011**, *76*, I33–I47. [CrossRef]
28. Siqueira, F.C.L.; Oliveira, V.C., Jr.; Barbosa, V.C.F. Fast iterative equivalent-layer technique for gravity data processing: A method grounded on excess mass constraint. *Geophysics* **2017**, *82*, G57–G69. [CrossRef]
29. Jirigalatu; Ebbing, J. A fast equivalent source method for airborne gravity gradient data. *Geophysics* **2019**, *84*, G75–G82. [CrossRef]
30. Mendonça, C.A. Subspace method for solving large-scale equivalent layer and density mapping problems. *Geophysics* **2020**, *85*, G57–G68. [CrossRef]
31. Soler, S.R.; Uieda, L. Gradient-boosted equivalent sources. *Geophys. J. Int.* **2021**, *227*, 1768–1783. [CrossRef]
32. Bu, J.; Ye, Z.; Liang, X.; Liu, L.; Liu, J.; Fu, T.; Zhang, C.; Zhu, Y. Marine gravity gradient model calculation based on wavelet numerical integration and CUDA parallel. *Comput. Geosci.* **2025**, *196*, 105852. [CrossRef]
33. Takahashi, D.; Oliveira, V.C., Jr.; Barbosa, V.C.F. Convolutional equivalent layer for gravity data processing. *Geophysics* **2020**, *85*, G129–G141. [CrossRef]
34. Takahashi, D.; Oliveira, V.C., Jr.; Barbosa, V.C.F. Convolutional equivalent layer for magnetic data processing. *Geophysics* **2022**, *87*, 1–59. [CrossRef]
35. Pedersen, L.B.; Rasmussen, T.M. The gradient tensor of potential field anomalies: Some implications on data collection and data processing of maps. *Geophysics* **1990**, *55*, 1558–1566. [CrossRef]
36. Dransfield, M.H. Airborne gravity gradiometry in the search for mineral deposits. In *Exploration in the New Millenium: Proceedings of Exploration 07, Fifth Decennial International Conference on Mineral Exploration, Toronto, ON, Canada, 9–12 September 2007*; Milkereit, B., Ed.; Decennial Mineral Exploration Conferences: Toronto, ON, Canada, 2007; pp. 341–354.
37. Aster, R.; Borchers, B.; Thurber, C. *Parameter Estimation and Inverse Problems*; Elsevier: Amsterdam, The Netherlands, 2018.
38. Golub, G.H.; Van Loan, C.F. *Matrix Computations*; JHU Press: Baltimore, MD, USA, 2013.
39. Oliveira Junior, V.C.; Takahashi, D.; Reis, A.L.; Barbosa, V.C. Computational aspects of the equivalent-layer technique. *Front. Earth Sci.* **2023**, *11*, 1253148. [CrossRef]
40. Martinez, C.; Li, Y. Gravity gradiometry processing and interpretation at the Kauring Test Site. In Proceedings of the 2012 SEG Annual Meeting, Las Vegas, Nevada, 4–9 November 2012; OnePetro: Richardson, TX, USA, 2012. [CrossRef]
41. Ennen, C.; Hall, S. Structural mapping of the Vinton salt dome, Louisiana, using gravity gradiometry data. In *SEG Technical Program Expanded Abstracts 2011*; Society of Exploration Geophysicists: Tulsa, OK, USA, 2011; pp. 830–835. [CrossRef]

42. Howard, D.; Grujic, M.; Lane, R. The Kauring airborne gravity and airborne gravity gradiometer test site, Western Australia. In *Proceedings of the Airborne Gravity 2010—Abstracts from the ASEG-PESA Airborne Gravity 2010 Workshop*; Lane, R., Ed.; Number 2010/23 in Geoscience Australia Record; Geoscience Australia: Canberra, Australia, 2010; pp. 107–114.
43. Howard, D.; Brett, J. *The Kauring Airborne Geophysical Test Range, Western Australia: A Non-Technical Overview*; Report; Geological Survey of Western Australia: Perth, Australia, 2010.
44. Liu, J.Z.; Li, H.L. Separation and interpretation of gravity field data based on two dimensional normal space-scale transform (NSST2D) algorithm: A case study of Kauring airborne gravity test site, western Australia. *Pure Appl. Geophys.* **2019**, *176*, 2513–2528. [CrossRef]
45. Coker, M.O.; Bhattacharya, J.P.; Marfurt, K.J. Fracture Patterns Within Mudstones on the Flanks of a Salt Dome: Syneresis or Slumping? 2007. Available online: https://mcee.ou.edu/aaspi/publications/2007/coker_etal_GCAGST2007.pdf (accessed on 23 July 2025).
46. Niu, T.; Zhang, G.; Zhang, M.; Zhang, G. Joint inversion of gravity and gravity gradient data using smoothed L 0 norm regularization algorithm with sensitivity matrix compression. *Front. Earth Sci.* **2023**, *11*, 1283238. [CrossRef]
47. Geoscience Australia. Canobie Airborne Gravity Gradiometer Survey, 2021. Dataset, 2021. Project Number: P5021. Includes Point-Located Data, Gridded Products, and QA Report. Available online : https://ecat.ga.gov.au/geonetwork/srv/eng/catalog.search#/metadata/146760 (accessed on 23 July 2025).
48. Reid, N.; Thorne, R.L.; Tang, J.; Purdy, D.; Brown, D.; Plet, C.; Kamber, B.; Pinchand, T. *Hydrogeochemistry of Eastern Isa Block*; Technical Report 2021/03, CSIRO Report EP2021-1969; Geological Survey of Queensland: Brisbane, Australia, 2021.
49. Geological Survey of Queensland. *North-West Queensland Mineral and Energy Province Report and Data Package*; Technical Report 1; CD-ROM; Queensland Department of Employment, Economic Development and Innovation: Brisbane, Australia, 2011; Volume 1.

Disclaimer/Publisher's Note: The statements, opinions and data contained in all publications are solely those of the individual author(s) and contributor(s) and not of MDPI and/or the editor(s). MDPI and/or the editor(s) disclaim responsibility for any injury to people or property resulting from any ideas, methods, instructions or products referred to in the content.

applied sciences

MDPI

Article

An Aeromagnetic Compensation Algorithm Based on a Temporal Convolutional Network

Han Wang and Boxin Zuo *

School of Computer Science, China University of Geosciences (Wuhan), Wuhan 430078, China; 1202211215@cug.edu.cn
* Correspondence: boxzuo@cug.edu.cn

Abstract: Aeromagnetic compensation is the main method for eliminating magnetic interference on flight platforms. With the improved sensitivity of aeromagnetic sensors, the generalization and compensation accuracy of existing aeromagnetic compensation methods have become insufficient to meet the needs of current aeromagnetic survey applications. In this article, we propose an aeromagnetic compensation method based on temporal convolutional networks to improve both generalization and compensation accuracy. In the proposed method, a neural network based on separable convolution with a residual connection is employed to improve the compensation accuracy and convergence stability, and a gradient correction loss function based on the Tolles–Lawson model is used to improve the generalization ability. We conducted experiments based on both simulated and real datasets and compared typical neural network compensation methods proposed by previous researchers. The results indicate that the method proposed in this article can achieve a lower standard deviation of residual magnetic interference than other neural network methods, demonstrating a better compensation performance and generalization ability.

Keywords: aeromagnetic compensation; deep learning; temporal convolutional network

1. Introduction

Aeromagnetic surveys are widely used to aid in the production of geological maps and are also commonly used during mineral exploration and petroleum exploration. The signals collected during aeromagnetic surveys include various interferences, which affect the survey's accuracy. Therefore, it is necessary to study aeromagnetic compensation to eliminate aeromagnetic interference.

In the early 1950s, Tolles and Lawson first analyzed the main sources of magnetic interference on flight platforms [1–4], including those from intrinsic magnetic fields, induced magnetic fields, and eddy-current magnetic fields. In 1961, Leliak further proposed a mathematical model for describing aeromagnetic interference: the Tolles–Lawson model (T–L model) [5]. In 1980, Leach described the solution of the T–L model as a linear regression problem and proposed a ridge regression method, which reduces the multicollinearity of the solving process [6]. In 2022, Feng proposed a compensation method to resist the interference of geomagnetic gradients in aeromagnetic surveys [7]. In 2023, Chen proposed a joint compensation method based on an inertial navigation system and a three-axis fluxgate magnetometer to improve the compensation accuracy [8] and then proposed a compensation method for airborne electronic equipment interference [9].

Considering the insufficient accuracy of traditional linear regression methods, more researchers have started studying neural-network-based aeromagnetic compensation methods. In 1993, Williams first used neural networks for aeromagnetic compensation and tested the effects of fully connected neural networks at different layers. The results showed that the neural network compensation method had a similar accuracy to the T–L model compensation method and could be used as an alternative [10].

In the study on maneuvering interference and low-frequency interference compensation methods (usually within 0.1–0.6 Hz frequency [11]), Ma proposed a dual estimation method based on an unscented Kalman filter in 2017, which combines the T–L model with a neural network to improve the compensation accuracy, and noise was added manually to solve the overfitting problem of neural networks [12]. In 2020, Xu proposed an end-to-end neural network model for magnetic anomaly detection and aeromagnetic compensation. The model consisted of two parts: a binary classification network for magnetic anomaly detection and a regression network for geomagnetic noise suppression [13]. In 2021, Yu proposed a neural network compensation method based on a generalized regression neural network to solve the overfitting problem of neural networks [14]. In the same year, Li designed a neural network compensation method based on physical features with the Levenberg–Marquardt backpropagation algorithm, which displayed an improved compensation accuracy [15]. In 2022, Yu proposed a neural network compensation method that combines a backpropagation neural network and a residual connection to address the vanishing-gradient problem during neural network training [16]. In the same year, Zhou proposed a compensation method based on a radial basis function artificial neural network, which uses the radial basis function as the activation function to improve the neural network's generalization [17]. Jiao proposed a model compression method to optimize the backpropagation neural network to solve the problem of insufficient computing resources during the training process on unmanned aerial vehicles [18]. In 2024, Wang proposed an incremental learning neural network compensation method that reduces the time required for neural network retraining [19].

In the study of high-frequency interference compensation algorithms, Albert used open flight data from the Massachusetts Institute of Technology and the Department of the Air Force Artificial Intelligence Accelerator Program (hereafter DAF-MIT AIA Open Flight Data) [20] and a simple, fully connected neural network to analyze the impact of different inputs on compensation accuracy in 2022 [21]. The results show that the current and voltage data input of some airborne electronic equipment can significantly improve compensation accuracy. In the same year, Wang proposed a compensation method combining wavelet decomposition and a long short-term memory neural network to reduce interference with airborne electronic equipment [22].

In recent years, time convolutional networks have been widely applied in various temporal tasks [23–25]. In order to establish connections between features and aeromagnetic interference in time series, improve both generalization and compensation accuracy, prevent exploding gradients, and achieve fast convergence, in this article, we propose an aeromagnetic compensation method based on temporal convolutional networks (hereafter MagTCN). In the proposed method, a neural network based on separable convolution with a residual connection is employed to improve the compensation accuracy and convergence stability, and a gradient correction loss function based on the T–L model is used to improve the generalization ability. MagTCN was applied to simulation datasets and DAF-MIT AIA Open Flight Data to verify its effectiveness.

The structure of this article is as follows: Section 2 describes the proposed method. Section 3 discusses the experimental details, including dataset preparation, model parameters, and evaluation metrics. Section 4 provides the results and discussion. Section 5 summarizes the conclusions.

2. Materials and Methods

2.1. Tolles–Lawson Model

The Tolles–Lawson model (T–L model) is widely used in aeromagnetic compensation of fixed-wing aircraft. The basic idea is to use the magnetic measurements of the vector magnetometer to calibrate the scalar magnetometer.

The main components of aeromagnetic survey data can be expressed as follows:

$$\mathbf{B}_{\text{tmi}} = \mathbf{B}_{\text{geo}} + \mathbf{B}_{\text{tgt}} + \mathbf{B}_{\text{a}} + \mathbf{B}_{\text{tv}} \tag{1}$$

where $\mathbf{B}_{\text{tmi}}$ represents the total magnetic field acquired by the magnetic sensor; $\mathbf{B}_{\text{geo}}$ represents the geomagnetic field; $\mathbf{B}_{\text{tgt}}$ represents the magnetic anomaly of the target; $\mathbf{B}_{\text{a}}$ represents the magnetic interference caused by the flight platform; and $\mathbf{B}_{\text{tv}}$ represents the diurnal variation in the geomagnetic field.

During short-term navigation, the diurnal variation in the geomagnetic field $\mathbf{B}_{\text{tv}}$ can be removed through monitoring stations. When there is no target, it can be assumed that the total magnetic field $\mathbf{B}_{\text{tmi}}$ only includes the geomagnetic field $\mathbf{B}_{\text{geo}}$ and the magnetic interference of the flight platform $\mathbf{B}_{\text{a}}$, which can be expressed as follows:

$$\mathbf{B}_{\text{tmi}} = \mathbf{B}_{\text{geo}} + \mathbf{B}_{\text{a}} \tag{2}$$

The above equation can be further derived as follows:

$$\left|\mathbf{B}_{\text{geo}}\right|^2 = \mathbf{B}_{\text{geo}} \cdot \mathbf{B}_{\text{geo}} = (\mathbf{B}_{\text{tmi}} - \mathbf{B}_{\text{a}}) \cdot (\mathbf{B}_{\text{tmi}} - \mathbf{B}_{\text{a}}) \tag{3}$$

$$\left|\mathbf{B}_{\text{geo}}\right|^2 = \mathbf{B}_{\text{tmi}} \cdot \mathbf{B}_{\text{tmi}} - 2\mathbf{B}_{\text{tmi}} \cdot \mathbf{B}_{\text{a}} + \mathbf{B}_{\text{a}} \cdot \mathbf{B}_{\text{a}} \tag{4}$$

$$\left|\mathbf{B}_{\text{geo}}\right| = \sqrt{\left|\mathbf{B}_{\text{tmi}}\right|^2 - 2\mathbf{B}_{\text{tmi}} \cdot \mathbf{B}_{\text{a}} + \left|\mathbf{B}_{\text{a}}\right|^2} \tag{5}$$

$$\left|\mathbf{B}_{\text{geo}}\right| = \left|\mathbf{B}_{\text{tmi}}\right| \sqrt{1 - 2\frac{\mathbf{B}_{\text{tmi}} \cdot \mathbf{B}_{\text{a}}}{\left|\mathbf{B}_{\text{tmi}}\right|^2} + \frac{\left|\mathbf{B}_{\text{a}}\right|^2}{\left|\mathbf{B}_{\text{tmi}}\right|^2}} \tag{6}$$

The magnetic interference of a flight platform, with a capacity of about 100 nT, is much smaller than that of the geomagnetic field, with a capacity of about 50,000 nT, so the $\left|\mathbf{B}_{\text{a}}\right|^2 / \left|\mathbf{B}_{\text{tmi}}\right|^2$ term can be ignored. Therefore, the above equation can be simplified as follows:

$$\left|\mathbf{B}_{\text{geo}}\right| \approx \left|\mathbf{B}_{\text{tmi}}\right| \sqrt{1 - 2\frac{\mathbf{B}_{\text{tmi}} \cdot \mathbf{B}_{\text{a}}}{\left|\mathbf{B}_{\text{tmi}}\right|^2}} \tag{7}$$

Using the first-order form of the series expansion formula, $\left|\mathbf{B}_{\text{geo}}\right|$ can be expressed as follows:

$$\left|\mathbf{B}_{\text{geo}}\right| \approx \left|\mathbf{B}_{\text{tmi}}\right| - \frac{\mathbf{B}_{\text{tmi}} \cdot \mathbf{B}_{\text{a}}}{\left|\mathbf{B}_{\text{tmi}}\right|} \tag{8}$$

By using a three-axis fluxgate magnetometer, the direction of the total magnetic field can be measured:

$$\hat{\mathbf{B}}_{\textbf{tmi}} \hat{\mathbf{B}}_{\text{tmi}} = \frac{\mathbf{B}_{\text{tmi}}}{\left|\mathbf{B}_{\text{tmi}}\right|} \tag{9}$$

Substituting Equation (9) into Equation (8) yields the following:

$$\left|\mathbf{B}_{\text{geo}}\right| \approx \left|\mathbf{B}_{\text{tmi}}\right| - \mathbf{B}_{\text{a}} \cdot \hat{\mathbf{B}}_{\text{tmi}} \tag{10}$$

$\mathbf{B}_{\text{a}} \cdot \hat{\mathbf{B}}_{\text{tmi}}$ represents the projection of the magnetic interference of the flight platform onto the total magnetic field, as shown in Figure 1.

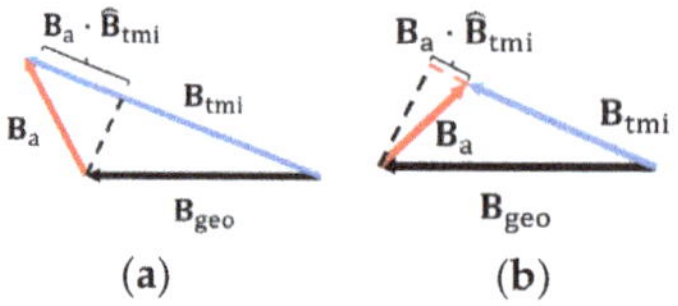

Figure 1. Modulus variation in the total magnetic field. (**a**) The situation where magnetic interference is in the same direction as the background geomagnetic field; (**b**) the situation where the magnetic interference is opposite to the background geomagnetic field.

If we can restore the geomagnetic field strength $|\mathbf{B}_{geo}|$, we can further obtain $\mathbf{B}_a \cdot \hat{\mathbf{B}}_{tmi}$ and establish the relationship between the flight platform attitude angle and $\mathbf{B}_a \cdot \hat{\mathbf{B}}_{tmi}$. Then, in the aeromagnetic survey, we can obtain the magnetic interference through the attitude angle and restore the magnetic anomaly of the target $\mathbf{B}_{tgt}$.

From the quasi-static electrodynamics model developed by Darwin [26], magnetic induction $\mathbf{H}$ can be expressed as follows:

$$\nabla \times \mathbf{H} = \mathbf{J} - \epsilon_0 \frac{d}{dt}\mathbf{E}_c \tag{11}$$

where $\mathbf{J}$ represents the current density; ϵ_0 represents the permittivity of free space; and $\mathbf{E}_c$ represents the electric field generated by Coulomb charges.

According to Ohm's law, Equation (11) can be further expressed as follows:

$$\nabla \times \mathbf{H} = \mathbf{J}_{perm} - \epsilon_0 \sigma^{-1} \frac{d}{dt}\mathbf{J}_{eddy} \tag{12}$$

where $\mathbf{J}_{perm}$ represents any unchanging currents; $-\epsilon_0 \sigma^{-1} \frac{d}{dt}\mathbf{J}_{eddy}$ represents the eddy currents.

Assume that $\mathbf{J}$ is permanent magnetization. Therefore, the time-varying field can be expressed as

$$\nabla \times \mathbf{H}_{eddy} = -\epsilon_0 \sigma^{-1} \frac{d}{dt}\mathbf{J}_{eddy} \tag{13}$$

A magnetization density $\mathbf{M}$ can be expressed as a bound current $\mathbf{J}$. Then, Equation (13) can be further expressed as

$$\mathbf{H}_{eddy} = -\epsilon_0 \sigma^{-1} \frac{d}{dt}\mathbf{M}_{eddy} \tag{14}$$

Further assuming that $\mathbf{M}_{eddy}$ is proportional to the external field $\mathbf{H}_{external}$, the relationship between the $\mathbf{B}$-field and the $\mathbf{H}$-field can be expressed as

$$\mathbf{B} = \mu_0(\mathbf{H} + \mathbf{M}) \tag{15}$$

where μ_0 represents the vacuum magnetic permeability.

In general, the magnetization model can be expressed as

$$\mathbf{M} = \mathbf{M}_{perm} + \mu \mathbf{H}_{external} \tag{16}$$

where μ represents a constant magnetic permeability.

Permanent magnetization $\mathbf{M}_{perm}$ and induced magnetization are both related to $\mathbf{H}_{external}$. Considering the contribution of eddy currents, the total magnetic field $\mathbf{B}_{tmi}$ can be expressed as

$$\mathbf{B}_{tmi} = \mu_0 \mathbf{H}_{external} + \mu_0 \left(\mathbf{M}_{perm} + \mu \mathbf{H}_{external} - \epsilon_0 \sigma^{-1} \frac{d}{dt}\mathbf{M}_{eddy} \right) \tag{17}$$

The second part on the right side of the above equation is the magnetic interference $\mathbf{B}_a$, which can be expressed as

$$\mathbf{B}_a = \mu_0 \mathbf{M}_{perm} + \mu_0 \boldsymbol{\mu} \mathbf{H}_{external} - \mu_0 \epsilon_0 \sigma^{-1} \frac{d}{dt} \mathbf{M}_{eddy} \quad (18)$$

The above equation can be simplified as follows:

$$\mathbf{B}_a = \mathbf{a} + \mathbf{b}\mathbf{B}_{tmi} + \mathbf{c}\dot{\mathbf{B}}_{tmi} \quad (19)$$

where **a**, **b**, and **c** represent the unknown coefficient vector; $\dot{()}$ represents a time-derivative.

Based on the above equation, the T–L model identifies the magnetic interference of the flight platform $\mathbf{B}_a$ in three parts: the constant magnetic interference field $\mathbf{B}_p$, the induced magnetic interference field $\mathbf{B}_{ind}$, and the eddy-current magnetic interference field $\mathbf{B}_{eddy}$. It can be expressed as follows:

$$\mathbf{B}_a = \mathbf{B}_p + \mathbf{B}_{ind} + \mathbf{B}_{eddy} \quad (20)$$

The inherent magnetic field of the flight platform is generated by permanent magnets, such as engines and motor cores, and its direction remains unchanged relative to the axial direction of the flight platform. The contribution of multiple intrinsic magnetic field sources to the load of the optical pump magnetic sensor is formed by vector superposition to form a fixed magnetic field bias, which can be expressed as follows:

$$\mathbf{B}_p = \begin{pmatrix} a_1 & a_2 & a_3 \end{pmatrix}^T \quad (21)$$

where **a** represents the intrinsic magnetic field coefficient composed of a_i.

The induced magnetic field of the flight platform is the result of the ferromagnetic components of the aircraft being excited by the background magnetic field, and its magnitude and direction are influenced by the strength, direction, and material magnetization of the components. The contribution of the induced magnetic field of the flight platform can be expressed as follows:

$$\mathbf{B}_{ind} = |\mathbf{B}_{tmi}| \begin{pmatrix} b_{11} & b_{12} & b_{13} \\ b_{21} & b_{22} & b_{23} \\ b_{31} & b_{32} & b_{33} \end{pmatrix} \hat{\mathbf{B}}_{tmi} \quad (22)$$

where **b** represents the induced magnetic field coefficient composed of b_{ij}.

The eddy-current magnetic field of the flight platform is generated by the change in magnetic flux caused by the cutting of magnetic induction lines on the fuselage metal. Eddy currents generate eddy magnetic fields through electromagnetic induction, and their magnitude and direction are influenced by the properties of conductive materials and the relative rate of change in external magnetic fields. They can be expressed as follows:

$$\mathbf{B}_{eddy} = |\mathbf{B}_{tmi}| \begin{pmatrix} c_{11} & c_{12} & c_{13} \\ c_{21} & c_{22} & c_{23} \\ c_{31} & c_{32} & c_{33} \end{pmatrix} \dot{\hat{\mathbf{B}}}_{tmi} \quad (23)$$

where **c** represents the eddy-current magnetic field coefficient composed of c_{ij}; $\dot{\hat{\mathbf{B}}}_{tmi}$ represents the rate of change in direction of $\mathbf{B}_{tmi}$.

Based on the above Equations (20)–(23), $\mathbf{B}_a$ can be expressed as follows:

$$\mathbf{B}_a = \left(\mathbf{a} + |\mathbf{B}_{tmi}|\, \mathbf{b}\, \hat{\mathbf{B}}_{tmi} + |\mathbf{B}_{tmi}|\, \mathbf{c}\, \dot{\hat{\mathbf{B}}}_{tmi} \right) \cdot \hat{\mathbf{B}}_{tmi} \quad (24)$$

Let cos X, cos Y, and cos Z represent the direction cosine of the angle between the airborne coordinate system and the geomagnetic field vector, and let $u_1 = \cos X$, $u_2 = \cos Y$, and $u_3 = \cos Z$. The relationship between u_1, u_2, and u_3 is as follows:

$$\begin{cases} u_1^2 + u_2^2 + u_3^2 = 1 \\ u_1 u_1' + u_2 u_2' + u_3 u_3' = 0 \end{cases} \tag{25}$$

$\mathbf{B}_\mathrm{a}$ can be simplified as follows:

$$\mathbf{B}_\mathrm{a} = B_\mathrm{geo} \begin{bmatrix} p_1 \\ p_2 \\ p_3 \\ p_4 \\ p_5 \\ p_6 \\ p_7 \\ p_8 \\ p_9 \\ p_{10} \\ p_{11} \\ p_{12} \\ p_{13} \\ p_{14} \\ p_{15} \\ p_{16} \end{bmatrix}^T \begin{bmatrix} u_1/B_\mathrm{geo} \\ u_2/B_\mathrm{geo} \\ u_3/B_\mathrm{geo} \\ u_1^2 \\ u_1 u_2 \\ u_1 u_3 \\ u_2 u_3 \\ u_3^2 \\ u_1 u_1' \\ u_2 u_1' \\ u_3 u_1' \\ u_1 u_3' \\ u_2 u_3' \\ u_3 u_3' \\ u_1 u_2' \\ u_3' u_2' \end{bmatrix} \tag{26}$$

where $p_1 \sim p_{16}$ represent the 16 coefficients to be calculated in the T–L model; B_geo represents the background magnetic field strength.

The background magnetic field term B_geo can be considered a constant and calculated together with the coefficient, so the value of $\mathbf{B}_\mathrm{a}$ can be further simplified as follows:

$$\mathbf{B}_\mathrm{a} = \mathbf{AC} \tag{27}$$

where $\mathbf{A}$ represents the coefficient matrix to be calculated in the T–L model; $\mathbf{C}$ represents the direction cosine matrix of the angle between the airborne coordinate system and the geomagnetic field vector.

During the calibration flight phase, we first collect $\mathbf{C}$ and $\mathbf{B}_\mathrm{a}$ and then calculate the coefficient matrix $\mathbf{A}$. During the survey phase, we calculate $\mathbf{B}_\mathrm{a}$ through $\mathbf{A}$ and the collected $\mathbf{C}$.

2.2. *MagTCN*

2.2.1. Separable Convolution Module

In MagTCN, we introduce separable convolutions, including depth-wise convolutions (DWConv) and point-wise convolutions (PWConv) [27]. Compared with the traditional convolution that mixes multi-dimensional information, the decoupled design makes it easier for MagTCN to model magnetic interference and reduces computational complexity.

As shown in Figure 2, for an input with a sequence length of T and K channels, DWConv simultaneously uses a $1 \times n \times 1$ filter for each channel, resulting in an output dimension of $1 \times Tfm \times K$.

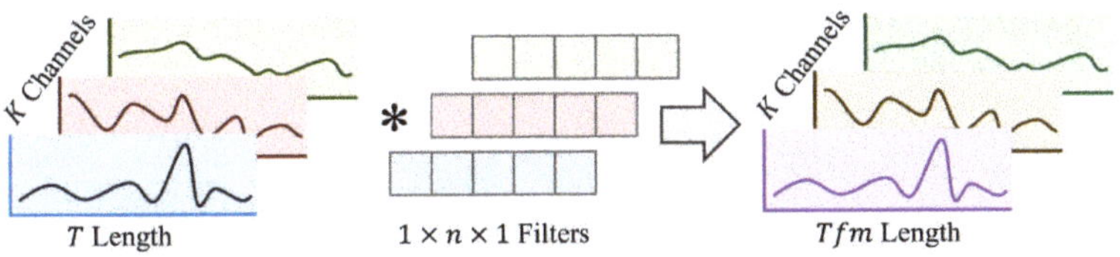

Figure 2. The convolution process of depth-wise convolutions. (*) indicates the operation of convolution on the input.

As shown in Figure 3, for an input with a sequence length of T and K channels, PWConv uses a filter with a dimension of $1 \times 1 \times M$ to cover all of the K channels. This performs stationary convolution on multiple variables at each time point, resulting in an output dimension of $1 \times T \times M$.

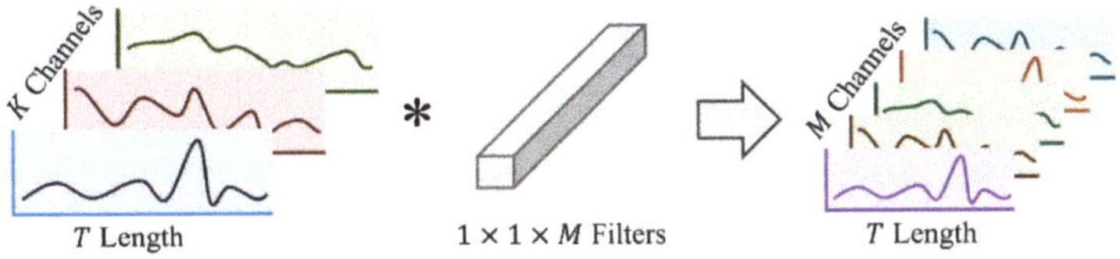

Figure 3. The convolution process of point-wise convolutions. (*) indicates the operation of convolution on the input.

2.2.2. Gaussian Error Linear Unit

The Gaussian Error Linear Unit (GELU) is a nonlinear activation function that provides random regularity to the activation function by inputting its own probability statistics to improve the generalization of neural networks [28]. The specific form can be expressed as follows:

$$\mathrm{GELU}(x) = x \times P(X \leq x) = x \times \mathbf{\Phi}(x) \tag{28}$$

where $\mathbf{\Phi}(x)$ represents the cumulative function of the Gaussian normal distribution of x, which can be expressed as follows:

$$x \times P(X \leq x) = x \int_{-\infty}^{x} \frac{e^{-\frac{(X-\mu)^2}{2\sigma^2}}}{\sqrt{2\pi\sigma}} dX \tag{29}$$

where μ and σ represent the mean and standard deviation of a normal distribution, respectively.

For input $x > 0$, the GELU has approximately linear output; for input $x < 0$, the output of the GELU is 0; when input x approaches 0, the GELU has nonlinear output, maintaining the continuity of the entire activation function.

2.2.3. Reversible Instance Normalization Module

In time-domain aeromagnetic compensation tasks, the difference in the mean or variance between the training and testing datasets leads to a decrease in compensation accuracy. To address the issue of distribution differences between different survey data, we introduced a reversible instance normalization module (RevIN) [29], on both sides of MagTCN, consisting of a normalization module and a denormalization module.

The input of the MagTCN backbone is represented as $X = \left\{x^{(i)}\right\}_{i=1}^{N}$, and the corresponding output is represented as $Y = \left\{y^{(i)}\right\}_{i=1}^{N}$. If K is the number of channels, T_x is the length of the input sequence, and T_y is the length of the output sequence, then there exists an input $x^{(i)} \in R^{K \times T_x}$ and an output $y^{(i)} \in R^{K \times T_y}$.

For the normalization part of the MagTCN backbone input, the mean and variance are used for normalization. The mean can be expressed as follows:

$$\mathbb{E}_t\left[x_{kt}^{(i)}\right] = \frac{1}{T_x}\sum_{j=1}^{T_x} x_{kj}^{(i)} \tag{30}$$

The variance can be expressed as follows:

$$\mathrm{Var}\left[x_{kt}^{(i)}\right] = \frac{1}{T_x}\sum_{j=1}^{T_x}\left(x_{kj}^{(i)} - \mathbb{E}_t\left[x_{kt}^{(i)}\right]\right)^2 \tag{31}$$

The normalization process can be expressed as follows:

$$\hat{x}_{kt}^{(i)} = \gamma_k\left(\frac{x_{kt}^{(i)} - \mathbb{E}_t\left[x_{kt}^{(i)}\right]}{\sqrt{\mathrm{Var}\left[x_{kt}^{(i)}\right] + \epsilon}}\right) + \beta_k \tag{32}$$

where $\gamma,\ \beta \in R^K$ represent learnable affine transformation parameters.

The denormalization process can be expressed as follows:

$$\hat{y}_{kt}^{(i)} = \sqrt{\mathrm{Var}\left[x_{kt}^{(i)}\right] + \epsilon}\cdot\left(\frac{\tilde{y}_{kt}^{(i)} - \beta_k}{\gamma_k}\right) + \mathbb{E}_t\left[x_{kt}^{(i)}\right] \tag{33}$$

2.2.4. The Structure of MagTCN

The structure of MagTCN is shown in Figure 4, where M represents the variable dimension, D represents the time dimension, and N represents the feature dimension. In MagTCN, we refer to the design of the Transformer [30]. DWConv is used to learn the time information between tokens for each feature, and a large kernel is used in DWConv to increase the receptive field and improve the temporal modeling ability, which has the same function as the self-attention module of the Transformer. Each point-wise convolutional block (PWBlock) consists of two PWConvs and one GELU. The first PWConv in PWBlock increases the dimension of the input vector, and the second one reduces the dimension of the vector again, which has the same function as the feedforward layer of the Transformer. The GELU in PWBlock is used to improve the generalization of MagTCN. Reshape and permute are used between the two PWBlocks to allow MagTCN to learn the temporal dependency and cross-variable dependency of features at the same time. The input of MagTCN is first normalized by RevIN's normalization module and then fed into the MagTCN backbone. After backbone processing, the data are fed into RevIN's denormalization module and residual connection. Finally, the output is formed by a linear layer and a flat layer.

The residual connection in MagTCN can be expressed as follows:

$$\mathbf{Z}^{\mathrm{r}} = f\left(\mathbf{Z}^{\mathrm{in}}\right) + \mathbf{Z}^{\mathrm{in}} \tag{34}$$

where $\mathbf{Z}^{\mathrm{r}}$ represents the output of the residual connection; $\mathbf{Z}^{\mathrm{in}}$ represents the input of MagTCN; and $f(\cdot)$ represents the backbone of MagTCN.

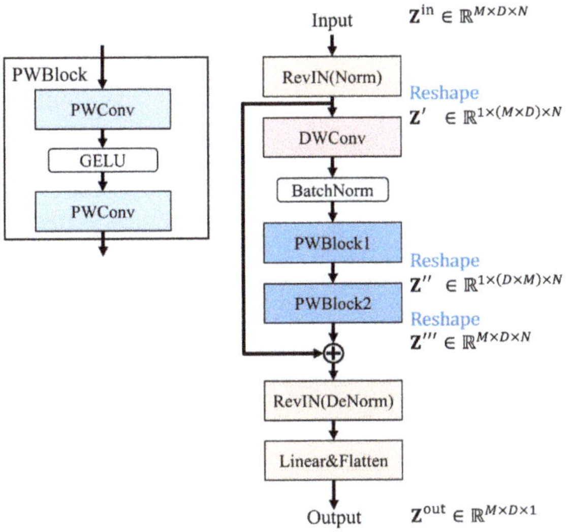

Figure 4. The neural network structure of the proposed method.

2.3. Loss Function

In order to ensure the steady improvement in the compensation accuracy and the generalization of MagTCN during neural network training, two types of losses are introduced: loss $\mathcal{L}_{TL}$, based on the T–L model, and truth loss $\mathcal{L}_{true}$. These two types of losses are based on the mean square error (MSE), which can be expressed as follows:

$$\text{MSE} = \frac{1}{n}\sum_{i=1}^{n}(y_i - \hat{y}_i)^2 \tag{35}$$

where y_i is the true label of the i-th sample, and $\hat{y}_i$ is the predicted value of the neural network model. When calculating $\mathcal{L}_{TL}$, y_i is the signal after aeromagnetic compensation via the T–L model; when calculating $\mathcal{L}_{true}$, y_i is the signal without interference.

In addition, the above two types of losses are combined and applied to neural network training through weights: at the kth iteration, the weight decay rate is v, the initial value of the weight decay coefficient is ρ_0, and the total weight value is W. The loss function $\mathcal{L}$ can be expressed as follows:

$$\mathcal{L} = \mathcal{L}_{TL} \times e^{-vk} + \mathcal{L}_{true} \times \left(W - e^{-vk}\right) \tag{36}$$

To prevent vanishing gradients, the gradient correction algorithm $f_{gc}(\cdot)$ is introduced for the above two losses.

The cosine similarity $\Phi(g_i, g_j)$ between the gradients g_i and g_j of the loss terms $\mathcal{L}_{TL} \times \rho_k$ and $\mathcal{L}_{true} \times (W - \rho_k)$ is calculated as follows:

$$\Phi(g_i, g_j) = \frac{2\|g_i\|_2\|g_j\|_2}{\|g_i\|_2^{\,2} + \|g_j\|_2^{\,2}} \tag{37}$$

The gradient g_i is updated to the following:

$$g_i = g_i - \frac{g_i \cdot g_j}{\|g_j\|^2} g_j \text{ when } \Phi < 0, \text{ else } g_i \tag{38}$$

The final loss function $\mathcal{L}$ can be expressed as follows:

$$\mathcal{L} = f_{gc}\left(\mathcal{L}_{TL} \times e^{-vk}, \mathcal{L}_{true} \times \left(W - e^{-vk}\right)\right) \tag{39}$$

3. Experimental Details

3.1. Data Preparation

3.1.1. Simulation Dataset

Based on the ArduPilot simulation flight control software, we simulated the aeromagnetic compensation calibration flight process for the fixed-wing aircraft Cessna-208B. The simulated flight path and maneuvering angle changes are shown in Figure 5, where the start recording position is 112.729° E, 16.720° N, the stop recording position is 112.734° E, 16.779° N (excluding takeoff and landing), and the average flight altitude is 3000 m. The simulated flight consists of two maneuvering circles, each consisting of four orthogonal flight segments with a length of 10 km on one side. The aircraft performs pitch, yaw, and roll movements of 5° to 15° in each of the four orthogonal directions.

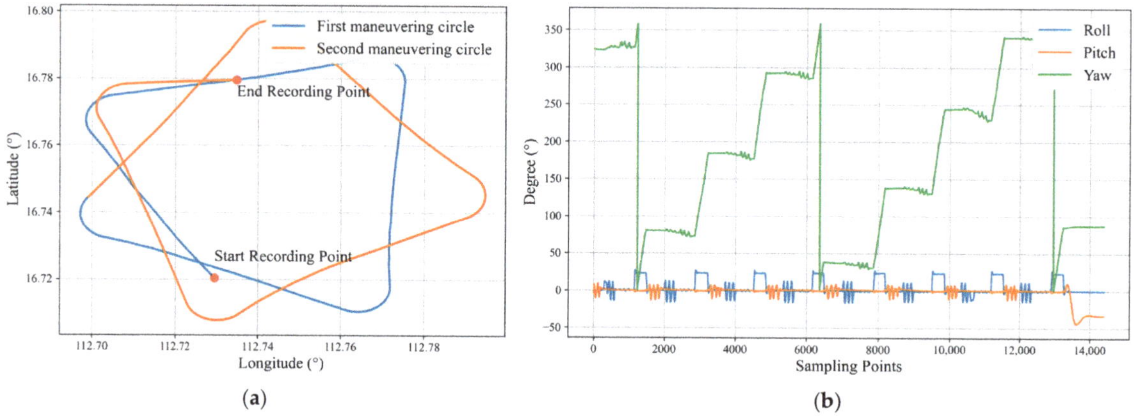

Figure 5. The flight maneuver simulation results: (**a**) the longitude and latitude of the flight path; (**b**) flight attitude changes during the route.

We simulated the main magnetic interference components of the flight platform based on the magnetic inclination, magnetic declination, and geomagnetic field strength at an altitude of 3000 m in the South China Sea (112.7° E, 17.6° N). The intrinsic magnetic field was simulated through the magnetic dipoles of motors and engines located on the aircraft's flaps, elevators, and rudders as field sources, as shown in Figure 6a. The induced magnetic field was simulated through finite element models of the aircraft's aluminum fuselage, steel fuel tank, and engine compartment, as shown in Figure 6b. In order to simulate the eddy-current magnetic field, the aircraft skin was regarded as equivalent to multiple coils located on the wings, nose, and tail, as shown in Figure 6c. Changes in the background magnetic field will cause changes in the magnetic flux passing through the coils, thereby generating eddy currents. The current will continue to change to form the corresponding eddy-current magnetic field. An optical pump magnetic sensor and a three-axis fluxgate sensor are installed in the center of the cabin to calculate the three types of magnetic interference at the sensor location.

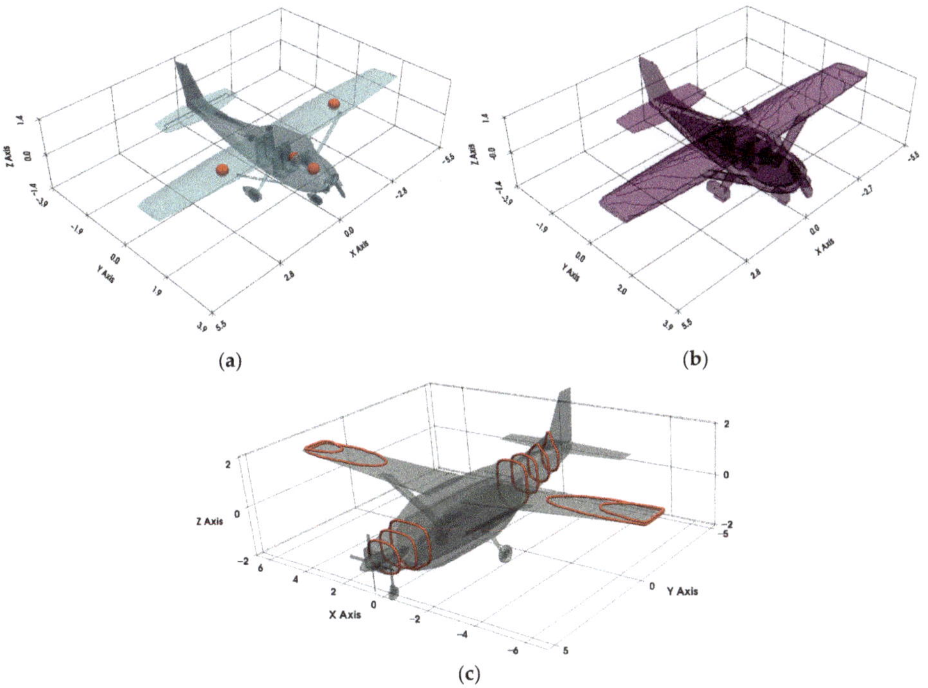

Figure 6. Illustration of flight platform interference simulation, where the gray part represents the aircraft model: (**a**) position of magnetic dipoles in intrinsic magnetic field simulation (indicated by red dots); (**b**) finite element division in induced magnetic field simulation (indicated by the purple square); (**c**) position of equivalent coils in eddy-current magnetic field simulation (indicated by red lines).

The magnetic signal on the route was collected at a frequency of 10 Hz based on the above simulation strategy, as shown in Figure 7.

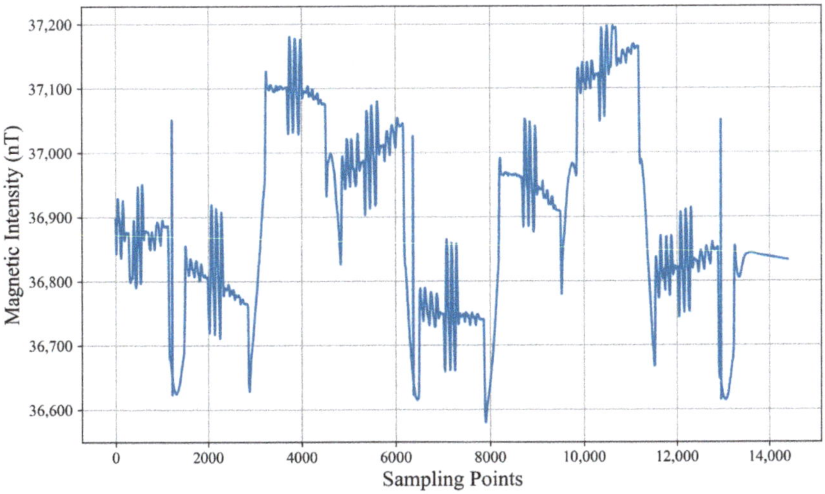

Figure 7. Total magnetic signal collected on the simulated route.

3.1.2. Real Dataset

The real dataset is based on flights 1002.02, 1002.20, and 1006.06 of the DAF-MIT AIA Open Flight Data [20], with 1002.02 used for training and the rest used for testing. The data collected include the aircraft attitude angle, magnetic sensors, accelerometers, longitude, latitude, altitude, current, and voltage. Among the five optical pump magnetic sensors, sensor 1 is located on the tail of the aircraft; sensors 2 and 5 are located at the front and rear of the cabin, respectively; and sensors 3 and 4 are located in the middle of the cabin. Fluxgate sensor B is located in the rear of the cabin, while sensors A, C, and D are located in the middle of the cabin. In this article, we will use data from optical pump sensor 5 and fluxgate sensor B.

3.2. Model Parameters

In terms of the input dimension of the neural network, based on a 10 Hz sampling rate, the input sequence length was set to 320 sampling points (32 s). The different datasets have different input features. In the simulation dataset experiment, 16 direction cosine coefficients (mentioned in Section 2.1) and three attitude angles (roll, pitch, and yaw) were selected as input features. In the real dataset experiment, nine additional current and voltage terms were added.

In terms of the convolutional layer parameters, the convolution kernel size of DWConv is set to 51, and the stride is set to 1; the convolution kernel size of PWConv is set to 1, and the stride is set to 1. For model training, the batch size is set to 256, the neural network optimizer is Adam, the learning rate is set to 0.001, and the number of training epochs is set to 200. In terms of the hardware environment, we used a computer with an Intel 12th Core Gen i5-13600KF CPU and a GeForce RTX 4060Ti GPU.

3.3. Evaluation Metrics

The standard deviation (STD) of the residual magnetic interference after aeromagnetic compensation and the improvement ratio (IR) are usually used to evaluate compensation accuracy. The standard deviation can be expressed as follows:

$$\mathrm{STD} = \sqrt{\frac{1}{n}\sum_{i=1}^{n}(x_i - \mu)^2} \tag{40}$$

where x represents the residual magnetic interference of the signal after aeromagnetic compensation, which is obtained by bandpass filtering (usually within 0.1–0.6 Hz frequency [11]); i represents the i-th sampling point; μ represents the arithmetic mean within the bandpass; and n represents the number of sampling points, which is consistent with the length of the original signal before aeromagnetic compensation.

The improvement rate is used to evaluate the degree of reduction in magnetic interference. It is calculated as the ratio of the standard deviation of magnetic interference in the bandpass before and after compensation and can be expressed as follows:

$$\mathrm{IR} = \frac{\mathrm{STD}_u}{\mathrm{STD}_c} \tag{41}$$

where STD_u represents the standard deviation of the magnetic interference in the bandpass of the original signal; STD_c represents the standard deviation of the magnetic interference in the bandpass after aeromagnetic compensation. The higher the improvement ratio, the higher the compensation accuracy.

4. Results and Discussion

4.1. Simulation Dataset Test Results

In this section, we compare MagTCN with several typical aeromagnetic compensation methods, including the three-branch neural network method proposed by Williams (hereafter WilliamsNN) [10], the simple backpropagation neural network method without a convolution structure (hereafter BPNN) [19], the convolutional residual neural network method using normal convolution (hereafter ResNet) [16], and the traditional linear regression T–L model method (hereafter T–L model) [6].

During the neural network training phase, we compared the convergence capabilities of several methods based on the data from the first maneuvering circle, as shown in Figure 8 and Table 1. The x-axis (Epochs) represents the number of times the entire training dataset is propagated forward and backward in the neural network. The y-axis (Loss) represents the error between the model's predicted values and the actual target values, as described in Section 2.3. Curves of different colors represent the loss between the predicted and label values of the corresponding neural network during the training process as the number of epochs increases. The faster the loss drops to a flat state, the faster the convergence. In Figure 8, WilliamsNN and BPNN do not fully converge at the 200th epoch. WilliamsNN has the slowest convergence speed and maintains a stable decline rate throughout the training phase. BPNN has a slightly better convergence speed, maintaining a faster decline rate before the sixth epoch and then maintaining a stable decline rate. ResNet and MagTCN both have faster convergence speeds and can achieve lower losses in the first few epochs, reaching full convergence in the 150th and 110th epochs, respectively. However, ResNet's loss fluctuates throughout the training phase and is relatively unstable. In contrast, MagTCN's loss is the fastest to reach a stable state with the same hyperparameters, demonstrating its fast and stable convergence ability.

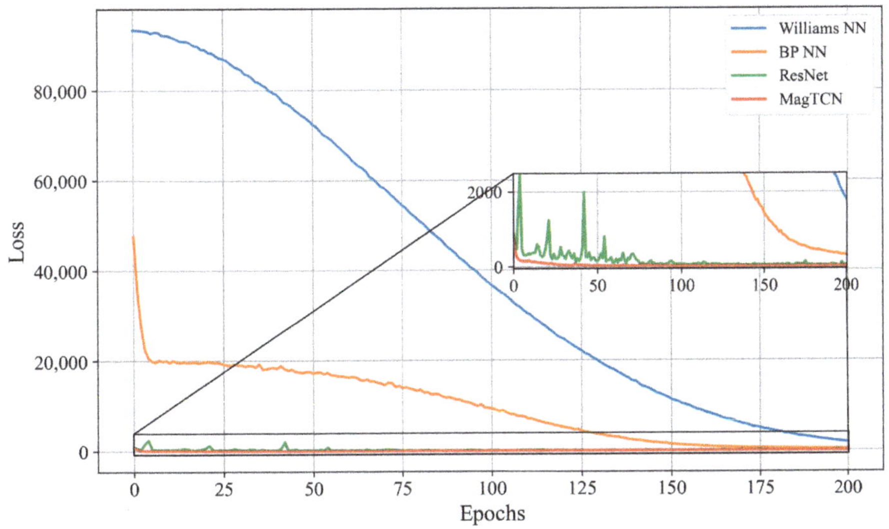

Figure 8. The losses of neural networks using simulation datasets.

Table 1. Statistics of neural network losses using simulation datasets.

Epoch	WilliamsNN Loss	BPNN Loss	ResNet Loss	MagTCN Loss
0	93,392.16	47,583.50	969.29	826.76
50	72,419.97	17,197.35	154.81	39.44
100	36,749.54	9321.64	47.71	16.32
150	11,337.10	1408.63	48.01	9.84
200	1792.37	303.06	42.36	13.76

Next, we compared the compensation accuracy and generalization of several compensation methods based on the data from the second maneuvering circle. Figure 9 shows the residual interference in the bandpass (0.1–0.6 Hz) of the magnetic signal on the route, where the curve named "origin" represents the interference before compensation, and the other curves represent the residual interference after compensation by the corresponding method. Large maneuvers during flight can cause large interferences (such as in the interval around the 1000th sampling point). The purpose of the compensation method is to eliminate these interferences as much as possible. The less residual interference there is after compensation, the better the compensation method. WilliamsNN has the worst compensation effect, and there is still more than 10 nT residual interference in some intervals (such as in the intervals from 4000 to 5000 sampling points and from 1500 to 2000 sampling points). BPNN has a slightly better compensation effect and can keep the residual interference within 1 nT in most intervals. However, in some intervals, residual interference is greater with BPNN than with WilliamsNN (as shown in the local enlarged figure of Figure 9). The T–L model, ResNet, and MagTCN all maintain low residual interference in the entire interval, making them much better than BPNN and WilliamsNN. In order to further compare the performance of several compensation methods, we calculated their STDs and IRs after compensation (as defined in Section 3.3, the lower the STD and the higher the IR, the better the compensation performance). As shown in Table 2, WilliamsNN has the worst STD and IR, which are 1.320 nT and 12.276, respectively. Those of BPNN are slightly better: 0.551 nT and 29.401, respectively. The compensation performance of the T–L model is better than that of BPNN and WilliamsNN, with an STD and IR of 0.037 nT and 427.641, respectively, which shows that the neural network compensation method that has not fully converged performs worse than the traditional T–L model compensation method. ResNet and MagTCN achieve full convergence in the training phase due to the convolutional structure and residual connection. Therefore, they are better than the T–L model in compensation performance. Among them, the STD and IR of ResNet are 0.033 nT and 480.082, respectively, and those of MagTCN are 0.024 nT and 651.61, respectively. The above results show that, based on the separable convolutional neural network structure and loss function proposed in this article, MagTCN is better than ResNet in compensation performance and has achieved the best compensation performance and generalization ability among the above compensation methods.

Table 2. The statistics of compensation results within the bandpass from simulation datasets.

	WilliamsNN	BPNN	ResNet	MagTCN	T–L Model
STD (nT)	1.320	0.551	0.033	0.024	0.037
IR	12.276	29.401	480.082	651.61	427.641

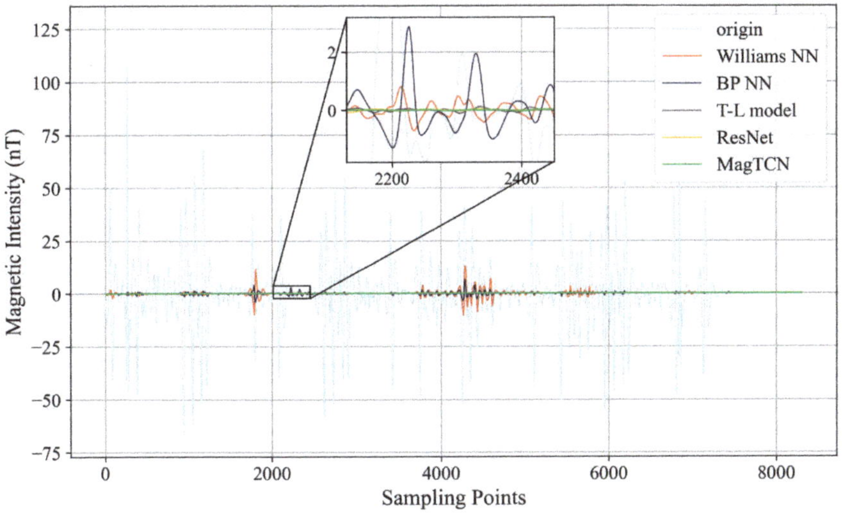

Figure 9. Compensation results within the bandpass from simulation datasets.

4.2. DAF-MIT AIA Open Flight Data Test Results for Flight 1002.20

In this section, we compare MagTCN with several typical aeromagnetic compensation methods, including WilliamsNN, BPNN, and ResNet.

During the neural network training phase, we compared the convergence capabilities of several methods based on flight 1002.02 of the DAF-MIT AIA Open Flight Data mentioned in Section 3.1.2, as shown in Figure 10 and Table 3. The x-axis (Epochs) represents the number of times the entire training dataset is propagated forward and backward in the neural network. The y-axis (Loss) represents the error between the model's predicted values and the actual target values, as described in Section 2.3. Curves of different colors represent the loss between the predicted and label values of the corresponding neural network during the training process as the number of epochs increases. The faster the loss drops to a flat state, the faster the convergence. In Figure 10, WilliamsNN and BPNN do not fully converge at the 200th epoch. WilliamsNN has the slowest convergence speed and maintains a stable decline rate throughout the training phase. BPNN has a slightly better convergence speed, maintaining a faster decline rate before the sixth epoch and then maintaining a stable decline rate. ResNet and MagTCN both have faster convergence speeds and can achieve lower losses in the first few epochs, reaching full convergence in the 150th and 100th epochs, respectively. However, ResNet's loss fluctuates throughout the training phase and is relatively unstable. In contrast, MagTCN's loss is the fastest to reach a stable state with the same hyperparameters, demonstrating its fast and stable convergence ability.

Next, we compared the compensation accuracy and generalization of several compensation methods based on flight 1002.20 of the DAF-MIT AIA Open Flight Data. Figure 11 shows the residual interference in the bandpass (0.1–0.6 Hz) of the magnetic signal on the route, where the curve named "origin" represents the interference before compensation, and the other curves represent the residual interference after compensation by the corresponding method. Large maneuvers during flight can cause large interference (such as around the 1000th sampling point). The purpose of the compensation method is to eliminate these interferences as much as possible. The less residual interference there is after compensation, the better the compensation method. WilliamsNN has the worst compensation effect, and there is still more than 5 nT residual interference in some intervals (such as in the intervals from 5000 to 5500 sampling points, from 7500 to 8000 sampling points, and from 12000 to 12500 sampling points). BPNN has a slightly better compensation effect and can keep the residual interference within 5 nT in most intervals. However, in some intervals, residual interference is greater with BPNN than with WilliamsNN (as shown in the local enlarged figure in Figure 11). ResNet and MagTCN all maintain low residual interference in the entire interval, making them much better than BPNN and WilliamsNN. In order to further compare the performance of several compensation methods, we calculated their STDs and IRs after compensation (as defined in Section 3.3, the lower the STD and the higher the IR, the better the compensation performance). As shown in Table 4, WilliamsNN has the worst STD and IR, which are 1.399 nT and 8.342, respectively. Those of BPNN are slightly better: 1.098 nT and 10.629, respectively. ResNet and MagTCN are better than WilliamsNN and BPNN in compensation performance due to their convolutional structures and residual connections. The STD and IR of ResNet are 0.028 nT and 402.786, respectively, and those of MagTCN are 0.025 nT and 455.920, respectively. The above results show that MagTCN has achieved the best compensation performance and generalization ability among the above compensation methods.

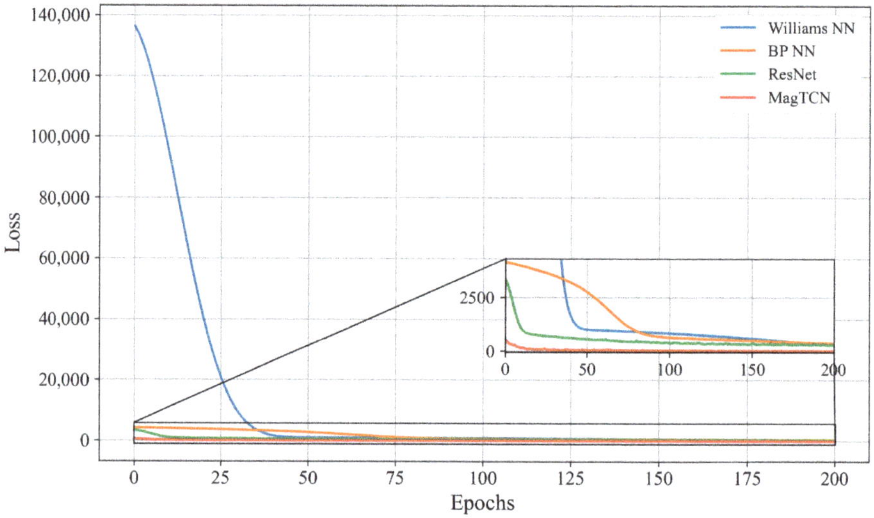

Figure 10. The losses of neural networks using real datasets.

Table 3. The statistics of neural network losses using real datasets.

Epoch	WilliamsNN Loss	BPNN Loss	ResNet Loss	MagTCN Loss
0	136,413.27	4169.42	3383.46	545.77
50	1003.05	2757.40	547.61	48.16
100	827.44	634.97	381.90	24.51
150	581.94	498.35	369.87	19.68
200	301.49	366.99	288.07	11.99

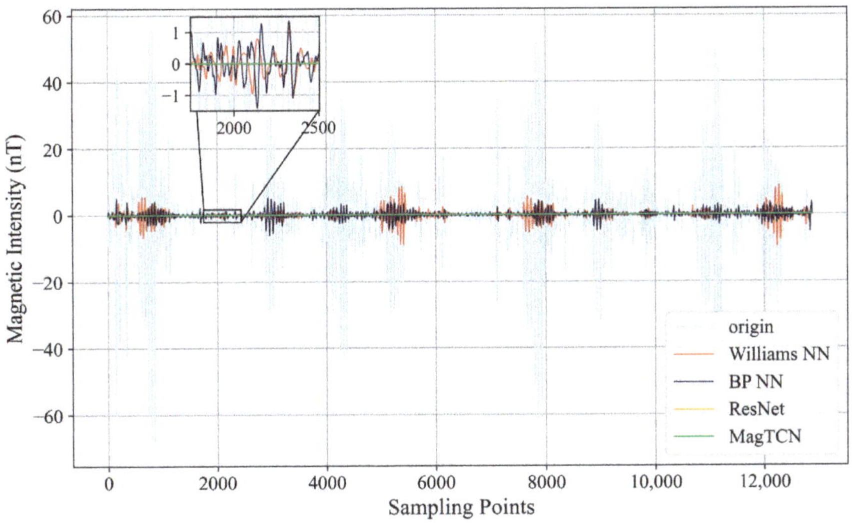

Figure 11. Compensation results within the bandpass of the flight 1002.20 signal.

Table 4. The statistics of compensation results within the bandpass of the flight 1002.20 signal.

	WilliamsNN	BPNN	ResNet	MagTCN
STD (nT)	1.399	1.098	0.028	0.025
IR	8.342	10.629	402.786	455.920

4.3. DAF-MIT AIA Open Flight Data Test Results for Flight 1006.06

In this section, we compare the compensation accuracy and generalization of several compensation methods based on flight 1006.06 of the DAF-MIT AIA Open Flight Data. Figure 12 shows the residual interference in the bandpass (0.1–0.6 Hz) of the magnetic signal on the route, where the curve named "origin" represents the interference before compensation, and the other curves represent the residual interference after compensation by the corresponding method. Large maneuvers during flight can cause large interference (such as around the 5000th sampling point). The purpose of the compensation method is to eliminate these interferences as much as possible. The less residual interference there is after compensation, the better the compensation method. BPNN has the worst compensation effect due to its poor generalization ability, and the residual interference is close to that before compensation in most intervals. Only some intervals (such as in the interval from 4000 to 6000 sampling points) have compensation effects. WilliamsNN has a better compensation effect and can keep the residual interference within 10 nT in most intervals. However, in some intervals (such as around the 500th and 6000th sampling points), there are still obvious residual interferences. ResNet and MagTCN all maintain low residual interference in the entire interval, which is much better than BPNN and WilliamsNN. In order to further compare the performance of several compensation methods, we calculated their STDs and IRs after compensation (as defined in Section 3.3, the lower the STD and the higher the IR, the better the compensation performance). As shown in Table 5, BPNN has the worst STD and IR, which are 10.664 nT and 1.436, respectively. Those of WilliamsNN are better: 2.099 nT and 7.298, respectively. ResNet and MagTCN are better than WilliamsNN and BPNN in compensation performance due to their convolutional structures and residual connections. The STD and IR of ResNet are 0.049 nT and 308.679, respectively, and those of MagTCN are 0.037 nT and 414.090, respectively. The above results show that MagTCN has achieved the best compensation performance and generalization ability among the above compensation methods.

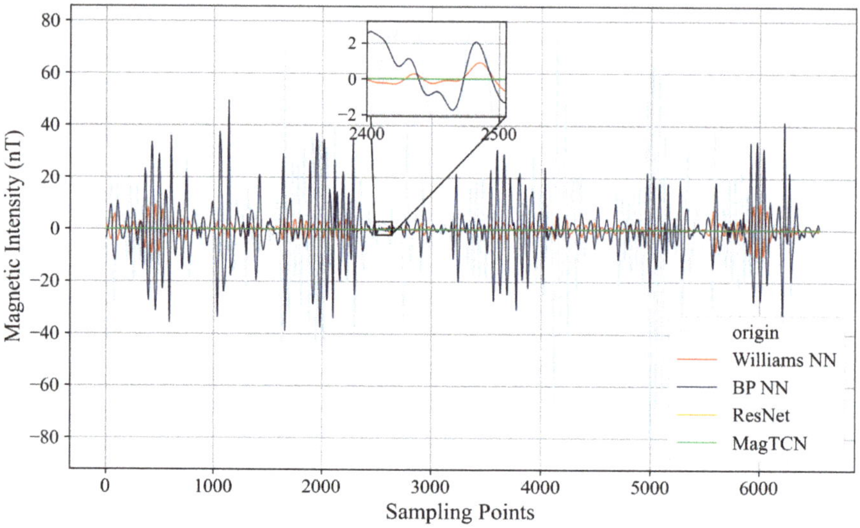

Figure 12. Compensation results within the bandpass of the flight 1006.06 signal.

Table 5. The statistics of compensation results within the bandpass of the flight 1006.06 signal.

	WilliamsNN	BPNN	ResNet	MagTCN
STD (nT)	2.099	10.664	0.049	0.037
IR	7.298	1.436	308.679	414.090

4.4. Compensation Algorithm Resource Consumption Test Results

In this section, we compare the time cost and floating-point operations (FLOPs) of MagTCN and the other compensation algorithms mentioned above to ensure that the proposed method can meet the requirements of real-time aeromagnetic compensation. The hardware and hyperparameters used in the experiment are basically the same as those in Section 3.2. In addition, we set the batch size to 1 to simulate the real-time environment and selected 30,000 inputs to test the average time cost.

As shown in Table 6, BPNN and WilliamsNN have lower average time costs and FLOPs due to their simple structures. The average time cost and FLOPs of BPNN are 0.428 ms and 0.238 MB, respectively, and those of WilliamsNN are 0.428 ms and 0.238 MB, respectively. ResNet and MagTCN take more time, 1.240 ms and 1.837 ms, respectively, but still meet the real-time requirements at a sampling frequency of 10 Hz (<100 ms). Their FLOPs can also be achieved by most graphics processing units [31].

Table 6. Resource consumption of several compensation algorithms.

	WilliamsNN	BPNN	ResNet	MagTCN
Average Time Cost (ms)	0.428	0.046	1.240	1.837
FLOPs (MB)	0.238	0.684	105.541	95.722

5. Conclusions

Past research into neural-network-based aeromagnetic compensation methods usually employed normal convolution structures, resulting in low generalization of the algorithm.

In this article, we propose an aeromagnetic compensation method, MagTCN, based on temporal convolutional networks, to improve both generalization and compensation accuracy. In the proposed method, a neural network based on separable convolution with a residual connection is employed to improve the compensation accuracy and convergence stability, and a gradient correction loss function based on the Tolles–Lawson model is used to improve the generalization ability. We conducted experiments based on both simulated and real datasets and compared typical neural network compensation methods proposed by previous researchers. The results indicate that the method proposed in this article can achieve a lower standard deviation of residual magnetic interference than other neural network methods, demonstrating a better compensation performance and generalization ability. In future work, we will further study the relationship between the input sequence length, convolution kernel receptive field, and compensation accuracy in temporal convolutional networks to achieve higher compensation performance.

Author Contributions: Conceptualization, B.Z.; methodology, B.Z. and H.W.; software, H.W.; validation, B.Z. and H.W.; formal analysis, B.Z.; investigation, B.Z. and H.W.; resources, B.Z.; data curation, B.Z.; writing—original draft preparation, H.W.; writing—review and editing, B.Z. and H.W.; visualization, H.W.; supervision, B.Z.; project administration, B.Z.; funding acquisition, B.Z. All authors have read and agreed to the published version of the manuscript.

Funding: This research received no external funding.

Institutional Review Board Statement: Not applicable.

Informed Consent Statement: Not applicable.

Data Availability Statement: The original contributions presented in the study are included in the article, further inquiries can be directed to the corresponding author.

Acknowledgments: The authors would like to thank the School of Computer Science, China University of Geosciences (Wuhan), for providing the deep learning servers.

Conflicts of Interest: The authors declare no conflicts of interest.

References

1. Tolles, W.E.; Lawson, J.D. *Magnetic Compensation of MAD Equipped Aircraft*; Airborne Instruments Lab. Inc.: Mineola, NY, USA, 1950; p. 201.
2. Tolles, W.E. Compensation of Aircraft Magnetic Fields. U.S. Patent 2,692,970, 26 October 1954.
3. Tolles, W.E. Magnetic Field Compensation System. U.S. Patent 2,706,801, 19 April 1955.
4. Tolles, W.E. Eddy-Current Compensation. U.S. Patent 2,802,983, 13 August 1957.
5. Leliak, P. Identification and evaluation of magnetic-field sources of magnetic airborne detector equipped aircraft. *IRE Trans. Aerosp. Navig. Electron.* **1961**, *ANE-8*, 95–105. [CrossRef]
6. Leach, B.W. Aeromagnetic compensation as a linear regression problem. In *Information Linkage Between Applied Mathematics and Industry*; Elsevier: Amsterdam, The Netherlands, 1980; pp. 139–161.
7. Feng, Y.; Zhang, Q.; Zheng, Y.; Qu, X.; Wu, F.; Fang, G. An improved aeromagnetic compensation method robust to geomagnetic gradient. *Appl. Sci.* **2022**, *12*, 1490. [CrossRef]
8. Chen, B.; Zhang, K.; Yan, B.; Zhu, W. The Conjunctive Compensation Method Based on Inertial Navigation System and Fluxgate Magnetometer. *Appl. Sci.* **2023**, *13*, 5138. [CrossRef]
9. Chen, B.; Huang, L.; Zhang, K.; Hu, J.; Zhu, W. Magnetic interference analysis and compensation method of airborne electronic equipment in an unmanned aerial vehicle. *Appl. Sci.* **2023**, *13*, 7455. [CrossRef]
10. Williams, P.M. Aeromagnetic compensation using neural networks. *Neural Comput. Appl.* **1993**, *1*, 207–214. [CrossRef]
11. Bickel, S.H. Small signal compensation of magnetic fields resulting from aircraft maneuvers. *IEEE Trans. Aerosp. Electron. Syst.* **2007**, *AES-15*, 518–525. [CrossRef]
12. Ma, M.; Zhou, Z.; Cheng, D. A dual estimate method for aeromagnetic compensation. *Meas. Sci. Technol.* **2017**, *28*, 115904. [CrossRef]

13. Xu, X.; Huang, L.; Liu, X.; Fang, G. Deepmad: Deep learning for magnetic anomaly detection and denoising. *IEEE Access* **2020**, *8*, 121257–121266. [CrossRef]
14. Yu, P.; Zhao, X.; Jia, J.; Zhou, S. An improved neural network method for aeromagnetic compensation. *Meas. Sci. Technol.* **2021**, *32*, 045106. [CrossRef]
15. Li, L.; Xu, Q.; Gu, H.; Zhou, L.; Liu, Z.; Cao, L. Aeromagnetic compensation algorithm based on levenberg-marquard neural network. *J. Geod. Geoinf. Sci.* **2021**, *4*, 74–83.
16. Yu, P.; Bi, F.; Jiao, J.; Zhao, X.; Zhou, S.; Su, Z. An Aeromagnetic Compensation Algorithm Based on a Residual Neural Network. *Appl. Sci.* **2022**, *12*, 10759. [CrossRef]
17. Zhou, S.; Yang, C.; Su, Z.; Yu, P.; Jiao, J. An aeromagnetic compensation algorithm based on radial basis function artificial neural network. *Appl. Sci.* **2022**, *13*, 136. [CrossRef]
18. Jiao, J.; Yu, P.; Zhao, X.; Bi, F. Real-time aeromagnetic compensation with compressed and accelerated neural networks. *IEEE Geosci. Remote Sens. Lett.* **2022**, *19*, 1–5. [CrossRef]
19. Wang, Q.; Song, J.; Zuo, C.; Dong, K.; Jin, F. An Aeromagnetic Compensation Algorithm Based on Broad Learning System. In Proceedings of the 2024 3rd International Conference on Robotics, Artificial Intelligence and Intelligent Control (RAIIC), Mianyang, China, 5–7 July 2024; pp. 373–377.
20. Gnadt, A.R.; Belarge, J.; Canciani, A.J.; Conger, L.; Curro II, J.A.; Edelman, A.; Morales, P.; O'Keeffe, M.F.; Taylor, J.; Jacobs, D. DAF-MIT AIA Open Flight Data for Magnetic Navigation Research. Available online: https://zenodo.org/records/12723700 (accessed on 7 February 2025).
21. Gnadt, A. Machine learning-enhanced magnetic calibration for airborne magnetic anomaly navigation. In Proceedings of the AIAA SciTech 2022 Forum, San Diego, CA, USA, 3–7 January 2022; p. 1760.
22. Wang, Y.; Han, Q.; Zhan, D.; Li, Q. A data-driven OBE magnetic interference compensation method. *Sensors* **2022**, *22*, 7732. [CrossRef] [PubMed]
23. Li, Q.; Zhang, L.; Liu, S. A-TCN: Attention-Based TCN for OSS Reliability Prediction on High-Dimensional Fault Dataset. In Proceedings of the 2024 IEEE 24th International Conference on Software Quality, Reliability, and Security Companion (QRS-C), Cambridge, UK,, 1–5 July 2024; pp. 187–195.
24. Xiang, X.; Li, X.; Zhang, Y.; Hu, J. A short-term forecasting method for photovoltaic power generation based on the TCN-ECANet-GRU hybrid model. *Sci. Rep.* **2024**, *14*, 6744. [CrossRef] [PubMed]
25. Wei, X.; Wang, Z. TCN-attention-HAR: Human activity recognition based on attention mechanism time convolutional network. *Sci. Rep.* **2024**, *14*, 7414. [CrossRef] [PubMed]
26. Larsson, J. Electromagnetics from a quasistatic perspective. *Am. J. Phys.* **2007**, *75*, 230–239. [CrossRef]
27. Chollet, F. Xception: Deep learning with depthwise separable convolutions. In Proceedings of the IEEE Conference on Computer Vision and Pattern Recognition, Honolulu, HI, USA, 21–26 July 2017; pp. 1251–1258.
28. Hendrycks, D.; Gimpel, K. Gaussian error linear units (gelus). *arXiv* **2016**, arXiv:1606.08415.
29. Kim, T.; Kim, J.; Tae, Y.; Park, C.; Choi, J.-H.; Choo, J. Reversible instance normalization for accurate time-series forecasting against distribution shift. In Proceedings of the International Conference on Learning Representations, Vienna, Austria, 4 May 2021.
30. Vaswani, A.; Shazeer, N.; Parmar, N.; Uszkoreit, J.; Jones, L.; Gomez, A.N.; Kaiser, Ł.; Polosukhin, I. Attention is all you need. *Adv. Neural Inf. Process. Syst.* **2017**, *30*, 5998–6008.
31. Vassilieva, N. Characterization and Benchmarking of Deep Learning. 2017. Available online: https://www.hpcuserforum.com/presentations/Wisconsin2017/HPDLCookbook4HPCUserForum.pdf (accessed on 7 February 2025).

Disclaimer/Publisher's Note: The statements, opinions and data contained in all publications are solely those of the individual author(s) and contributor(s) and not of MDPI and/or the editor(s). MDPI and/or the editor(s) disclaim responsibility for any injury to people or property resulting from any ideas, methods, instructions or products referred to in the content.

applied sciences

MDPI

Article

Susceptibility and Remanent Magnetization Estimates from Orientation Tools in Borehole Imaging Logs

Julio Cesar S. O. Lyrio *, Ana Patrícia C. C. Laier, Jorge Campos Junior †, Ana Natalia G. Rodrigues and Luciano dos Santos Martins

Petrobras, Av. Henrique Valadares 28, Rio de Janeiro 20231-030, Brazil; alaier@petrobras.com.br (A.P.C.C.L.); ana.rodrigues@petrobras.com.br (A.N.G.R.); luciano_martins@petrobras.com.br (L.d.S.M.)
* Correspondence: jlyrio@petrobras.com.br
† Retired.

Abstract: Orientation tools in borehole imaging logs acquire magnetic information that is currently used for spatial and geographical orientation of the images. We propose to use this magnetic field information to estimate both magnetic susceptibility and remanent magnetization of rocks inside wells. Measurements of these magnetic parameters are not often available in hydrocarbon exploration to support forward modeling of magnetic data, an interpretation tool that has played important role in the exploration risk reduction in the Pre-Salt prospects of Campos Basin, Brazil. The acquired magnetic data requires corrections for tool rotation and diurnal variation of the Earth's magnetic field before calculation. Then, using a set of simple equations and reasonable assumptions we were able to estimate the magnetic susceptibility of carbonates and basalts, as well as the remanent magnetization of the basalts, from a Pre-Salt well in Campos Basin. When compared to susceptibility values measured in laboratory for the same rock interval, our results show a significant match. This promising result shows the importance of our methodology in providing reliable information that can minimize uncertainties in forward modeling of magnetic data, which contributes to reduction of hydrocarbon exploration risks. Given that direct susceptibility and remanence measurements require oriented samples, a complex and expensive operation in wells, our results offer this rock information without any extra costs since imaging logs are commonly acquired in exploration wells. Besides its use in hydrocarbon exploration, our methodology can be applied to mineral exploration where magnetic susceptibility is an important property for rock identification.

Keywords: magnetic susceptibility; borehole imaging logs; hydrocarbon exploration

1. Introduction

Borehole imaging tools, like the Formation Micro-Imaging (FMI), for instance, carry triaxial magnetometers in an auxiliary orientation tool (also known as inclinometers) that acquire magnetic information used for spatial and geographical orientation of the images. Until recently, the use of the magnetic data acquired by these orientation tools was restricted to support the inclinometry data. According to [1], the perception that changes in the measured magnetic field were caused by the presence of magnetic rocks in the borehole has extended the use of this data to help identifying igneous rock intervals during analysis of imaging tools data. The fact that these magnetic data can detect the presence of igneous rocks rises the question if they are sensitive enough to allow estimates of some magnetic properties of these rocks, more specifically, the magnetic susceptibility.

Interpretation of magnetic data is traditionally used in petroleum exploration to map the geological distribution of magnetic rocks in subsurface. Among various interpretation techniques, the forward modeling of magnetic data is a typical approach for estimation of positions, depths, geometries and magnetic properties of the rocks in exploration areas. In general, the forward modeling technique searches for a reliable geological solution that fits the magnetic anomaly observed in the survey area. This interpretation process is mainly ruled by two variables: the geometry of the rocks and their magnetic susceptibility. Well data and other geophysical data, like seismic and gravity for instance, may provide the necessary control to the source's geometry variability, while measurements of magnetic susceptibility are rare specially in oil industry.

Forward modeling has played important role to the exploration of Pre-Salt prospects in offshore Brazil, where it has been used to help in the differentiation between carbonates and igneous rocks, whose distinction in seismic images are often difficult. Examples of such use can be found in [2], who discussed the utility and problems encountered when modeling in Santos Basin, offshore Brazil. Figure 1 shows a representative example from [2]. In both Figure 1a,b, the upper panels show the magnetic profiles where the blue dotted line represent the observed data, the continuous blue line is the data calculated from the model and the continuous red line accounts for the misfit. The models are shown in the lower panels where the blue region are the non-magnetic rocks (carbonates and other sedimentary rocks) while the pink regions represent the igneous rocks, whose magnetic susceptibility comes from literature values due to the absent of measurements. The green lines are the interface between the layers as interpreted from seismic data. The initial interpretation (Figure 1a) suggested the presence a thicker and shallower igneous package at the top of the structure. The modeling solution (exhibited in Figure 1b) shows that a thinner and deeper igneous package fits better the observed data and, therefore, should be more likely to be true [2]. Latter drilling has proved that the model proposed by the forward modeling was correct.

The importance of magnetic forward modeling to Pre-Salt exploration, specially where seismic interpretation is ambiguous, contrasts with the lack of susceptibility measurements in boreholes. Since magnetic susceptibility in rocks may vary in orders of magnitude, the use of literature values may not always provide reliable results. Estimates from borehole magnetic data collected in nearby wells may provide the necessary control for the forward modeling. The incorporation of more reliable magnetic susceptibility estimates into these models would make them more realistic contributing for exploration risks reduction.

Despite its utility for magnetic modeling, the development of tools for borehole magnetic susceptibility measurements was greatly driven by the magnetostratigraphy due to its application in determining the absolute age of geological formations. Examples of interpretation of magnetic logging data in sediments can be found in [3,4] who have studied the magnitude and variations of magnetic susceptibility in sedimentary and volcanoclastic rocks to understand their influence on induction logs.

Magnetic well logging tools have been developed by both oil and service companies since the middle of last century. Two magnetic well logging instruments capable of acquiring magnetic susceptibility as well as the total magnetic field were developed by Magnolia Petroleum according to [5]. The development of the GHMT (Geological High Sensibility Magnetic Tool) by Total with the help of Schlumberger was described by [6,7]. The GHMT is a magnetic logging tool developed for recording the total magnetic field and the magnetic susceptibility of rocks in oil wells. Besides oil companies, the mineral industry was also researching borehole magnetic susceptibility tools. A magnetic susceptibility log system for uranium exploration whose development was supported by the Department of Energy of United States was described by [8,9]. The basic theory and results of an inductive magnetic susceptibility logger (Geonics EM39S) for shallow lithological applications was presented by [10] who have used it to study unconsolidated sediments and rocks in Canada. A magnetic tensor gradiometer, developed by [11] in Australia for iron ore exploration, is an instrument capable to measure magnetic field gradients in all directions within a borehole, however it does not provide susceptibility measurements.

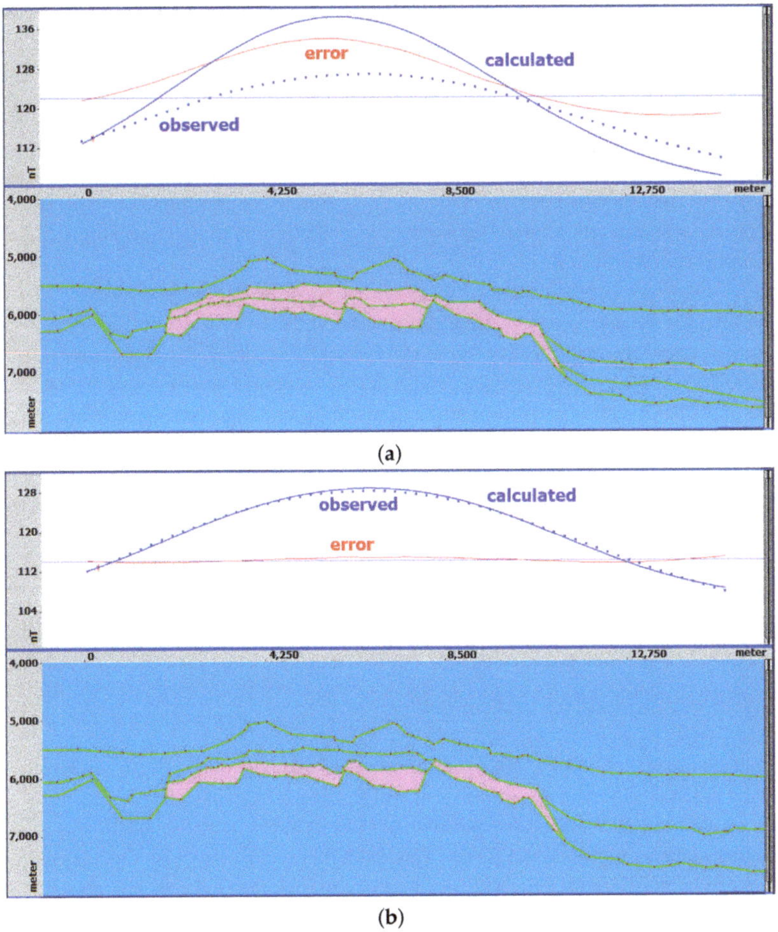

Figure 1. Forward modeling example of Brazilian Pre-Salt exploration. The upper panels show the magnetic profiles. The models are shown in the lower panels where the blue regions are the non-magnetic rocks while the pink regions represent the igneous rocks. The green lines are the interface between layers as interpreted from seismic data. (**a**) The initial model with a thicker igneous package. (**b**) The final model suggesting less igneous rocks, which was confirmed by drilling (modified from [2]).

Alternatively, methods of getting indirect measurements of the magnetic susceptibility of rocks have been proposed as well. Such methods were based mainly on changes in acquisition methodology, choice of parameters and data processing of induction and NMR (Nuclear Magnetic Resonance) tools. The use of induction logs to determine relative magnetic susceptibility with focus on the perturbations in induction results caused by the magnetic formations was proposed by [12]. Apparent magnetic susceptibility contrast estimates between rock and a saturating liquid was performed by [13] from measures of the free induction decay in core plugs, powdered rocks and drilled cuttings, but not in well logs.

The developments and applications of the logging magnetic tools have yielded promising results. Examples of the application of borehole magnetic susceptibility studies in different fields can be found in [14], who evaluates the use of high-resolution magnetic susceptibility logs to characterize climate-forced sedimentation evolution and in [15] who has used high-sensitivity induction magnetic susceptibility probe to infer mineralogical changes and lithology variations in shallow mapping applied to groundwater studies. Environmental studies has also made use of magnetic susceptibility to identify variations in sediment composition as in [16], but in most of the cases, the logging is carried out on a core instead of inside the borehole. Magnetic susceptibility logs are also common in ocean drilling programs, as in [17] for instance, but they usually log cores of recovered sediments and do not take borehole measurements. The use of magnetic susceptibility downhole finds applicability even in archaeology, as in [18] who reviews the role of magnetic susceptibility in archaeogeophysical studies.

Despite of the large applicability of magnetic susceptibility for sediment studies, little application of borehole magnetic logging have emerged in oil exploration industry. Even though [19] provides an extensive review of magnetic susceptibility measurements applications for reservoir characterization, most of the discussed results were based on laboratory rather than logging measurements. As a consequence, to our knowledge, this type of borehole logging tools can not be found for commercial use in oil industry nowadays. In this paper we propose to overcome this lack of direct measurements by estimating the magnetic susceptibility of the rocks along the borehole using the magnetic data acquired by the magnetometer included in the orientation tools of imaging logs.

2. Materials and Methods

The total field F measured inside a borehole can be related to the Earth's magnetic field H and to the total magnetization M_T of a rock as:

$$F = (H + M_T) = H + (M_H + M_R) = H + (\chi H + M_R), \quad (1)$$

where M_H is the component of magnetization produced in response to the Earth's magnetic field, M_R is the natural remanent magnetization and χ is the rock's magnetic susceptibility. The natural remanent magnetization is a permanent magnetization that remains in the rock even in the absence of an inducing field, recording its magnetization history. The magnetic susceptibility is a dimensionless physical property of the rock that in cases of sufficiently weak fields, like the Earth's magnetic field, represents a constant of proportionality relating the induced magnetization and the applied field. Since the induced magnetization is essentially parallel to the applied field, the magnetic susceptibility is a scalar quantity for most of rocks. Equation (1) is of fundamental importance because it represents the basis for all estimates carried in this study.

As Equation (1) is a vector sum, it can only be correctly done with the knowledge of the three parameters that describe a magnetic field vector: magnitude F, inclination I and declination D. Alternatively, the sum can be simplified by decomposing each vector in its components in the x-, y- and z-directions as defined in Figure 2a. However, the triaxial magnetometers used in image logs to determine tool orientation and/or well direction measure the three components of the magnetic field F with respect to the tool coordinate system instead of true geographic coordinate system. In order to correctly perform the vector sum in Equation (1) it is necessary to project the field components from the tool's system into the geographic system described in Figure 2a.

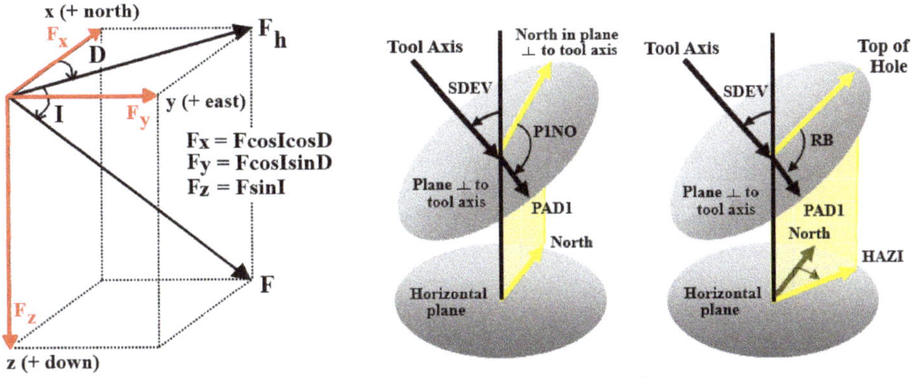

(a) (b)

Figure 2. Graphic illustration of some important definitions. (**a**) Spatial relationship between the components, F_x, F_y and F_z, of the magnetic field described in terms of three parameters: magnitude F, inclination I and declination D. The sign convention for field and its components is: x positive to north, y positive to east and z positive downward. (**b**) Representation of the meaning of four tool angles, SDEV, HAZI, RB and P1NO, which describes tool's orientation in the geographical coordinate system (modified from [20]).

In the tool orientation system, the x-axis points to the direction of a reference marker, called P1 or Pad #1, located in the plane perpendicular to the z-axis, which for vertical wells correspond to the borehole axis (Figure 2b). As the tool rotates during logging operations, the tool's x-axis points not to north as defined in Figure 2a, but to different directions. Consequently, the magnitude of acquired horizontal components do not represent the true amplitude in north and east direction. Figure 3a shows the magnetic field acquired in a typical well as an example of the variation in amplitude of the horizontal components caused by the tool's rotation. The strong fluctuations occurring in the horizontal components of the acquired magnetic field need to be corrected before calculations.

According to [20] the orientation of the tool in the geographical coordinate system can be described by four angles: Sonde Deviation (SDEV), Hole Azimuth (HAZI), Relative Bearing (RB) and Rotation of P1 from North (P1NO), as illustrated in Figure 2b. Because P1NO angle represents the rotation of P1 from north, its measurements provide the declination of the tool's x-axis along the logged interval. It is possible to compute the angle between tool's x-axis and the direction of measured field using the relations between magnetic field and its components shown in Figure 2a. The sum of these two angles provide the true magnetic declination of the total inducing field. The calculation of the true declination allows us to compute the correct F_x and F_y components by applying again the relations in Figure 2a using the correct angles this time. Figure 3b shows the field components after amplitude correction. As expected, the strong fluctuations present in the horizontal components of Figure 3a were significantly reduced.

Once the true amplitude of the field components has been recovered, estimates for the magnetic susceptibility χ can be made using two different approaches. The first and more simplistic case is that of sedimentary rocks. Since the majority of sedimentary rocks has very low content of ferromagnetic minerals, which are responsible for retaining magnetization, we assume that remanent magnetization is not preserved in these rocks. This assumption considers that the resolution of magnetometers used in wells are not fine enough to register the weak magnetic field disturbance caused by such low content of ferromagnetic minerals. In these cases, Equation (1) simplifies to:

$$F_\alpha = H_\alpha + \chi_\alpha H_\alpha \tag{2}$$

where α may represent any of the x, y and z directions. In this case, the magnetic susceptibility χ can be easily calculated by:

$$\chi_\alpha = \frac{F_\alpha}{H_\alpha} - 1. \tag{3}$$

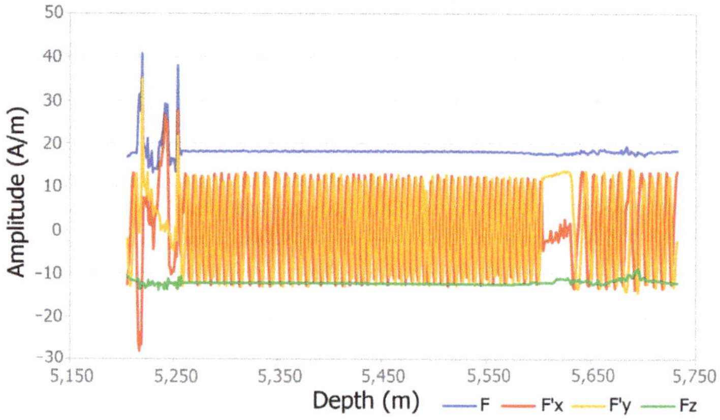

(**a**)

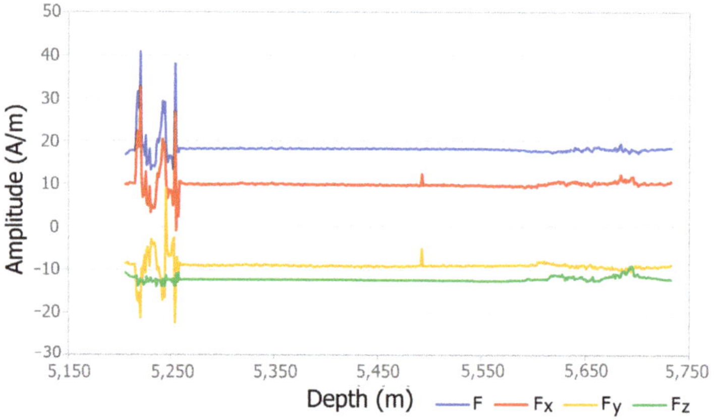

(**b**)

Figure 3. Example of the amplitude fluctuation caused by tool's rotation in a typical well. The strong amplitude fluctuations occurring in the horizontal components shown in (**a**) were greatly reduced after correction, as exhibit in (**b**).

The second approach assumes the presence of a natural remanent magnetization M_R caused by the existing magnetic field at the time that rocks with high content of ferromagnetic minerals were formed. This case leads to a situation where there is only one equation and two unknowns χ and M_R as:

$$F_\alpha = H_\alpha + \chi_\alpha H_\alpha + M_{R_\alpha}. \tag{4}$$

However, if we assume χ and M_R as constant along some measured interval, we will have n measurements of F_α and can also get n values for H_α in the same interval. Although fundamental to get the estimates, this assumption seems geologically reasonable as you can make your measured interval n as small as two samples (0.3 m). This allows us to construct a system of n equations and two unknowns χ and M_R:

$$\begin{cases} F_{\alpha_1} - H_{\alpha_1} = \chi_\alpha H_{\alpha_1} + M_{R_\alpha} \\ \quad\vdots \\ F_{\alpha_n} - H_{\alpha_n} = \chi_\alpha H_{\alpha_n} + M_{R_\alpha} \end{cases}, \tag{5}$$

which in matrix form becomes:

$$\begin{bmatrix} H_{\alpha_1} & 1 \\ & \vdots \\ H_{\alpha_n} & 1 \end{bmatrix} \begin{bmatrix} \chi_\alpha \\ M_{R_\alpha} \end{bmatrix} = \begin{bmatrix} F_{\alpha_1} - H_{\alpha_1} \\ \vdots \\ F_{\alpha_n} - H_{\alpha_n} \end{bmatrix}. \tag{6}$$

This system can be solved for χ_α and M_{R_α} using the least squares method. So, the mean value of χ and the magnitude of vector M_R will be given by:

$$\chi = \frac{\chi_x + \chi_y + \chi_z}{3} \quad \text{and} \quad M_R = \sqrt{M_{R_x}^2 + M_{R_y}^2 + M_{R_z}^2}. \tag{7}$$

Additionally, the inclination I_R and declination D_R of the remanent magnetization may be estimated by rearranging Equation (4) in terms of the remanent magnetization, and making use of the spatial relations shown in Figure 2a. Therefore, we can define the remanent magnetization components as:

$$\begin{aligned} M_{R_x} &= F_x - H_x - \chi H_x = M_R \cos I_R \cos D_R, \\ M_{R_y} &= F_y - H_y - \chi H_y = M_R \cos I_R \sin D_R, \\ M_{R_z} &= F_z - H_z - \chi H_z = M_R \sin I_R. \end{aligned} \tag{8}$$

The vertical components of remanent magnetization in Equation (8) allows to estimate the inducing remanent field inclination I_R with the help of previously estimated remanent magnetization magnitude M_R,

$$I_R = \arcsin\left(\frac{F_z - H_z - \chi H_z}{M_R}\right). \tag{9}$$

Similarly, remanent magnetization declination D_R can be estimated from the horizontal components of remanent field (M_{R_x} and M_{R_y}) as:

$$D_R = \arccos\left(\frac{F_x - H_x - \chi H_x}{M_R \cos I_R}\right) \quad \text{or} \quad D_R = \arcsin\left(\frac{F_y - H_y - \chi H_y}{M_R \cos I_R}\right). \tag{10}$$

The estimates of magnitude M_R, inclination I_R and declination D_R completely define the remanent magnetization.

We have shown that estimates of magnetic susceptibility and remanence are possible through simple calculations when magnetic field data from orientation tools in borehole imaging logs are available. The proposed methodology was applied to real data example from Campos Basin, Brazil.

3. Results

As an example of application to a real problem, we have applied the proposed method to estimate both the magnetic susceptibility and the remanent magnetization of a sequence that includes Pre-Salt carbonates and igneous rocks from Campos Basin, Brazil. The Campos Basin is one of the most important and prolific offshore Brazilian oil provinces, currently accounting for about 18% of the Brazilian total hydrocarbon production which is dominated (about 79%) by the Pre-Salt sequence [21]. The magnetic data used in this estimate was acquired in well 1-RJS-755-RJ, located at 2950 m water depth about 200 km off the coast of Rio de Janeiro state (see Figure 4). This well has identified the presence of oil in carbonate reservoirs of the Pre-Salt section.

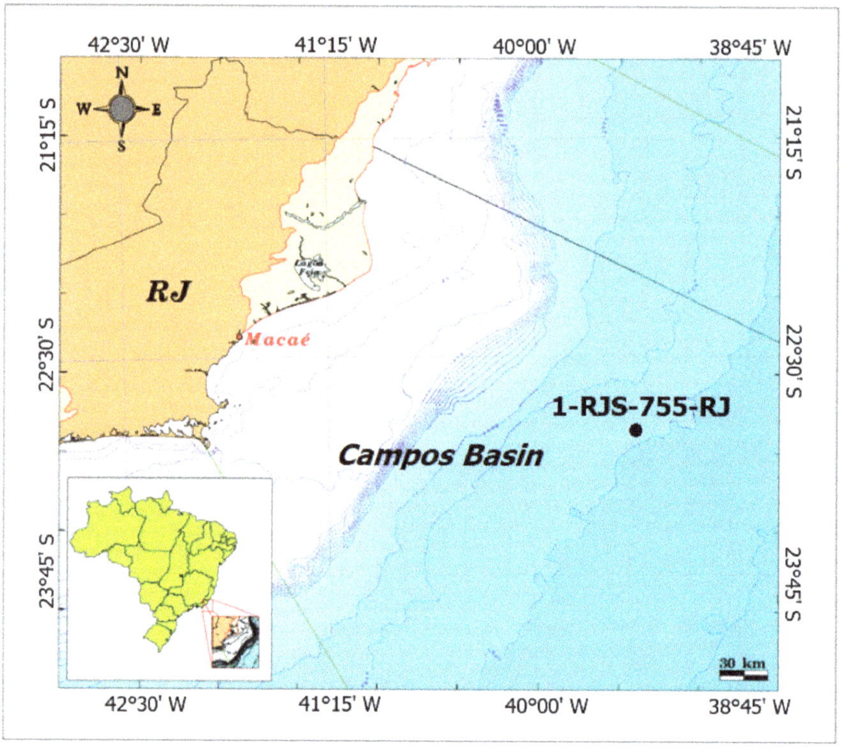

Figure 4. Geographic location of well 1-RJS-755-RJ at 2950 m water depth in the Campos Basin, Brazil. Contour lines represent the bathymetry (modified from [22]).

3.1. Campos Basin Geologic Settings

The Campos Basin, located at the South Atlantic continental margin along the southeastern coast of Brazil, is part of the Central Segment of the South Atlantic Ocean. The development of Campos Basin was initiated during the Late-Jurassic/Early-Cretaceous break-up of Gondwana super continent, followed by the opening of the South Atlantic Ocean. Early mechanical subsidence was accomplished by episodic events that reactivated the Proterozoic basement fabric under regional E–W extension. Its stratigraphy includes Early Cretaceous to recent sediments. According to [23], the stratigraphic record of the Campos Basin may be divided into three tectonostratigraphic mega-sequences: Non Marine Rift, Transitional Post-Rift, and Marine Drift, as shown in Figure 5. The Non-Marine Rift is associated with synthetic and antithetic faults forming several half-grabens filled with fluvial-deltaic sediments. The Transitional Post-Rift is characterized by evaporite deposition, where salt movements play important role by controlling stratigraphic and structural features responsible for several exploratory plays. The Marine Drift may be subdivided into a carbonate Restricted Marine sequence (Albian to Turonian) and a siliciclastic Open Marine sequence (Late Cretaceous to Quaternary).

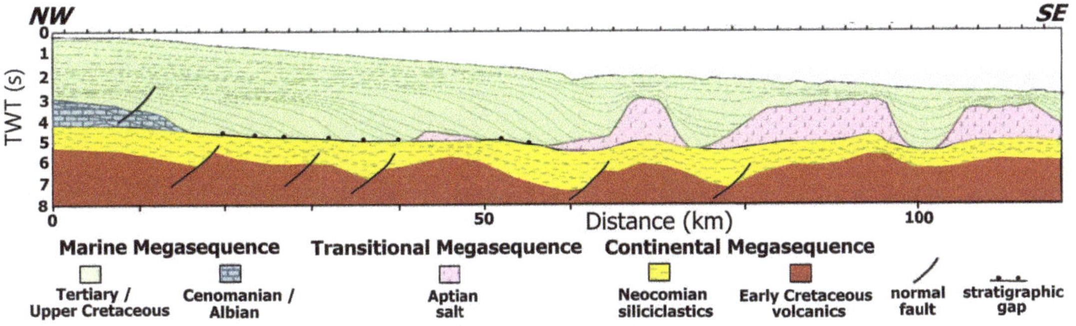

Figure 5. Schematic dip-oriented geological section in the Campos showing the tectonostratigraphic mega-sequences (modified from [23]).

The overall structural style in Campos Basin, exhibited in Figure 6, is controlled by a detachment surface at the base of Aptian salt as proposed by [24]. Above the Aptian detachment the thin-skinned salt tectonics with extensional–proximal and compressional–distal domains have been active during divergent margin phase since the Albian, and is characterized by listric fault-rollover systems, turtle-back structures, raft tectonics, down building diapirs, salt walls, and compressional salt tongues [25,26]. Bellow the detachment, the basement is characterized by a system of horsts and grabens limited by steep normal faults that were active during the rift phase in the Early Cretaceous [27]. As suggested by [24] the basement structure controls the salt tectonic domains, with extensional structures to the West, and diapirs, salt walls, and compressional structures to the East of the main internal basement high.

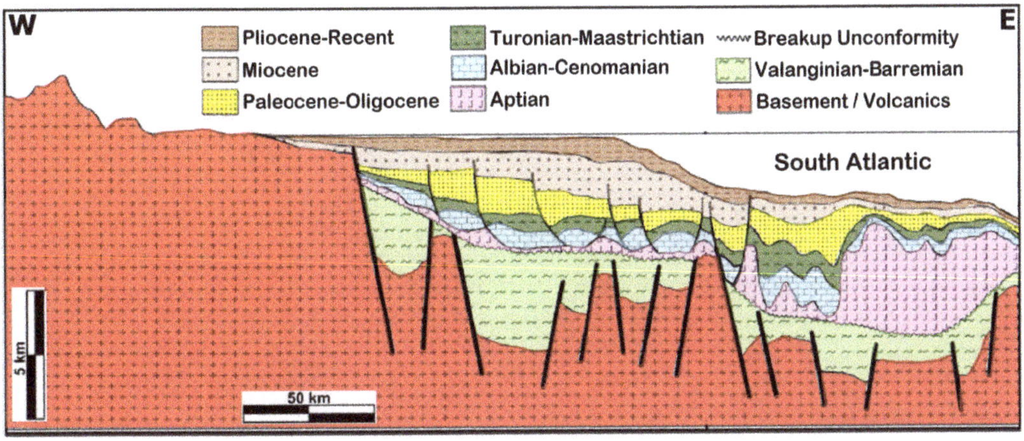

Figure 6. Structural section through central Campos Basin showing the dominant detached structural style (modified from [24]).

In a recent paper, [28] has proposed that the Pre-Salt units bellow the detachment are constituted by four depositional sequences that nearly corresponds to the tectonic stages of the rifting: rift initiation, rift climax, transition and post-rift, as shown in Figure 7a. The rift initiation stage is represented by intense tholeiitic volcanic activity. According to [24], the rift climax stage is defined by the development, in the basement, of an extensive NE–SW and NNW–SSE system of horsts and grabens limited by steep normal faults. The rift transition stage is characterized by a carbonate section whose basal portions is associated to the rift climax stage. Carbonate deposits including argillaceous, laminites and spherulitic facies represent the post rift stage as mentioned by [28]. This stage is of great exploratory importance because the entire carbonate section constitutes the main hydrocarbon reservoirs of the Pre-Salt sequence and will be the focus of our magnetic susceptibility and remanence estimates.

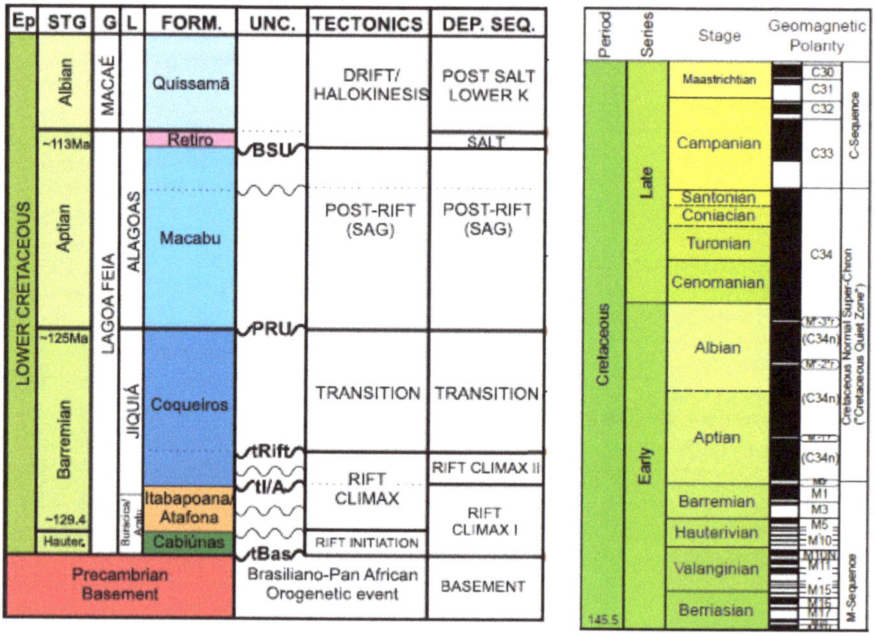

(a) (b)

Figure 7. Simplified stratigraphic and tectonic framework of Campos Basin. (**a**) The main unconformities are: BSU (Base-Salt Unconformity), PRU (Post-Rift Unconformity), tRift (top Rift), tI/A (top Itabapoana/Atafona fms.), tBas (top basement), modified from [28]. (**b**) Geomagnetic polarity time scale for Cretaceous, modified from Concise Geologic Time Scale [29].

3.2. *Magnetic Dataset*

In Campos Basin, there is a large number of wells where borehole imaging logs and, consequently, magnetic field data from their orientation tools were acquired. Our choice for well 1-RJS-755-RJ is due to the fact that 58 core samples were collected and analyzed in laboratory for magnetic susceptibility along about 550 m of Pre-Salt rocks. The sampled interval includes carbonates on the top of an igneous sequence composed by basalts, diabases, hyaloclastites and breccias. From the total, 47 samples were identified as igneous rocks and 11 as carbonates, all showing high alteration degree [22]. This large number of laboratory measurements provides the necessary control to validate our susceptibility estimates, but since samples were not oriented, the directions of the remanent magnetization were not determined, which makes comparison impossible.

3.2.1. Magnetic Sample Analysis

The laboratory analysis of these samples was performed in the MFK1-FA Multifunction Frequency Kappabridge (AGICO), one of the most sensitive laboratory instruments for measuring both anisotropy of magnetic susceptibility (AMS) and bulk magnetic susceptibility (Figure 8). In addiction to magnetic susceptibility, analysis have also included magnetic hysteresis measurements, IRM (Isothermal Remanent Magnetization) acquisition curves, FORCs (First Order Reversal Curves) and thermomagnetic curves. This set of analysis allows the understanding of the magnetic behavior of these rocks as well as to characterize the main magnetic minerals present in the samples [22].

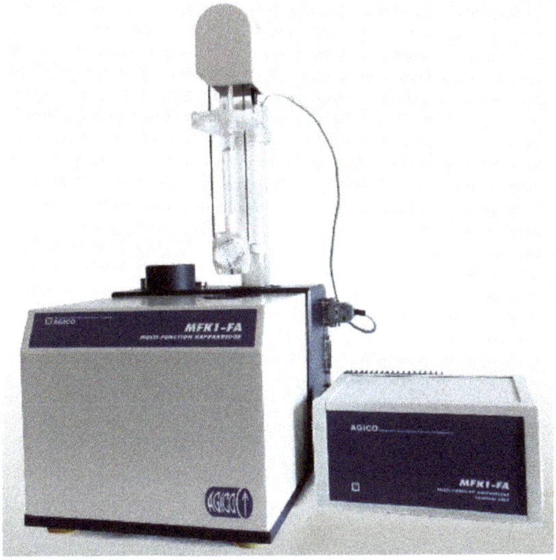

Figure 8. The MFK1-FA instrument used to measure the magnetic susceptibility in 58 core samples collected in well 1-RJS-755-RJ (https://www.agico.cz/text/products/olddev/olddev.php, (accessed on 31 January 2025)).

According to [22], the results of the magnetic susceptibility analysis for the carbonate samples show low values due to the prevailing presence of the diamagnetic minerals (i.e. calcite) in their composition. Measurements for magnetic hysteresis, IRM acquisition, FORCs and thermomagnetic curves for the carbonates have shown extreme noise level due to the low intensity of the magnetic signal. Despite the noise level, the magnetic hysteresis results show curves typical of low coercivity ferromagnetic minerals, while results for both IRM acquisition and thermomagnetic curves were considered as inconclusive. Although weak, the FORC signal of some samples show characteristics of multidomain grains which suggests the presence of detrital magnetite.

For igneous samples, the magnetic susceptibility values were considered lower than would be expected for these kind of rocks. This was interpreted by [22] as consequence of the high alteration degree of the samples. However, the intensity of the magnetic signal was strong enough to allow the magnetic hysteresis and the IRM acquisition to show curves typical of low coercivity ferromagnetic minerals for most of the samples. Two samples, one of basalt and the other of diabase, have shown curves with characteristic behavior of high coercivity ferromagnetic minerals. Analysis of the FORC diagrams have exhibited bi-dimensional data indicating the presence of low coercivity multidomain grains. Thermomagnetic results have indicated that magnetite is the main mineral phase formed but maghemite, greigite, titanomagnetite and possibly pyrite were also identified.

3.2.2. Magnetic Field Data

The magnetic data of well 1-RJS-755-RJ was acquired in Feb/2021 by Schlumberger during a microresistivity imaging survey using a Oil-Base Microimager (OBMI) tool along a depth interval extending from 6125 to 6681 m. The logged segment was composed by a carbonate section extending from 6125 to 6210 m on top of a large basalt section located between 6211 and 6681 m. We have included the lithologic log in Figure 9 and others to allow visual correlation between the measured fields and the corresponding rocks.

The acquired data consists of a set of multiple measurements taken at a 0.15 m interval. However, for the proposed method, only the measurements of depth, magnetic field inclination, magnetic field intensity in x, y and z directions and the PAD1 (see Figure 2b) azimuth angle are strictly necessary. The behavior of the acquired magnetic data, exhibited in Figure 9a, is characterized by fluctuation in the amplitudes of the horizontal components F_x and F_y caused by the tool rotation during acquisition. Components F and F_z are not affected by the tool's rotation, with F showing an almost constant amplitude as F_z exhibits a slight amplitude decrease with depth.

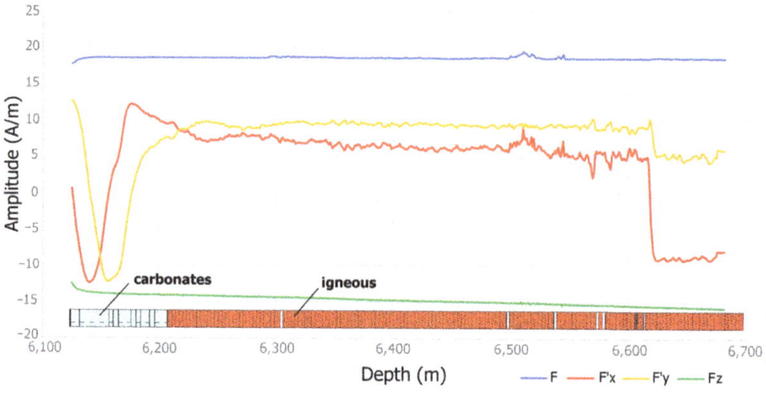

(**a**)

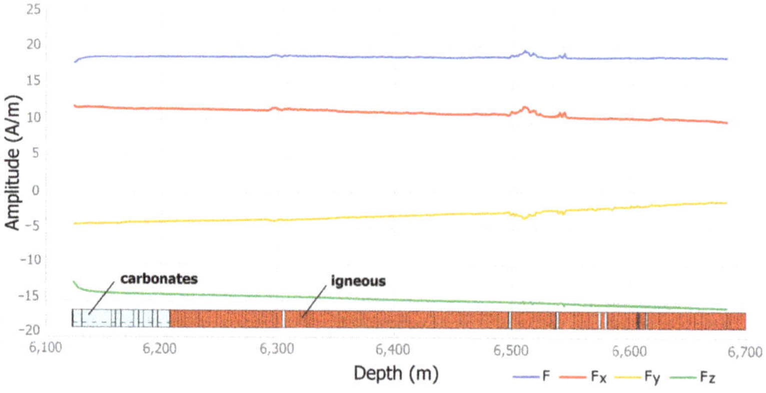

(**b**)

Figure 9. Magnetic field and its components as acquired in well 1-RJS-755-RJ before (**a**) and after amplitude correction (**b**). Notice the significant change in the magnitude of the horizontal components after correction of the disturbance caused by tool rotation. The lithologic log is included for illustration.

3.3. Magnetic Corrections

3.3.1. Magnetic Amplitude Correction

The magnetic amplitude correction is the most important correction in this methodology because it recovers the true field amplitudes before doing any estimate. Since the amplitudes of components are the fundamental parameters for all estimates, the precision of this correction directly controls the quality of estimates.

The amplitude fluctuations seen in F_x and F_y components need to be corrected to recover their true values in north and east directions before susceptibility and remanence estimation. Figure 9b exhibits the magnetic field and its components after applying amplitude correction. Besides the expressive reduction in the amplitude fluctuations, there were also significant changes in the magnitudes of the horizontal components after correction. The amplitude of the F_y component was significantly reduced and now it is possible to see a slight increase with depth. On the other hand, F_x component has suffered a small increase in amplitude unveiling a smooth decreases with depth. Total field F and component F_z do not need correction and their form remains unchanged. In terms of behavior, component F_z shows a smooth decrease with depth while the total field F exhibits a flat behavior except by small fluctuations located around depth 6500 m. Even though these localized amplitude variations seem to be directly related to remaining fluctuations in the horizontal components, they also are present in the F_z component, although with a much smaller amplitude that is difficult to see in the scale used in Figure 9 to represent all field components together.

3.3.2. Magnetic Diurnal Variation Correction

It is well known that the Earth's magnetic field suffers temporal variations during the day caused by interactions between the Sun and electrical charged particles in the ionosphere. Such diurnal variation is characterized by a slow but progressive increase in the field intensity as the solar incidence increases, decreasing towards the background value as the solar incidence reduces. As the logging operations can last for hours, Earth's magnetic field diurnal variations may occur during such periods of magnetic data acquisition. Once estimates are based mainly on field amplitudes, any temporal variation in the field introduces localized errors in the estimates. If the logging time interval is known, it is possible to remove such variations using for that the magnetic field behavior registered by the closest magnetic observatory. In the case of well 1-RJS-755-RJ, the magnetic field variations during logging operation (from 02:26 to 07:58 AM) was corrected using records obtained by Vassouras Magnetic Observatory (VSS) in Rio de Janeiro, Brazil, the closest to the well location. Figure 10 shows the variations of the Earth's magnetic field and its components along the day of logging (10 February 2021). Because the amplitude of field variations were not large during operation period, corrections were small and visually imperceptible in the scale used to represent the field and its components together and will not be shown here.

After all corrections the measured magnetic field and its components can be used in the magnetic susceptibility and remanence estimates. Due to the visualization difficulties caused by the large scale necessary to show all components together, we exhibit in Figure 11 only the F field, which allows better evaluation of rock's magnetic response in depth. Notice that the field response to the igneous rocks becomes more evident only below the base of the first interval (around 6300 m). It is important to notice that there is no significant change in the transition between the carbonates and the first igneous interval, who shows only a slight increase in field amplitude with depth. Since the first interval represents the top of the igneous sequence, it seems possible that it has been exposed to a higher alteration degree, which may have reduced its contribution to the measured magnetic field.

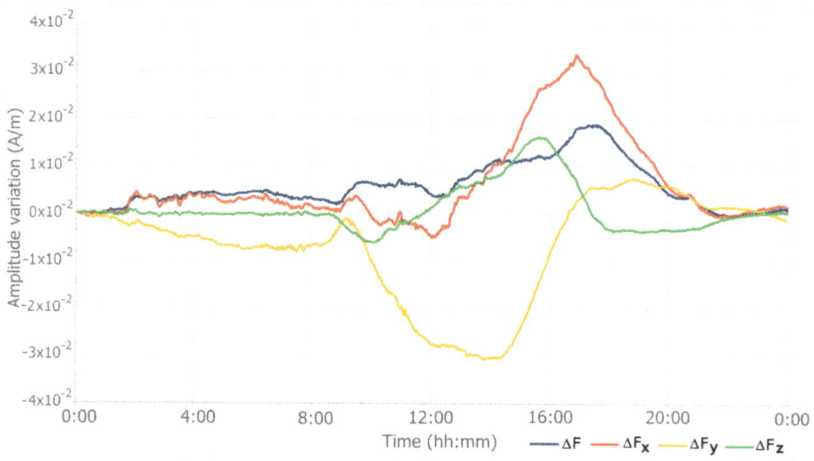

Figure 10. Diurnal variation of Earth's magnetic field and its components on 10 February 2021 according to Vassouras Magnetic Observatory (VSS) in Rio de Janeiro, Brazil. The logging operation was from 02:26 to 7:58 AM, a period of relatively small variations.

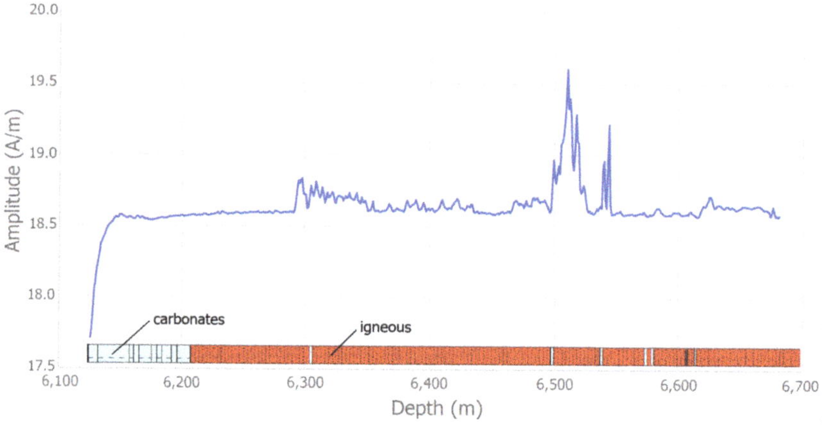

Figure 11. Measured magnetic field F after corrections. The field components were omitted to allow better visualization of the variation caused by the presence of igneous rocks. The lithologic log is included for illustration.

3.4. Magnetic Susceptibility Estimates

According to Equation (4), estimates of magnetic susceptibility and remanence require knowing the behavior of the Earth's magnetic field H in depth at the location of well 1-RJS-755-RJ on the date of logging (10 February 2021). One way of doing this is considering the IGRF (International Geomagnetic Reference Field) as a good representation of field H. However, the IGRF only represents the dipolar portion of the Earth's field and does not take into account any regional field. Another option is the Enhanced Magnetic Model 2017 (EMM2017), a mathematical model also considered to satisfactorily represent the Earth's magnetic field. The EMM2017 is a compilation of satellite, aerial, marine and terrestrial data, which uses spherical harmonics of the scalar magnetic potential of degree and order 790 to represent the field [30]. As it includes regional data into its model, it seems to be a better choice to represent H, assuming the differences to the measured field F are just the local field changes in depth caused by the presence of igneous rocks. Although this is a strong assumption, the fact that the magnetic field decays rapidly with distance allows us to assume that the contribution to the magnetic field caused by rocks close to the sensor will by far overcome those of deeper portions of the basin. Therefore, we chose to use the EMM2017 to represent the Earth's magnetic field H in our estimates. Figure 12a shows the variation in depth of the Earth's magnetic field according to the EMM2017 model. The small amplitude variations in depth of Earth's field model are not visible in the scale of Figure 12a and the curves looks almost constant. Moreover, visual comparison with the field amplitudes measured in well 1-RJS-755-RJ after correction (Figure 9b) confirms EMM2017 model as a good approximation to Earth's magnetic field. The difference between measured (F) and model field (H) is shown in Figure 12b. The small differences in amplitude result from the local geologic variations as can be seen by the correlation with the lithologic log.

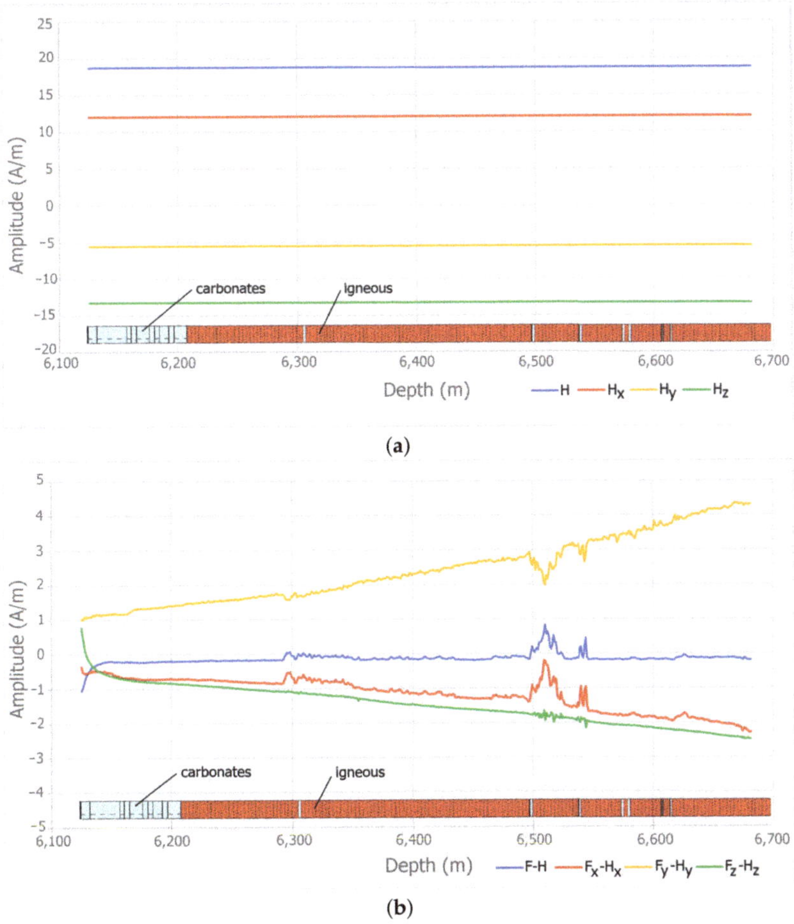

Figure 12. (**a**) The Earth's magnetic field in depth at position of well 1-RJS-755-RJ according to the EMM2017 model. (**b**) Differences in depth between the field measured in well 1-RJS-755-RJ and Earth's magnetic field model. These small differences in amplitude are caused by the local geologic variation in depth. The lithologic log is included for illustration.

Estimates of magnetic susceptibility χ for the carbonate interval (6125 to 6210 m) is simpler because carbonates are assumed do not acquire remanent magnetization and can be achieved applying Equation (3). The statistics for magnetic susceptibility estimates for the carbonate interval in the different directions is shown in Table 1. Analysis of the statistics shows results dominated by negative values, which means that the measured field is smaller than the predicted by the Earth's model field. Since carbonates are mainly constituted by calcite, a diamagnetic mineral, i.e., a mineral whose magnetic response is in the opposite direction to the applied field, this result seems reasonable. In terms of absolute amplitude, estimated susceptibility are, at least, one order of magnitude higher than the average values proposed for carbonates in the literature (Figure 13) as shown by [31]. However, we must consider that literature values are world averages. It should include several types of carbonates in different stratigraphic positions. On the other hand, our carbonates are located directly on the top of igneous rocks whose strong magnetic field may be influencing the measurements taken on the carbonates right above and inducing higher susceptibility values.

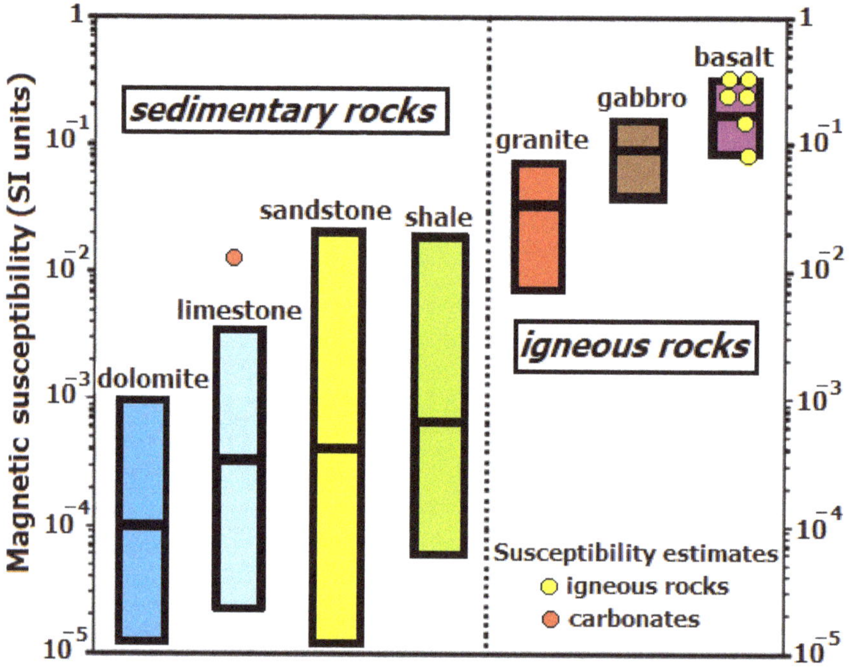

Figure 13. Ranges and mean values of the magnetic susceptibility of the most common rock types (modified from [31]). For comparison, susceptibility estimates from our methodology are plotted as yellow circles (igneous rocks) and red circle (carbonates).

Table 1. Statistics for magnetic susceptibility estimates in different directions for the carbonate interval (6125 to 6210 m) of well 1-RJS-755-RJ. The χ value represents the mean susceptibility according to Equation (7).

	χ	χ_x	χ_y	χ_z
Minimum	-5.6×10^{-2}	-5.9×10^{-2}	-2.6×10^{-1}	-5.8×10^{-2}
Maximum	-1.0×10^{-2}	-2.9×10^{-2}	-1.8×10^{-1}	6.4×10^{-2}
Mean	-1.4×10^{-2}	-5.2×10^{-2}	-2.3×10^{-1}	4.8×10^{-2}
Standard deviation	8.7×10^{-3}	8.0×10^{-3}	2.2×10^{-2}	2.3×10^{-2}

The interval extending from 6211 to 6681 m is composed by igneous rocks. The lithologic profile included in Figure 9 shows that the igneous interval can be divided into six smaller segments separated by tiny carbonate layers. These intervals extend from 6211 to 6300 m, 6305 to 6495 m, 6500 to 6535 m, 6540 to 6570 m, 6585 to 6605 m and 6620 to 6681 m. Because both susceptibility and remanence can be different in these igneous intervals, estimates were be done separately using Equation (7) after solving the system proposed in Equation (5) for each interval. The estimates for the six igneous intervals are exhibited in Table 2. Since Equation (5) assumes both susceptibility and remanence as constant along these intervals the estimates for each component are represented by single values.

Table 2. Magnetic susceptibility (χ) and remanence (M_R) estimates for the six igneous intervals of well 1-RJS-755-RJ.

	Interval (m)					
Estimates	**6211–6300**	**6305–6495**	**6500–6535**	**6540–6570**	**6585–6605**	**6620–6681**
χ_x	2.3×10^{-1}	8.7×10^{-1}	5.5×10^{-1}	8.6×10^{-1}	5.6×10^{-1}	7.2×10^{-1}
χ_y	-8.2×10^{-1}	-9.3×10^{-1}	-7.0×10^{-1}	-5.9×10^{-1}	-5.0×10^{-1}	-9.3×10^{-1}
χ_z	1.0×10^{0}	9.9×10^{-1}	3.8×10^{-1}	5.0×10^{-1}	9.2×10^{-1}	1.0×10^{0}
χ	1.4×10^{-1}	3.1×10^{-1}	7.8×10^{-2}	2.6×10^{-1}	3.3×10^{-1}	2.6×10^{-1}
M_{R_x} (A/m)	-7.9×10^{-1}	-1.4×10^{0}	-1.2×10^{0}	-1.8×10^{0}	-1.9×10^{0}	-2.1×10^{0}
M_{R_y} (A/m)	1.7×10^{0}	2.8×10^{0}	3.1×10^{0}	3.4×10^{0}	3.7×10^{0}	4.3×10^{0}
M_{R_z} (A/m)	-1.1×10^{0}	-1.7×10^{0}	-1.9×10^{0}	-2.1×10^{0}	-2.1×10^{0}	-2.5×10^{0}
M_R (A/m)	2.2×10^{0}	3.6×10^{0}	3.8×10^{0}	4.4×10^{0}	4.7×10^{0}	5.4×10^{0}

The results presented in Table 2 show magnitude values of susceptibility around 1×10^{-1} SI unit, which is within the expected range for igneous rocks according to [31] (Figure 13). In terms of remanence, the estimated values ranging from around 2.0 to over 6.0 Am^{-1} are lower in amplitude when compared to the strength of actual Earth's magnetic field, whose lower values are around 16 Am^{-1}. Low values of the Earth's magnetic field may be related to geomagnetic field reversals, as proposed by [32]. In a paper describing numerical simulations of geomagnetic field reversals, Ref. [32] has found that just before each reversal the intensity of the magnetic field became very weak. So, it seems possible that when those igneous rocks acquired magnetization, the Earth's field was in a reversal process. To reinforce this hypothesis, the geomagnetic polarity scale shown in Figure 7b indicates that the Hauterivian, the time stage when these rocks were formed, was undergoing intense polarity changes. Another possibility is that the intensity of the remanent field was not fully preserved inside these rocks along the time. In addiction, the fact that both χ and M_R estimates do not show widely disperse values gives confidence to the results.

Equations (8)–(10) allow estimates of both inclination I_R and declination D_R of the remanent magnetic field at the time those igneous rocks were formed. The results for the six igneous intervals of well 1-RJS-755-RJ are very consistent as shown in Table 3, which gives credibility to the estimates. In terms of remanence inclination the results point to an inclination angle between $-27°$ and $-30°$ pointing to the same direction of the actual Earth's field but with a smaller angle since the actual field inclination at well position is about $-45°$. The declination estimates show small values dispersion ranging from around 112° to 118°, values that are very different from today's field at well location, which is around 24°. However, the geomagnetic field reversal simulations of [32] indicates that during a reversal the field not only becomes weaker, but also multiple poles may appear in different parts of the globe. The low field intensity associated to the unusual declination angle may support the hypothesis that the igneous rocks may have acquired magnetization during a period of intense magnetic field changes, as illustrated by the geomagnetic polarity scale of Figure 7b (Hauterivian stage). As no remanent inclination and declination measures are available to compare, it is difficult to extend this discussion since, in the worst hypothesis, we may be speculating upon not valid results.

Table 3. Inclination I_R and declination D_R estimates for remanence of the six igneous intervals in well 1-RJS-755-RJ.

Interval (m)	I_R (°)	D_R (°)
6211–6300	−29.98	114.63
6305–6495	−29.14	115.96
6500–6535	−29.43	112.05
6540–6570	−28.20	117.68
6585–6605	−27.53	116.66
6620–6681	−26.96	116.32

As mentioned in Section 3.2, magnetic susceptibility laboratory measurements were made in 58 rock samples distributed along the analyzed interval. The laboratory values are plotted together with the susceptibility estimates in Figure 14. For the carbonate interval estimate, results show values one order of magnitude higher than those measured in the lab. It is possible that the close presence of igneous rocks may be influencing the measured field and, consequently, the susceptibility estimates. Estimates for the igneous rocks presented better results when compared with the lab analysis. The first igneous interval, for instance, shows estimates that are very close to the measurements. For second and third interval estimates are a little lower than the measures while the predictions for the remaining intervals show higher values than the measurements. Taken into account that the process uses only the magnetic field acquired inside the well for its estimates, we consider these results as satisfactory to provide susceptibility and remanence values for forward magnetic modeling purposes.

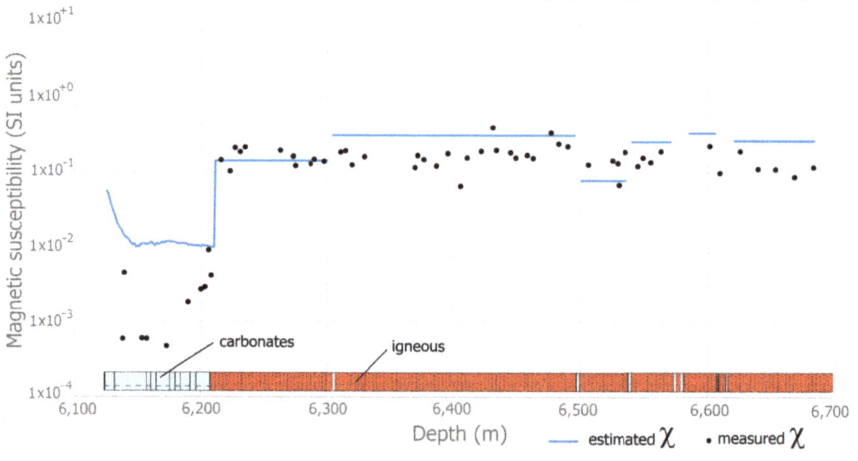

Figure 14. Magnetic susceptibility estimates for well 1-RJS-755-RJ and the result of 58 susceptibility laboratory measurements made in rock samples. The lithologic log along the well is exhibited in the base of the chart for illustration.

4. Conclusions

Results achieved with this research lead to the conclusion that both magnetic susceptibility and remanent magnetic field can be satisfactorily estimated for forward magnetic modeling purposes from the magnetic field acquired by orientation tools in borehole imaging logs. The methodology can also be applied in mining boreholes if they have acquired the magnetic field. The use for magnetostratigraphy, however, depends on a resolution improvement in the acquired magnetic data.

When applied to real data the proposed methodology has estimated magnetic susceptibility values inside the range of values published in the literature for igneous rocks, and very close to laboratory measurements taken from rock samples collected in the well. The methodology can be improved by doing adjustments in the correction of the rotation tool effect, the most important correction. The choice of the Earth's magnetic field model can also be improved by taking always the most complete model. Finally, the use of a base magnetometer near the well may provide better data to the correction of diurnal field variation.

Although measurements of intensity, inclination and declination for magnetic remanence were not available to validate our estimates, the small values dispersion of each variable and the consistence of the results in terms of what we would expect for a magnetic field, gives confidence to the process. The promising results achieved here has proved that the proposed methodology is able to provide reliable susceptibility values for magnetic forward modeling and inversion, specially in the case of hydrocarbon exploration where direct measures are rare. Applications in mining and magnetostratigraphy may also be possible but require further investigations.

Author Contributions: Conceptualization, J.C.S.O.L., A.P.C.C.L. and J.C.J.; methodology, J.C.S.O.L.; software, J.C.S.O.L.; validation, A.P.C.C.L. and J.C.J.; formal analysis, A.N.G.R.; investigation, J.C.S.O.L.; resources, A.P.C.C.L., J.C.J. and L.d.S.M.; writing—original draft preparation, J.C.S.O.L.; writing—review and editing, J.C.S.O.L.; visualization, A.N.G.R.; supervision, L.d.S.M.; project administration, L.d.S.M. All authors have read and agreed to the published version of the manuscript.

Funding: This research received no external funding.

Institutional Review Board Statement: Not applicable.

Informed Consent Statement: Not applicable.

Data Availability Statement: The data used in this research may be requested to the Brazilian National Petroleum Agency-ANP. Instructions to request data can be found at https://www.gov.br/anp/pt-br/assuntos/exploracao-e-producao-de-oleo-e-gas/dados-tecnicos/acesso-aos-dados-tecnicos, accessed on 31 January 2025.

Acknowledgments: The authors are thankful to the colleague I. O. Carmo for the valuable suggestions. We also would like to thank others colleagues from PETROBRAS for the encouragement and for many useful discussions. We thank PETROBRAS for permission to publish this work.

Conflicts of Interest: The authors declare that they have no known competing financial interests or personal relationships that could have appeared to influence the work reported in this paper. Julio Cesar S. O. Lyrio, Ana Patrícia C. C. Laier, Ana Natalia G. Rodrigues and Luciano dos Santos Martins are employees of PETROBRAS, who provided funding and technical support for the work. Jorge Campos Junior was employee of PETROBRAS by the time this work was developed, but is now retired.

References

1. Laier, A.P.; Jesus, C.M.; Ferraz, E.K.P.; Caires, A.C.B.; Júnior, J.C. Implicações Exploratórias da Orientação de Perfis de Imagem. In Proceedings of the XI Seminário de Interpretação Exploratória, Petrobras, Rio de Janeiro, Brazil, 12 March 2018.
2. Oliveira, J.A.B. *Modelagens Magnéticas Como Suporte ao Mapeamento de Rochas Ígneas;* Nota técnica; Petrobras: Rio de Janeiro, Brazil, 2017.
3. Pozzi, J.P.; Martin, J.P.; Pocachard, J.; Feinberg, H.; Galdeano, A. In-situ magnetostratigraphy: Interpretation of magnetic logging in sediments. *Earth Planet. Sci. Lett.* **1988**, *88*, 357–373. [CrossRef]

4. Nelson, P.H. Magnetic susceptibility logs from sedimentary and volcanic environments. In Proceedings of the SPWLA 34th Annual Logging Symposium, Calgary, AB, Canada, 13–16 June 1993.
5. Broding, R.; Zimmerman, C.; Somers, E.; Wilhelm, E.; Stripling, A. Magnetic well logging. *Geophysics* **1952**, *17*, 1–26. [CrossRef]
6. Lalanne, B.; Bouisset, P.; Pages, G.; Pocachard, J. Magnetic Logging: Borehole Magnetostratigraphy and Absolute Datation in Sedimentary Rocks. In Proceedings of the Middle East Oil Show, Manama, Bahrain, 16–19 November 1991; SPE-21437-MS, pp. 841–850.
7. Pages, G.; Boutemy, Y.; Barthies, V.; Pocachard, J. Wireline Magnetostratigraphy Principles And Field Results. In Proceedings of the SPWLA 35th Annual Logging Symposium, Tulsa, OK, USA, 19–22 June 1994.
8. Emilia, D.A.; Allen, J.W.; Chessmore, R.B.; Wilson, R.B. The DOE/Simplec magnetic susceptibility logging system. In Proceedings of the SPWLA 22nd Annual Logging Symposium, Mexico City, Mexico, 23–26 June 1981.
9. Scott, J.H.; Seeley, R.L.; Barth, J.J. A magnetic susceptibility well-logging system for mineral exploration. In Proceedings of the SPWLA 22nd Annual Logging Symposium, Mexico City, Mexico, 23–26 June 1981.
10. McNeill, J.D.; Hunter, J.A.; Bosnar, M. Application of a borehole induction magnetic susceptibility logger to shallow lithological mapping. *J. Environ. Eng. Geophys.* **1996**, *1*, 77–90. [CrossRef]
11. Leslie, K.; Foss, C.A.; Hillan, D.; Blay, K. A downhole magnetic tensor gradiometer for developing robust magnetisation models from magnetic anomalies. In Proceedings of the Iron Ore, Perth, WA, Australia, 13–15 July 2015; pp. 1–11.
12. Barber, T.; Anderson, B.; Mowat, G. Using induction tools to identify magnetic formations and to determine relative magnetic susceptibility and dielectric constant. *Log Anal.* **1995**, *36*, 16–26.
13. Mitchell, J. Monitoring Lithology Variations in Drilled Rock Formations Using NMR Apparent Magnetic Susceptibility Contrast. *Appl. Magn. Reson.* **2020**, *51*, 205–219. [CrossRef]
14. Huret, E.; Thiesson, J.; Tabbagh, A.; Galbrun, B.; Collin, P. Improvement of cyclostratigraphic studies by processing of high-resolution magnetic susceptibility logging: Example of PEP1002 borehole (Bure, Meuse, France). *Comptes Rendus Geosci.* **2011**, *343*, 379–386. [CrossRef]
15. Crow, H.L.; Hunter, J.A.; Olson, L.C.; Pugin, A.J.M.; Russell, H.A.J. Borehole geophysical log signatures and stratigraphic assessment in a glacial basin, southern Ontario. *Can. J. Earth Sci.* **2018**, *55*, 829–845. [CrossRef]
16. Nowaczyk, N.R. Logging of Magnetic Susceptibility. In *Tracking Environmental Change Using Lake Sediments: Basin Analysis, Coring, and Chronological Techniques*; Last, W.M., Smol, J.P., Eds.; Springer: Dordrecht, The Netherlands, 2001; pp. 155–170. [CrossRef]
17. Bloemendal, J.; Tauxe, L.; Valet, J.P. High-resolution, whole-core magnetic susceptibility logs from LEG 1081. In *Ocean Drilling Program: Initial Report*; Ocean Drilling Program; Texas A&M University: College Station, TX, USA, 1986; Volume 108, p. 1005.
18. Dalan, R.A. A review of the role of magnetic susceptibility in archaeogeophysical studies in the USA: Recent developments and prospects. *Archaeol. Prospect.* **2008**, *15*, 1–31. [CrossRef]
19. To, T.H.; Nguyen, T.V.; Nguyen, L.K.; Dinh, H.D. A review of the applications of magnetic susceptibility measurements for improved reservoir characterization. *J. Min. Earth Sci.* **2022**, *63*, 25–34. [CrossRef] [PubMed]
20. Cheung, P.; Hayman, A.; Vessereau, P.; Verges, P.; Laval, L.; Yerbes, M.; Rylander, E. In Situ Correction of Wireline Triaxial Accelerometer and Magnetometer Measurements. In Proceedings of the SPWLA 48th Annual Logging Symposium, Austin, TX, USA, 3–6 June 2007.
21. ANP—Agência Nacional de Petróleo, Gás Natural e Biocombustíveis. Boletim da Produção de Petróleo e Gás Natural—06/2024. 2024. Available online: https://www.gov.br/anp/pt-br/centrais-de-conteudo/publicacoes/boletins-anp/boletins/arquivos-bmppgn/2024/junho.pdf (accessed on 6 August 2024).
22. Rodrigues, A.N.G. *Propriedades Magnéticas do poço 1-RJS-755-RJ (Urissanê), Intervalo 6137.70–6683.00 m, Bacia de Campos (Brasil)*; Relatório Interno; Petrobras: Rio de Janeiro, Brazil, 2021.
23. Cainelli, C.; Mohriak, W.U. Some remarks on the evolution of sedimentary basins along the Eastern Brazilian continental margin. *Episodes J. Int. Geosci.* **1999**, *22*, 206–216. [CrossRef] [PubMed]
24. Fetter, M. The role of basement tectonic reactivation on the structural evolution of Campos Basin, offshore Brazil: Evidence from 3D seismic analysis and section restoration. *Mar. Pet. Geol.* **2009**, *26*, 873–886. [CrossRef]
25. Cobbold, P.R.; Szatmari, P. Radial gravitational gliding on passive margins. *Tectonophysics* **1991**, *188*, 249–289. [CrossRef]
26. Demercian, S.; Szatmari, P.; Cobbold, P.R. Style and pattern of salt diapirs due to thin-skinned gravitational gliding, Campos and Santos basins, offshore Brazil. *Tectonophysics* **1993**, *228*, 393–433. [CrossRef]
27. Chang, H.K.; Kowsmann, R.O.; Figueiredo, A.M.F.; Bender, A.A. Tectonics and stratigraphy of the East Brazil Rift system: An overview. *Tectonophysics* **1992**, *213*, 97–138. [CrossRef]
28. Strugale, M.; Schmitt, R.S.; Cartwright, J. Basement geology and its controls on the nucleation and growth of rift faults in the northern Campos Basin, offshore Brazil. *Basin Res.* **2021**, *33*, 1906–1933. [CrossRef]
29. Ogg, J.G.; Ogg, G.M.; Gradstein, F.M. *The Concise Geologic Time Scale*; Cambridge University Press: Cambridge, UK, 2008.
30. NOAA/NCEI—National Center for Environmental Information. Enhanced Magnetic Model (EMM). 2022. Available online: https://www.ngdc.noaa.gov/geomag/EMM (accessed on 6 August 2024).

31. Lowrie, W. *Fundamentals of Geophysics;* Cambridge University Press: Cambridge, UK, 1997.
32. Glatzmaiers, G.A.; Roberts, P.H. A three-dimensional self-consistent computer simulation of a geomagnetic field reversal. *Nature* **1995**, *377*, 203–209. [CrossRef]

Disclaimer/Publisher's Note: The statements, opinions and data contained in all publications are solely those of the individual author(s) and contributor(s) and not of MDPI and/or the editor(s). MDPI and/or the editor(s) disclaim responsibility for any injury to people or property resulting from any ideas, methods, instructions or products referred to in the content.

applied sciences

MDPI

Article

Inversion of Elastic and Fracture Parameters in Tilted Transverse Isotropic Media with Parameter Standardization

Guangzhi Zhang [1,2,*], Shengzhao Dai [1,2], Han Li [1,2], Hongjian Hao [1,2] and Tengfei Chen [1,2]

1 Key Laboratory of Deep Oil and Gas, China University of Petroleum (East China), Qingdao 266580, China; daishengzhao0402@163.com (S.D.); b24010007@s.upc.edu.cn (H.L.); haohongjian12345@163.com (H.H.); fei1368054261@163.com (T.C.)
2 School of Geoscience, China University of Petroleum (East China), Qingdao 266580, China
* Correspondence: zhanggz@upc.edu.cn

Abstract: This study addresses inversion challenges in tilted transverse isotropic (TTI) media affected by inclined fractures. A new method is proposed to derive the reflection coefficient for such media, combining scattering theory with the steady-phase method. To enhance inversion accuracy and stability, a scale normalization technique is introduced. The approach improves parameter consistency during the inversion process. The results highlight the potential of this method to offer valuable technical support for fractured reservoir exploration and development.

Keywords: fracture; tilted transverse isotropic; reflection coefficient equation; parameter normalization; seismic inversion

1. Introduction

In the exploration and development of tight carbonate reservoirs, fracture development plays a crucial role in determining reservoir permeability and the formation of reservoir space. The presence of fractures induces anisotropy in the subsurface medium, which manifests as variations in elastic properties depending on the direction. To quantitatively characterize this anisotropy, specific anisotropy parameters for different types of media are proposed (Thomsen, 1986; Rüger, 1996; Tsvankin, 1997) [1–3]. The petrophysical characterization of fractured media is a critical aspect of understanding the behavior of reservoirs, particularly in terms of permeability and anisotropy. Several models have been proposed to represent the effects of fractures on the elastic properties of subsurface materials. These include Hudson's micro-fracture model (Hudson, 1980, 1981) [4,5], Schoenberg's linear slip model (Schoenberg, 1980) [6], and the moving mesh method (Ivančić and Solovchuk, 2022) [7]. Among the most widely used are Hudson's micro-fracture model and Schoenberg's linear slip model, both of which provide essential frameworks for representing fractured rock as equivalent anisotropic media. Schoenberg further introduced the concepts of normal weakness δ_N and tangential weakness δ_T to quantitatively assess the anisotropy induced by fractures, offering a means to characterize fracture-related anisotropy in terms of material behavior. Building upon this, Bakulin (2000) developed a relationship between the fracture weakness parameter [8], the Thomsen anisotropy parameters, and fracture density, providing a unified theoretical foundation for describing the anisotropic effects of fractures in a comprehensive manner. These advancements have significantly improved our ability to model and interpret the complex anisotropic behavior of fractured reservoirs, offering valuable insights into reservoir exploration and development (Figure 1) [9].

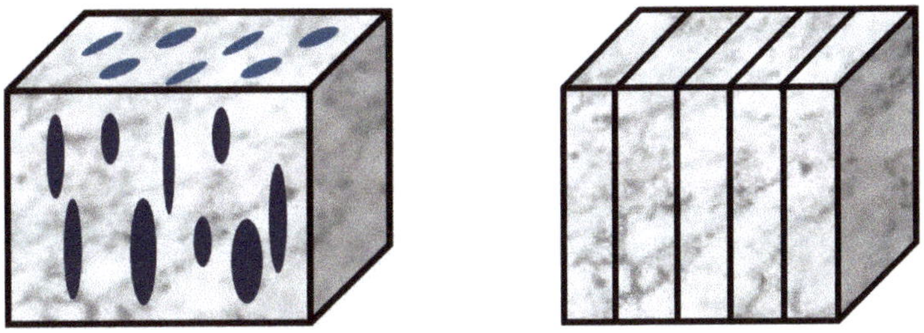

Figure 1. Classical fracture model (Chen, 2015) [9]. (**a**) Hudson's penny-shaped crack model and (**b**) Schoenberg's linear-slip model.

The azimuthal AVO (Amplitude Versus Offset) inversion has emerged as a powerful and reliable technique for predicting fracture parameters, playing a pivotal role in characterizing subsurface anisotropy and heterogeneity (Bachrach, 2015; Li, 2022; Chen, 2024) [10–12]. Central to the success of azimuthal AVO inversion is the accurate estimation of reflection coefficients, which are highly sensitive to fracture characteristics, such as orientation and density. The response of fractured media is strongly dependent on the orientation of the fractures relative to the seismic wave propagation direction. When fractures are vertically oriented, the medium can be effectively modeled as a horizontally layered transversely isotropic (HTI) medium. Numerous studies have focused on deriving and refining reflection coefficient models for HTI media to better understand the seismic response of such systems (Rüger, 1997; Chen et al., 2017) [13,14]. In contrast, when fractures are horizontally oriented, the medium can be treated as a vertically layered transversely isotropic (VTI) medium, and extensive research has been conducted on the reflection coefficients for VTI media, providing insights into the seismic behavior of such systems (Shaw and Sen, 2004; Zhang et al., 2019) [15,16]. When fractures are inclined, the resulting anisotropy becomes more intricate, and the medium is typically described using the tilted transversely isotropic (TTI) model. This model captures the more complex interactions between fracture geometry and seismic waves, and recent advancements have enhanced the understanding of fracture tilt effects on the elastic properties of the medium (Pan et al., 2021; Zhao et al., 2024) [17,18].

In this study, we derive the reflection coefficients for TTI media, considering the influence of a set of inclined fractures with rotational invariance. Our approach is grounded in the assumption of weak anisotropy and the theory of seismic scattering. In the context of seismic inversion, a major challenge lies in addressing the inconsistent scales of inversion parameters, which can lead to inaccuracies in the inversion process. To overcome this issue, we propose a novel inversion equation that transforms the parameters into a uniform scale, thereby improving the reliability of the inversion results. To demonstrate the effectiveness of our method, we perform validations using both synthetic model data and actual field data. The results underscore the practical applicability of our approach to enhance seismic inversion accuracy in anisotropic media.

2. Theory and Method

2.1. Derivation of the Reflection Coefficient Equation

In the investigation of the stiffness matrix for anisotropic media, we consider a specific case involving a group of horizontal, rotationally invariant fractures within a vertically transversely isotropic (VTI) medium. The stiffness matrix for this configuration is derived and expressed as follows:

$$C^{VTI} = C^{Iso} + C_f^H \tag{1}$$

where C^{Iso} represents the stiffness coefficient matrix of the background isotropic medium while C_f^H corresponds to the stiffness coefficient matrix induced by a single group of horizontal fractures. Similarly, the stiffness coefficient matrix of HTI media, which is caused by a set of vertical, rotationally invariant fractures, can be expressed as the sum of the background isotropic medium's stiffness matrix and the matrix resulting from the vertical fractures. Building on this principle of superposition, the stiffness matrix for TTI media can be formulated as the sum of the stiffness matrix for the background isotropic medium and the stiffness matrix induced by a set of inclined fractures. Furthermore, the stiffness matrix resulting from a single group of inclined fractures in TTI media can be decomposed into two components: one from the vertical fractures and the other from the horizontal fractures. Therefore, the total stiffness coefficient matrix for the TTI medium is given by the following:

$$C^{TTI} = C^{Iso} + C_f^T = C^{Iso} + C_{f'}^V \cos\xi + C_{f'}^H \sin\xi \quad (2)$$

where ξ represents the angle between the fracture and the vertical direction in the TTI medium. $C_{f'}^V \cos\xi$ refers to the difference in the stiffness matrix caused by the inclined fracture in the vertical direction, and the influence of this component on the seismic amplitude is solely dependent on the incident angle. On the other hand, $C_{f'}^H \sin\xi$ represents the difference in the stiffness matrix due to the inclined fracture in the horizontal direction, where its impact on the seismic amplitude is influenced by both the incident angle and the azimuth angle. Therefore, the stiffness coefficient matrix of the TTI medium, induced by a single group of inclined fractures, is given by the following:

$$C = \begin{bmatrix} C_{11} & C_{12} & C_{13} & 0 & 0 & 0 \\ C_{12} & C_{22} & C_{23} & 0 & 0 & 0 \\ C_{13} & C_{23} & C_{33} & 0 & 0 & 0 \\ 0 & 0 & 0 & C_{44} & 0 & 0 \\ 0 & 0 & 0 & 0 & C_{55} & 0 \\ 0 & 0 & 0 & 0 & 0 & C_{66} \end{bmatrix} \quad (3)$$

Its specific forms are as follows:

$$\begin{aligned}
C_{11} &= M\left[1-(1-2g)^2\delta_N\sin\xi-\delta_N\cos\xi\right],\\
C_{12} &= \lambda[1-(1-2g)\delta_N\sin\xi-\delta_N\cos\xi],\\
C_{13} &= \lambda(1-\delta_N\sin\xi-\delta_N\cos\xi),\\
C_{22} &= M\left[1-(1-2g)^2\delta_N\sin\xi-(1-2g)^2\delta_N\cos\xi\right],\\
C_{23} &= \lambda[1-\delta_N\sin\xi-(1-2g)\delta_N\cos\xi],\\
C_{33} &= M\left[1-\delta_N\sin\xi-(1-2g)^2\delta_N\cos\xi\right],\\
C_{44} &= \mu(1-\delta_T\sin\xi),\\
C_{55} &= \mu(1-\delta_T\sin\xi-\delta_T\cos\xi),\\
C_{66} &= \mu(1-\delta_T\cos\xi).
\end{aligned}$$

where λ and μ refer to the Lame coefficient of the background isotropic medium, while M denotes the longitudinal wave modulus, and its unit is Pa. g refers to the square of the ratio between the transverse and longitudinal wave velocities. δ_N represents the normal weakness of the fracture, which quantifies the difference in elastic parameters across the fracture surface and along its normal direction. δ_T refers to the tangential weakness of the fracture, capturing the difference in elastic parameters along the tangential direction of the fracture. For g and δ_N, δ_T, they are all dimensionless parameters.

Based on Shaw and Sen's (2004) theory of reflection coefficient equation derivation based on seismic scattering [15], the following equation is produced:

$$R_{\mathrm{PP}}(\theta, \phi) = \frac{1}{4\rho_b \cos^2\theta}(\Delta\rho_b \chi + \Delta C \eta) \tag{4}$$

where ρ_b represents the density of the medium, χ is a parameter associated with the polarization vector, η is the matrix that accounts for the incident angle and azimuth angle of the seismic data, and ΔC is the perturbation matrix of the medium. Thomsen (1986) identified that the majority of rocks demonstrate weak anisotropy. Furthermore, studies on TTI media by Shaw and Sen (2004) and Pan (2021) assume that such media are weakly anisotropic and have drawn reasonable conclusions [1,15,17]. When deriving the perturbation matrix for TTI media, we neglect the effects of $\Delta M\delta_N$, $\Delta M\delta_T$, $\Delta\mu\delta_N$ and $\Delta\mu\delta_T$ due to the assumption of weak anisotropy. Using this theoretical framework for the derivation of the reflection coefficient, the final expression for the reflection coefficient is as follows:

$$R_{\mathrm{PP}}(\theta, \phi) = a(\theta)\frac{\Delta M}{M} + b(\theta)\frac{\Delta\mu}{\mu} + c(\theta, \phi)\Delta\delta_N + d(\theta, \phi)\Delta\delta_T \tag{5}$$

where

$$\begin{aligned}
a(\theta) &= \tfrac{\sec^2\theta}{4}, \quad \mathrm{b}(\theta) = -2g\sin^2\theta \\
c(\theta, \phi) &= \left\{ \begin{array}{l} \frac{\sin\xi \sec^2\theta}{4}\left(1 - 2g\sin^2\theta\right)^2 \\ -\frac{\cos\xi \sec^2\theta}{4}\left[2g\left(\sin^2\theta \sin^2\phi + \cos^2\theta\right) - 1\right]^2 \end{array} \right\} \\
d(\theta, \phi) &= \left[g\sin\xi \sin^2\theta + g\cos\xi \sin^2\theta \cos^2\phi\left(1 - \tan^2\theta \sin^2\phi\right)\right]
\end{aligned}$$

2.2. Reflection Coefficient Equation Analysis

To assess the rationality and validity of the derived equation, we counted some logging data in the reservoir section of a basin in eastern China, and then we constructed a two-layer model with the specific parameters listed in Table 1. Using this model, we compared the reflection coefficient obtained in this study with the reflection coefficient derived by Pan (2021), Shaw, and Sen (2004) [15,17]. The results of this comparison are presented in Figure 2. To further illustrate the accuracy of the reflection coefficient derived in this paper, we calculated the difference between the reflection coefficient obtained by Shaw and Sen and the reflection coefficient derived by us and normalized this difference by dividing it by the value of the reflection coefficient. The resulting relative error is shown in Figure 3, providing a more intuitive measure of the accuracy of our derivation.

Table 1. Two-layer model parameter values.

Layer	M [Gpa]	μ [Gpa]	δ_N	δ_T
Upeer	1.85	1.15	0	0
Lower	2.48	1.32	0.12	0.07

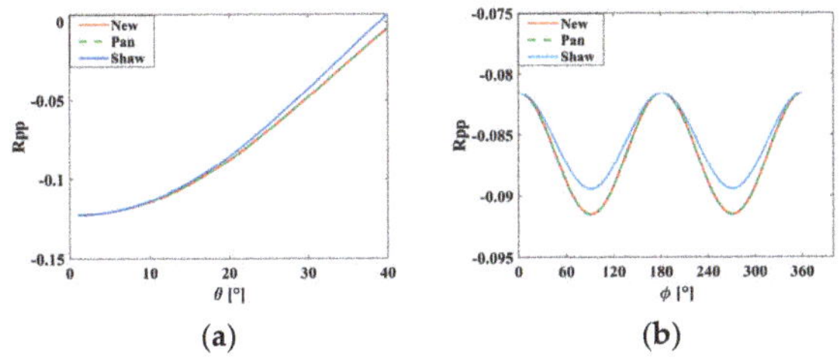

Figure 2. Comparison of reflection coefficients: (**a**) the azimuth angle is 60° and (**b**) the incidence angle is 30°.

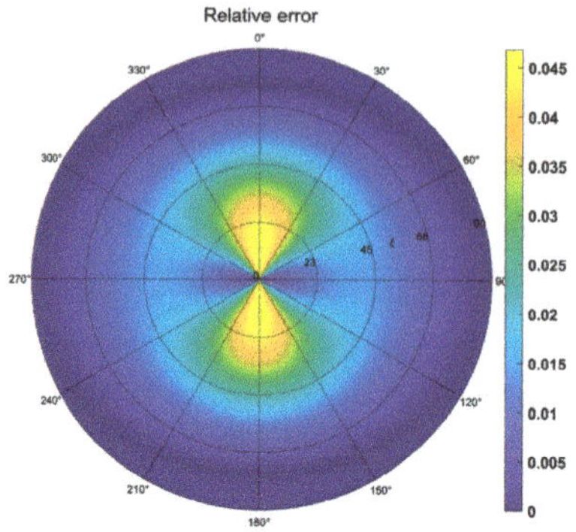

Figure 3. Relative error of reflection coefficient.

As illustrated in Figure 3, the largest error values are observed primarily around the 0° and 180° azimuths. This is due to the increased sensitivity of the reflection coefficient's azimuthal response as the fracture orientation becomes more perpendicular (i.e., as the value of ξ decreases). While significant errors are evident at certain azimuth angles, the overall error in the reflection coefficient derived remains well within 5%. This indicates that the derived reflection coefficient maintains a high degree of accuracy and consistency. Consequently, it can be concluded that the reflection coefficient derived in this study has considerable reference value and practical applicability for fracture analysis.

2.3. Derivation of the Inversion Equation

Azimuth AVO inversion is one of the important means of obtaining reservoir fracture parameters. In this process, the convolution model can be used to effectively establish the mathematical relationship between the reflection coefficient and the seismic data, as follows:

$$d_{nml\times 1} = G_{nml\times 4n} m_{4n\times 1} \tag{6}$$

where d is the seismic data and G is the observation matrix with the following equation:

$$G = \begin{bmatrix} W_{\theta_1}K_M(\theta_1) & W_{\theta_1}K_\mu(\theta_1) & W_{\theta_1}K_{\delta_N}(\theta_1,\varphi_1) & W_{\theta_1}K_{\delta_T}(\theta_1,\varphi_1) \\ \vdots & \vdots & \vdots & \vdots \\ W_{\theta_l}K_M(\theta_l) & W_{\theta_l}K_\mu(\theta_l) & W_{\theta_l}K_{\delta_N}(\theta_l,\varphi_m) & W_{\theta_L}K_{\delta_T}(\theta_l,\varphi_m) \end{bmatrix}$$

where W_{θ_i} is the matrix representing the mathematical relations of the wavelets in the convolution process as follows:

$$W_{\theta_i} = \begin{bmatrix} w_{\theta_i}^1 & 0 & 0 & \cdots \\ w_{\theta_i}^2 & w_{\theta_i}^1 & 0 & \ddots \\ w_{\theta_i}^3 & w_{\theta_i}^2 & w_{\theta_i}^1 & \ddots \\ \vdots & \ddots & \ddots & \ddots \end{bmatrix}$$

and

$$K_M(\theta_1) = \begin{bmatrix} a(\theta_1) & 0 & \cdots & 0 \\ 0 & a(\theta_1) & \cdots & 0 \\ \vdots & \vdots & \ddots & \vdots \\ 0 & 0 & 0 & a(\theta_1) \end{bmatrix}_{n\times n}, K_\mu(\theta_1) = \begin{bmatrix} b(\theta_1) & 0 & \cdots & 0 \\ 0 & b(\theta_1) & \cdots & 0 \\ \vdots & \vdots & \ddots & \vdots \\ 0 & 0 & 0 & b(\theta_1) \end{bmatrix}_{n\times n},$$

$$K_{\delta_N}(\theta_1,\varphi_1) = \begin{bmatrix} c(\theta_1,\varphi_1) & 0 & \cdots & 0 \\ 0 & c(\theta_1,\varphi_1) & \cdots & 0 \\ \vdots & \vdots & \ddots & \vdots \\ 0 & 0 & 0 & c(\theta_1,\varphi_1) \end{bmatrix}_{n\times n},$$

$$K_{\delta_T}(\theta_1,\varphi_1) = \begin{bmatrix} d(\theta_1,\varphi_1) & 0 & \cdots & 0 \\ 0 & d(\theta_1,\varphi_1) & \cdots & 0 \\ \vdots & \vdots & \ddots & \vdots \\ 0 & 0 & 0 & d(\theta_1,\varphi_1) \end{bmatrix}_{n\times n}$$

During the inversion process, significant differences in the order of magnitude among the parameters to be inverted can cause instability, as the large numerical disparity negatively impacts the convergence and accuracy of the inversion algorithm. To address this issue effectively, it is essential to normalize all inversion parameters to a common scale, thus mitigating any potential interference from these magnitude differences. In this study, seismic data from different azimuths were first superimposed to enhance the effect of the isotropic background. This approach allowed for the normalization of the background isotropic parameters by dividing them by the mean value of the isotropic background. For the normalization of fracture parameters, the seismic data at various azimuths were differentiated to remove the influence of the isotropic background. The fracture parameters were then normalized by dividing by the mean value of fracture weakness, ensuring consistent standardization. This methodology effectively controls the variability in the inversion process, leading to more reliable and accurate results.

By implementing this approach of standardization, the final inversion equation not only mitigates the instability caused by the significant order-of-magnitude differences between the various parameters but also enhances the overall accuracy of the inversion results. This allows the inversion model to more effectively capture the characteristics of fractures, ultimately ensuring the extraction of more reliable reservoir fracture parameters in practical applications. As a result, the normalized inversion equation can be formulated as follows:

$$d^*_{nml\times 1} = G_{nml\times 4n} m^*_{4n\times 1} \tag{7}$$

where

$$d^*_{nml\times 1} = \begin{bmatrix} \frac{S_2-S_1}{M^0} + \frac{S_2-S_1}{\delta_N^0} \\ \vdots \\ \frac{S_X+S_{X-1}}{M^0} + \frac{S_X-S_{X-1}}{\delta_N^0} \end{bmatrix} \quad m^*_{4n\times 1} = \begin{bmatrix} \frac{m_{[1,\ldots,2n]}}{M^0} \\ \frac{m_{[2n+1,\ldots,4n]}}{\delta_N^0} \end{bmatrix}$$

In the final inversion process, the correlation between the parameters to be inverted is first assessed by analyzing the logging data. This correlation is then accounted for and eliminated from the inversion equation to prevent solution instability arising from parameter interactions. Furthermore, a low-frequency model is incorporated to effectively smooth and constrain the inversion results, ensuring greater stability and reliability. Throughout these steps, the final inversion equation is expressed as follows:

$$F(m^*) = \|(d - Gm^*)\|^2 + 2v_n^2 \sum_{i=1}^{4N} \ln(1 + m_i^2/v_m^2) + \Xi \tag{8}$$

where

$$\Xi = \gamma_M(\eta_M - PL'_M)^T(\eta_M - PL'_M) + \gamma_\mu\left(\eta_\mu - PL'_\mu\right)^T\left(\eta_\mu - PL'_\mu\right)$$
$$+\gamma_{\delta_N}\left(\eta_{\delta_N} - PL'_{\delta_N}\right)^T\left(\eta_{\delta_N} - PL'_{\delta_N}\right) + \gamma_{\delta_T}\left(\eta_{\delta_T} - PL'_{\delta_T}\right)^T\left(\eta_{\delta_T} - PL'_{\delta_T}\right)$$
$$\eta_M = 1/2\cdot\ln(L_M/L_{M^0}), \eta_\mu = 1/2\cdot\ln(L_\mu/L_{\mu^0}),$$
$$\eta_{\delta_N} = 1/2\cdot\ln(L_{\delta_N}/L_{\delta_N^0}), \eta_{\delta_T} = 1/2\cdot\ln(L_{\delta_T}/L_{\delta_T^0})$$

The damped least squares algorithm is employed to solve Equation (8), facilitating the estimation of both the background elastic parameters and the fracture weakness parameters.

3. Examples

3.1. Synthetic Example

To assess the effectiveness and practicality of the proposed inversion method, a model test was conducted using data from well A, located in a basin in eastern China. Figure 4 shows the planar position of the well. Initially, key parameters such as the mineral content, pore characteristics, and anisotropy were thoroughly analyzed. Based on these parameters, a TTI petrophysical model tailored for the specific geological setting of the study area was constructed. This model was then used to compute the fracture parameter curve for the well. Next, a synthetic seismic record was generated by convolving a Ricker wavelet with a central frequency of 30 Hz and a reflection coefficient. To simulate real-world conditions and align with the methodology of Li (2022) [11], noise with a signal-to-noise ratio of two was added to the synthetic data, mimicking the interference typically observed in actual seismic recordings. Finally, the results obtained from the proposed inversion method were compared to those from a traditional inversion approach. This comparison allowed for a detailed evaluation of the advantages of the new method, particularly in terms of its ability to enhance the stability and accuracy of the inversion results.

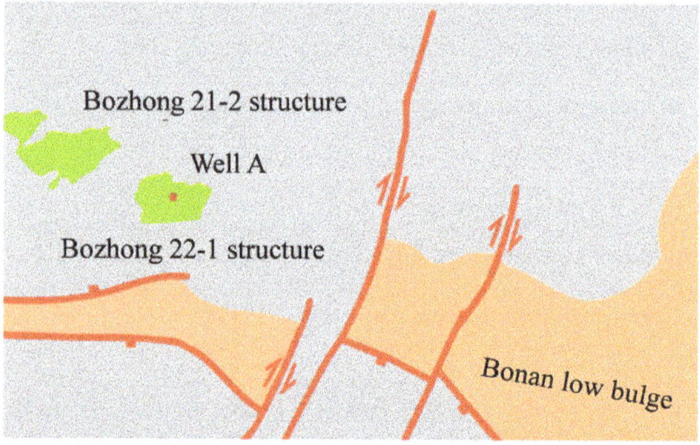

Figure 4. The location of well A.

Figure 5 presents the model test results for both the proposed inversion method and the traditional method under noise-free conditions. As shown, both methods yielded accurate inversion results without noise. Figure 6 further compares the accuracy of the two methods, with errors remaining within a reasonable range and the results appearing stable for both approaches. However, when noise is introduced, the performance diverges. Figure 7 illustrates the inversion results at a signal-to-noise ratio of two, while Figure 8 compares the associated errors. Under noisy conditions, the traditional method shows a significant decline in its fit with the logging curve, particularly for the fracture weakness parameter inversion, leading to increased errors. In contrast, the proposed method demonstrates superior stability and a notable reduction in inversion errors, especially in fracture parameter estimation, where the error is reduced by over 50%. These findings highlight the advantages of the proposed inversion method in improving the stability and accuracy of inversion, particularly for complex fractured reservoir predictions.

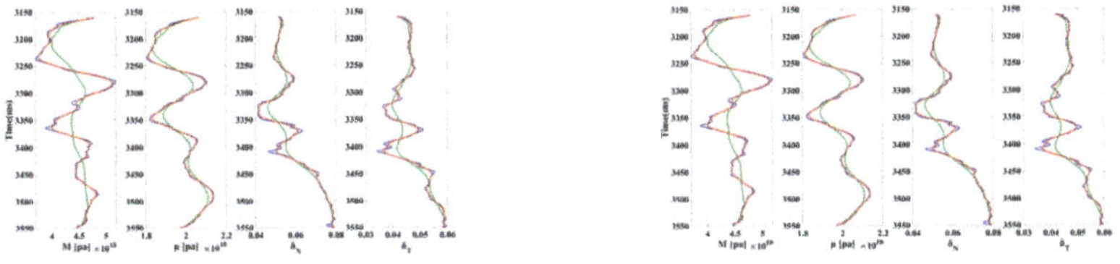

(a) (b)

Figure 5. Inversion results in noise-free conditions. (**a**) The inversion method in this paper, and (**b**) the traditional inversion method. (The green line is the initial value, the blue line is the true value, and the red line is the inverted value).

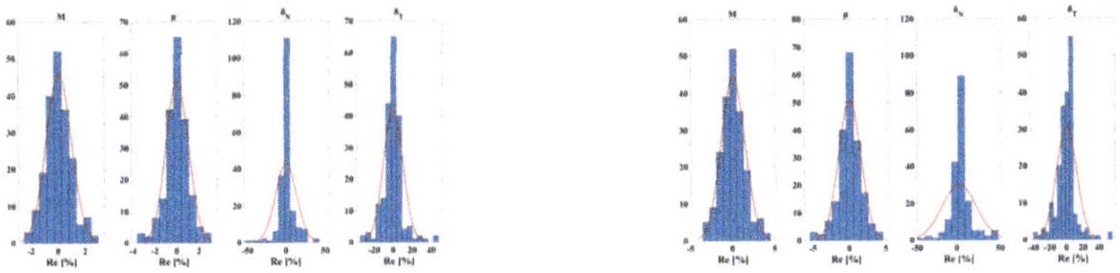

(a) (b)

Figure 6. Inversion error in noise-free conditions. (**a**) The inversion method in this paper, and (**b**) the traditional azimuth AVO inversion method.

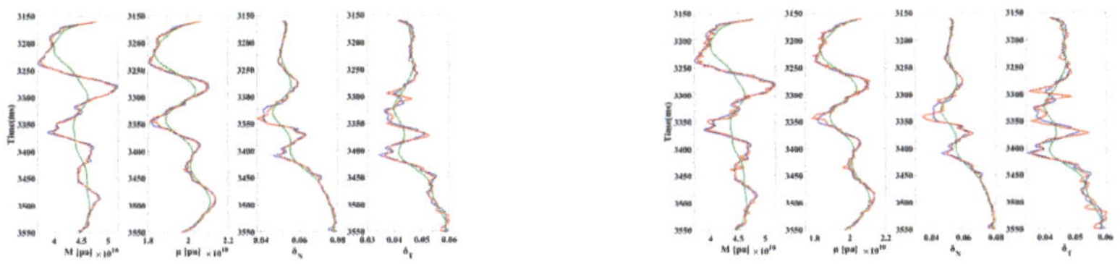

(a) (b)

Figure 7. Inversion results with S/N = 2. (**a**) The inversion method in this paper, and (**b**) the traditional azimuth AVO inversion method. (The green line is the initial value, the blue line is the true value, and the red line is the inverted value).

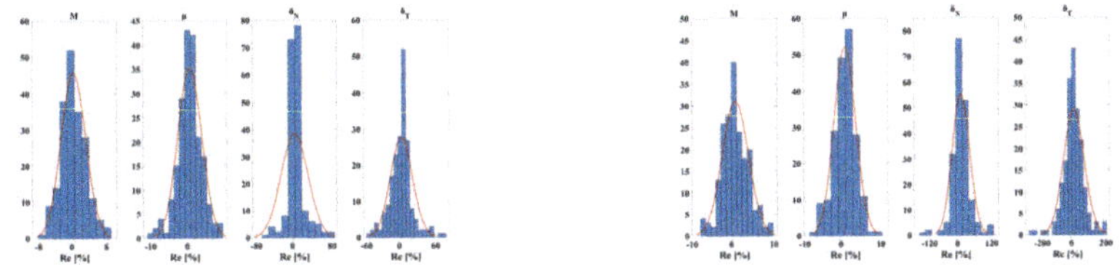

(a) (b)

Figure 8. Inversion error with S/N = 2. (**a**) The inversion method in this paper, and (**b**) the traditional azimuth AVO inversion method.

3.2. Field Data Example

The inversion method demonstrated high stability and reliability in model tests, and its application to the actual work area has yielded promising results. The violent tectonic movements in the Yanshan region, particularly since the Paleogene, have triggered frequent fractures, with the main basal faults significantly influencing fracture distribution. This has led to the development of a complex network of inclined fractures and numerous smaller fractures within the carbonate rocks, enhancing the reservoir quality and making it a high-potential target.

To improve the seismic data quality, the data in the work area were angle-stacked to increase the signal-to-noise ratio. Angle selection was based on the coverage frequency, with angles that exhibited high signal strength and extensive coverage being prioritized. The final selected incident angles were 12°, 20°, and 28°, and the azimuth angles were 32.5°, 77.5°, 122.5°, and 167.5°. The processed seismic data are shown in Figure 9.

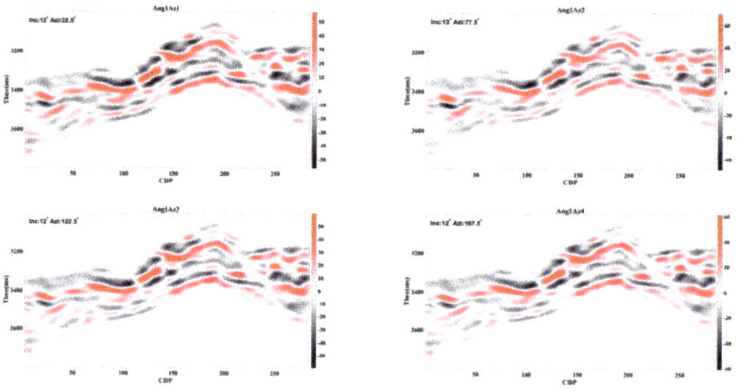

(**a**)

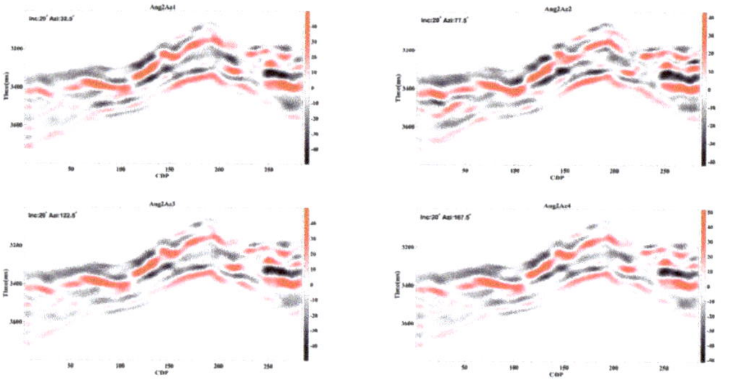

(**b**)

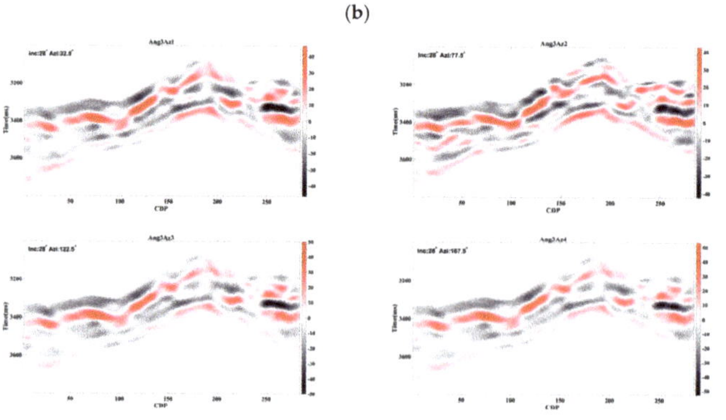

(**c**)

Figure 9. The seismic data (**a**) and the angle of incidence are equal to 12°. (**b**) The angle of incidence is equal to 20°. (**c**) The angle of incidence is equal to 28°.

In the data inversion process, the initial model for the parameters to be inverted was established using logging data, followed by seismic data inversion through the damping least squares method. Figure 10 presents the inversion results, which align well with the actual logging data. Specifically, the inversion results mirror the trends in the logging data for high values, which correspond to high inversion values, and low values, which correspond to low inversion values. This confirms the reliability of the inversion process.

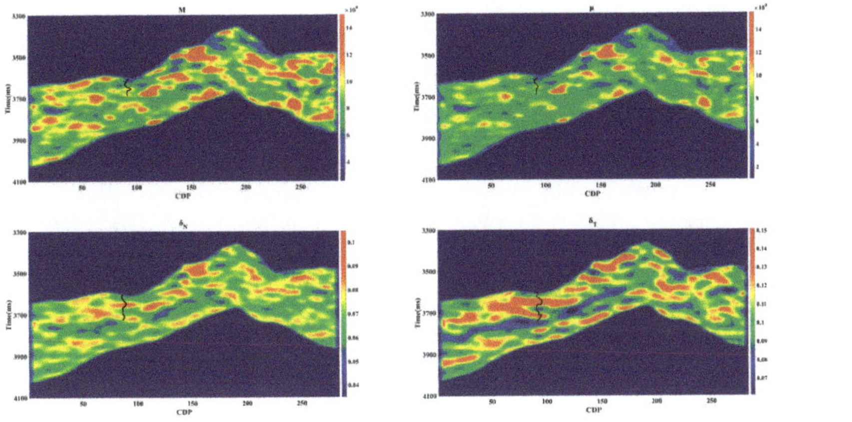

Figure 10. Inversion results of model parameters.

Furthermore, the inversion results were sliced at the target layer, as shown in Figure 11. These slices indicate that the inversion results for the target horizon at well A exhibit high values, supporting their consistency with logging data interpretation. Additionally, the fracture weakness map reveals more significant fracture development in the northwest area, which is consistent with existing geological understanding. These findings validate the accuracy and robustness of the inversion results, demonstrating that the proposed reflection coefficient and inversion method are both reasonable and effective for practical application.

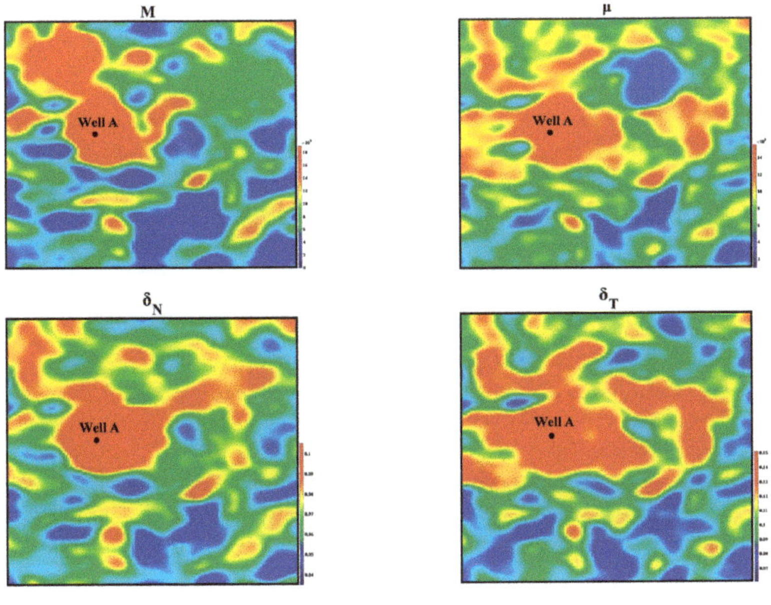

Figure 11. Time slice of the inversion result.

4. Discussions

In this study, we derive the stiffness matrix of a TTI medium with a single group of inclined fractures using the linear slip theory and the projection decomposition method. The resulting stiffness matrix consists of two components: one representing the stiffness of the background isotropic medium and the other capturing the stiffness changes induced by the inclined fractures. By applying the projection decomposition approach, the stiffness variation due to the fractures is further separated into two parts. The first component $C_{f'}^{H} \sin \xi$ reflects the stiffness change in the horizontal direction, where vertical forces are less resisted, making the medium more prone to deformation. The second component $C_{f'}^{V} \cos \xi$ represents the stiffness variation in the vertical direction, where horizontal forces lead to greater susceptibility to deformation.

In parameter inversion, the background isotropic medium parameters are typically expressed as the ratio of the difference to the sum of the upper and lower values, with magnitudes often around 10^{-2} and occasionally 10^{-3}. In contrast, fracture parameters are represented by the difference between the upper and lower values, with magnitudes generally on the order of 10^{-4} to 10^{-5} and sometimes as low as 10^{-6}. This significant difference, up to three or even four orders of magnitude, can lead to instability in the inversion process and reduce its accuracy. Therefore, it is crucial to normalize the inversion parameters prior to the inversion to mitigate these issues and improve stability and precision. In addition, the moving mesh method can also solve some of the stability problems of fracture parameter inversion (Ivančić and Solovchuk, 2022) [7]. Since the primary focus of our paper is not on the moving mesh method, we have not delved deeply into its intricacies.

This study focuses on the TTI media model induced by a single group of inclined fractures, which is primarily applicable to inclined fracture reservoirs. Based on the paper of Hsu and Schoenberg (1993) [19], the difference between fracture weakness parameters δ_N and δ_T can serve as an indicator for fracture fluid type. Based on the findings of Ponthot and Belytschko (1998) [20], the presence of fluid influences the elastic properties of fractures, thereby altering the fracture pressure, which subsequently leads to changes in the fracture's state. However, due to time and resource constraints, this paper does not address the fluid indicator factors. Future research can explore the inversion of these fluid factors in more detail.

5. Conclusions

This paper presents a new reflection coefficient equation for TTI media induced by a single group of inclined fractures, derived using the projection decomposition approach. Compared to previous models, the new equation demonstrates decent accuracy and rationality. To address the issue of inconsistent parameter scales during inversion, we normalized the parameters by adding and subtracting seismic data from different azimuths. One-dimensional testing using well models and actual data tests shows that the results align with both logging interpretations and geological insights from the study area, further validating its effectiveness. The findings of this study hold significant promise for practical applications in reservoir characterization, particularly in fractured reservoir exploration and development. The improved accuracy and stability of the inversion process have the potential to enhance decision-making in well placement. Finally, in the future, we will study fluid interpretation and large-scale data applications to address the shortcomings of this article.

Author Contributions: Methodology, S.D.; Software, G.Z.; Validation, T.C.; Investigation, H.L. and H.H.; Resources, H.H.; Data curation, H.L.; Writing—original draft, S.D.; Writing—review and editing, G.Z. and S.D.; Supervision, T.C.; Funding acquisition, G.Z. All authors have read and agreed to the published version of the manuscript.

Funding: This research was funded by the National Natural Science Foundation of China under the Special Fund for Basic Research Operations of the Central Universities (U23B6010) and the CNOOC "14th Five Year Plan" Major Science and Technology Project (KJGG2022-0304).

Data Availability Statement: The original contributions presented in this study are included in the article. Further inquiries can be directed to the corresponding author.

Acknowledgments: We sincerely thanks HampsonRussell Suite software version 12.1 for its support.

Conflicts of Interest: The authors declare no conflict of interest.

Nomenclature

M	Longitudinal wave modulus, which is equal to the product of the density and the square of the longitudinal wave velocity
μ	The transverse wave modulus, which is equal to the product of the density and the square of the shear wave velocity
δ_N	Normal weakness
δ_T	Tangential weakness
Re	Relative error
AngXazY	Seismic data for the X-th angle of incidence and the Y-th azimuth
Inc	The angle of incidence
Azi	The azimuth

References

1. Thomsen, L. Weak elastic anisotropy. *Geophysics* **1986**, *51*, 1954–1966. [CrossRef]
2. Rüger, A. Reflection Coefficient and Azimuthal AVO Analysis in Anisotropic Media. Ph.D. Thesis, Colorado School of Mines, Golden, CO, USA, 1996.
3. Tsvankin, I. Anisotropic parameters and P-wave velocity for orthorhombic media. *Geophysics* **1997**, *62*, 1292–1309. [CrossRef]
4. Hudson, J.A. Overall properties of a cracked solid. *Math. Proc. Camb. Philos. Soc.* **1980**, *88*, 371–384. [CrossRef]
5. Hudson, J.A. Wave speeds and attenuation of elastic waves in material containing cracks. *Geophysics* **1981**, *64*, 133–150. [CrossRef]
6. Schoenberg, M. Elastic wave behavior across linear slip interfaces. *J. Acoust. Soc. Am.* **1980**, *68*, 1516–1521. [CrossRef]
7. Ivančić, F.; Solovchuk, M. Moving mesh strategy for simulating sliding and rolling dynamics of droplets on inclined surfaces with finite element method. *Comput. Methods Appl. Mech. Eng.* **2022**, *400*, 115404. [CrossRef]
8. Bakulin, A.; Grechka, V.; Tsvankin, I. Estimation of fracture parameters from reflection seismic data—Part I: HTI model due to a single fracture set. *Geophysics* **2000**, *65*, 1788–1802. [CrossRef]
9. Chen, H.; Yin, X. *Study on Methodology of Pre-Stack Seismic Inversion for Fractured Reservoirs Based on Rock Physics*; China University of Petroleum: Qingdao, China, 2015.
10. Bachrach, R. Uncertainty and nonuniqueness in linearized AVAZ for orthorhombic media. *Lead. Edge* **2015**, *34*, 1048–1056. [CrossRef]
11. Li, L.; Zhang, G.; Pan, X.; Guo, X.; Zhang, J.; Zhou, Y.; Lin, Y. Anisotropic poroelasticity and AVAZ inversion for in situ stress estimate in fractured shale-gas reservoirs. *IEEE Trans. Geosci. Remote Sens.* **2022**, *60*, 5911113. [CrossRef]
12. Chen, F.; Zong, Z.; Lang, K.; Li, J.; Yin, X.; Miao, Z.; Xiao, W. Geofluid discrimination in stress-induced anisotropic porous reservoirs using seismic AVAZ inversion. *IEEE Trans. Geosci. Remote Sens.* **2024**, *62*, 5931714. [CrossRef]
13. Rüger, A. P-wave reflection coefficients for transversely isotropic models with vertical and horizontal axis of symmetry. *Geophysics* **1997**, *62*, 713–722. [CrossRef]
14. Chen, H.; Pan, X.; Ji, Y.; Zhang, G. Bayesian Markov Chain Monte Carlo inversion for weak anisotropy parameters and fracture weaknesses using azimuthal elastic impedance. *Geophys. J. Int.* **2017**, *210*, 801–818. [CrossRef]
15. Shaw, R.K.; Sen, M.K. Born integral, stationary phase and linearized reflection coefficients in weak anisotropic media. *Geophys. J. Int.* **2004**, *158*, 225–238. [CrossRef]
16. Zhang, F.; Zhang, T.; Li, X.Y. Seismic amplitude inversion for the transversely isotropic media with vertical axis of symmetry. *Geophys. Prospect.* **2019**, *67*, 2368–2385. [CrossRef]
17. Pan, X.; Li, L.; Zhou, S.; Zhang, G.; Liu, J. Azimuthal amplitude variation with offset parameterization and inversion for fracture weaknesses in tilted transversely isotropic media. *Geophysics* **2021**, *86*, C1–C18. [CrossRef]

18. Zhao, Y.; Wen, X.T.; Li, C.L.; Liu, Y.; Xie, C.L. Systematic prediction of the gas content, fractures, and brittleness in fractured shale reservoirs with TTI medium. *Pet. Sci.* **2024**, *158*, 225–238. [CrossRef]
19. Hsu, C.J.; Schoenberg, M. Elastic waves through a simulated fractured medium. *Geophysics* **1993**, *58*, 964–977. [CrossRef]
20. Ponthot, J.P.; Belytschko, T. Arbitrary Lagrangian-Eulerian formulation for element-free Galerkin method. *Comput. Methods Appl. Mech. Eng.* **1998**, *152*, 19–46. [CrossRef]

Disclaimer/Publisher's Note: The statements, opinions and data contained in all publications are solely those of the individual author(s) and contributor(s) and not of MDPI and/or the editor(s). MDPI and/or the editor(s) disclaim responsibility for any injury to people or property resulting from any ideas, methods, instructions or products referred to in the content.

applied sciences

MDPI

Article

Seismic Prediction of Porosity in the Norne Field: Utilizing Support Vector Regression and Empirical Models Driven by Bayesian Linearized Inversion

Jorge A. Teruya Monroe [1,†], Jose J. S. de Figueiredo [1,2,3,*,†] and Carlos E. S. Amanajas [1]

1 Programa de Pós-Graduação em Geofísica, Universidade Federal do Pará, Rua Augusto Correa, 01, Belém 66075-110, PA, Brazil; jteruya2541@gmail.com (J.A.T.M.); ceamanajas@gmail.com (C.E.S.A.)
2 Petrophysics, and Rock Physics Laboratory, Faculty of Geophysics, Federal University of Pará (UFPA), Belém 66075-110, PA, Brazil
3 National Institute of Science and Technology of Petroleum Geophysics, Room 206-E, Ondina University Campus, Salvador 40170-115, BA, Brazil
* Correspondence: jadsomjose@gmail.com
† These authors contributed equally to this work.

Abstract: This work aims to improve the characterization of petrophysical properties by accurately estimating subsurface porosity using seismic and well data. The study includes Bayesian Linearized Inversion to obtain elastic parameters (e.g., compressional e shear wave velocities and densities). This reduces processing uncertainty and provides a reliable substitute for the standard Amplitude versus Offset inversion method. Furthermore, incorporating sparse spike wavelets with Bayesian Linearized Inversion refines the inversion output, facilitating the extraction of petrophysical properties. Combined with log data from seventeen wells, these inverted parameters serve as inputs for two porosity prediction models: the empirical Han's equation and a more adaptable Support Vector Regression model, the latter demonstrating superior precision in most cases due to its flexible fitting and calibration capabilities. Results from the Norne field in the North Sea confirm the approach's viability, with the Support Vector Regression model achieving a significant Pearson correlation coefficient of 90% in porosity prediction, underscoring the potential of machine learning techniques in improving subsurface exploration results.

Keywords: Bayesian linearized inversion; porosity; machine learning

1. Introduction

Porosity has a significant impact on the economic feasibility of a deposit, as is widely known. Porosity can be quantified using a variety of approaches, including seismic data [1] (by indirect methods), well logs [2,3] (via direct measurements), and accurate laboratory analysis [4]. Seismic data are used to estimate elastic characteristics, which are then used to calculate petrophysical properties. However, it is critical to recognize the uncertainties that pervade both stages of this complex inversion process. If these uncertainties are not managed properly, they might lead to cumulative errors that propagate throughout the structure of the final output. The integrated technique, which harmonizes seismic data with well logs, combines rock physics and seismic inversions. This approach extracts elastic and petrophysical parameters directly from the initial data [5–9]. Beyond that, one essential component in the thorough characterization of siliciclastic reservoirs is the porosity attribute. It is essential for controlling fluid flow and storing fluid in these rock formations. Most of the published research has focused on using seismic and well-log measurements to estimate porosity, e.g., Su et al. [10] and Boateng et al. [11], presented a seismic petrophysical inversion technique (similar to [12]) to determine porosity from the seismic impedance. Guo et al. [13] presented a Monte Carlo-based seismic petrophysical inversion technique that simultaneously estimates porosity and pore aspect ratio using the Xu-White model and the Zoeppritz equation. These investigations demonstrate that the quantitative representations of petrophysical results can be more accurate by incorporating the changing porosity using inversion techniques. In regions with few well-log readings, Feng [14] calculated reservoir porosity using seismic data and deep learning to make precise predictions.

The main objective of the work was to reduce initial uncertainty by continuously modifying the models' input data and increasing the prediction models' accuracy. When using inversion models, the initial uncertainty of the input data is typically neglected, leading to values that underestimate this uncertainty [15]. In the current work, two distinct inversion models—the BLI model and the AVO model—were utilized to enhance the porosity results obtained from the prediction models to reduce the effects of uncertainty in the resulting values. The BLI model was used because it combines the data inversion methodology with the Bayesian process, widely used for its ability to use a priori information to minimize the effects of uncertainty effectively [16–22], thus improving the accuracy and reliability of the model results. Statistical rock physics modeling integrates statistical analyses with rock physics models to extrapolate information beyond well-log observations [23,24]. Ensuring initial input data distributions adhere to a Gaussian distribution is critical [25]. It is worth mentioning that meticulous pre-processing and analysis were conducted to control distributions and meet Gaussian distribution requirements. The model utilized, developed by [25], offers an analytical matrix comprising mean and covariance, representing Gaussian posterior distribution results within a Bayesian structural framework. This strategic selection notably enhances result quality by managing intrinsic dataset uncertainty alongside a priori information. The investigation begins with seismic and well data inversion, utilizing techniques such as Bayesian Linearized Inversion (BLI) and Amplitude Versus Offset (AVO).

At the start of the investigation, to effectively manage computational resources, predictive models were initially applied to a subset of the seismic data (a time window of 500 milliseconds versus 100 traces). Additionally, information from the primary records of the seventeen wells was utilized. Of the seventeen wells, two serve as primary wells (6608-10-02 and 6608-10-04) to implement various inversion and prediction processes, and fifteen secondary wells were used to train and evaluate the different models used in the study. Also, at the beginning, the Ricker wave is used as the initial wave, which is then replaced by a sparse spike wave; the latter acts as an optimization of the wavelet, thus improving the results obtained, and verifying that the choice of an appropriate wave influences significantly influences optimizing the results of the porosity model. The velocities of the P and S waves, as well as the density used to obtain the porosity values, are predicted by two different models: one based on support vector regression (SVR) and the other on Han's Equation [26].

The initial values used in the predictive model based on the empirical equation show improved dimensionality after the inversion process, going from 1D to 2D. This increase was sought to capture nonlinear relationships between inherent attributes of the data and associated uncertainties. Furthermore, this expansion aimed to strengthen the predictive capacity of the model by allowing it to encapsulate more complex relationships between the attributes and the target variable while mitigating potential modeling errors. The main objective of this study is to improve the accuracy of porosity predictions by optimizing inversion models and minimizing inherent uncertainties. Using a set of techniques such as

Bayesian Linear Inversion (BLI) in combination with Support Vector Regression (SVR) and Han's empirical model [26], we can optimize the characterization of the reservoir.

2. Theoretical Background

2.1. Bayesian Linearized Inversion (BLI)

Given their ability to infer the spatial distribution in the subsurface, inversion procedures are essential for defining and predicting the physical properties of hydrocarbon reservoirs [27]. Beyond figuring out the parameter values for the model that fits the data the best, these prediction and modeling techniques also try to model uncertainty to improve the accuracy of the results [28]. The seismic inversion solution in a Bayesian context, where the procedure creates a relationship between the observed data **d** and the modeling parameters, which are modified via a likelihood function $p(\mathbf{d}|\mathbf{m})$. A posterior distribution $p(\mathbf{m}|\mathbf{d})$ is used to derive this likelihood function. The equation can be written as follows:

$$p(\mathbf{m}|\mathbf{d}) \propto \alpha\, p(\mathbf{d}|\mathbf{m})p(\mathbf{m}), \tag{1}$$

where the posterior distribution is represented by $p(\mathbf{m}|\mathbf{d})$ and the likelihood function is $p(\mathbf{d}|\mathbf{m})$; this function indicates the probability of obtaining the observed data based on the model parameters. The model parameters are represented by **m**, the observed data by **d**, the prior distribution of the geological model is $p(\mathbf{m})$, and the normalization constant with values that tend to unity by α. Mathematically, this can be represented as:

$$\alpha = \frac{1}{\int p(\mathbf{d}|\mathbf{m}) \cdot p(\mathbf{m})}. \tag{2}$$

Bayes' theorem is utilized to incorporate prior knowledge regarding the model parameters, which imposes limitations on the solution and represents the uncertainty in the results. The likelihood function $p(\mathbf{d}|\mathbf{m})$ incorporates the forward convolutional model $g(\mathbf{m})$ inside the Bayesian framework. It serves as the initial reference point as established in the model described in [29]. The likelihood function quantifies the probability of observing the data **d** given the model parameters **m**. This is accomplished by revising the initial distribution using the information that has been observed. The prior distribution functions, denoted as $p(\mathbf{m})$, represent the initial knowledge or assumptions about the model parameters before any data are considered. Finally, the parameter α acts as a normalization factor for the probability distribution $p(\mathbf{m})$, guaranteeing that the resulting posterior distribution integrates with 1.

2.2. The Forward Model

The statistical model is characterized by its effective integration of both physical and statistical models, allowing for the inclusion of scenarios that may not be captured just through physical data such as well-logs. The computation of seismic data based on elastic parameters can be formulated as follows:

$$\mathbf{d} = g(\mathbf{m}) + \mathbf{e}_d, \tag{3}$$

where e_d represents the observed noise, derived from the data **d**, $g(\mathbf{m})$ represents the seismic wavelet extracted from the reflectivity function, and $\mathbf{e}_d$ presents a great initial uncertainty derived from the observed data **d**, which has a strong impact throughout the inversion process. Within Bayesian processes, the term **d** is incorporated into the probability function $p(\mathbf{d}|\mathbf{m})$ (see Equation (1)). This equation creates a connection that unites the model parameters with the observed data, establishing the basis for calculating the values of the goal distribution $p(\mathbf{m}|\mathbf{d})$. The values of $p(\mathbf{d}|\mathbf{m})$, which represent the likelihood function, show us the level of correspondence between **d** and $g(\mathbf{m})$, considering the parameters **m**. The analytical solution can be expressed based on BLI theory for the direct linearized model presented in Equation (3). A normal or Gaussian distribution with zero means $e \sim N(0, \Sigma_e)$ is assumed for the error term $\mathbf{e}_d$ in the data. The posterior distribution $p(\mathbf{s}|\mathbf{d})$ also follows a Gaussian Mixture (GM) with an analytical expression. In other words, the distribution remains consistent at any depth and can be described as a linear combination of Gaussian distributions with a fixed number of components.

2.2.1. Convolutional Model

Using Equation (6) to set up the convolution model in the discrete domain and the reflectivity formulation from [25], we arrive at:

$$\mathbf{d} = \mathbf{SADz} + \mathbf{e}_d, \tag{4}$$

where **d** is a vector that comprises concatenated seismic traces (acquired at different angles) with dimensions $n_\theta n_d \times 1$, **z** is the logarithmic vector of elastic properties with dimensions $3n \times 1$, **S** is the convolution matrix of size $n_\theta n_d \times n_\theta n_d$ incorporating the discretized wavelet for each angle, **A** is a matrix of size $n_\theta n_d \times n_\theta n_d$ containing coefficients $a_\alpha(t,\theta)$, $a_\beta(t,\theta)$, and $a_\rho(t,\theta)$, **D** is a first-order differential operator of dimension $n_\theta \times n_d \times 3n$ [30], and $\mathbf{e}_d$ represents the seismic noise.

2.2.2. Statistical Model

Assuming that the error term $\mathbf{e}_d$ follows a multivariate Gaussian distribution:

$$P(\mathbf{m}|\mathbf{d}) = exp\left(-\frac{1}{2}(\mathbf{d} - g(\mathbf{m}))^T \boldsymbol{\Sigma}_\mathbf{d}{}^{-1}(\mathbf{d} - g(\mathbf{m}))\right), \tag{5}$$

where $\boldsymbol{\Sigma}_d$ is the covariance matrix of the error and $g(\mathbf{m})$ is the forward model. Introducing low-frequency information, such as well-log data, into the model can be accomplished by introducing it as an extra constraint through regularization techniques such as Tikhonov regularization. The resultant solution will demonstrate enhanced accuracy, specificity, and less ambiguity. When comparing a solution that includes low-frequency information to one that does not, we have the following equation:

$$P(\mathbf{m}) = \exp\left(-\frac{1}{2}(\mathbf{m} - \mu_m)^T \boldsymbol{\Sigma}_\mathbf{m}{}^{-1}(\mathbf{m} - \mu_m)\right), \tag{6}$$

where μ_m represents a vector formed by combining the low-frequency models of the logarithmic elastic properties and $\boldsymbol{\Sigma}_m$ is the covariance matrix that describes the spatial relationship between the properties and quantifies the difference in the prior distribution.

2.2.3. Objective Function

Within the Bayesian inversion, the objective function calculates the divergence between the values of the previous model and the observed data (**m**). The objective function allows us to adjust the previous model to the observed data **m** to determine the subsequent model, with which we describe the physical structure of the model while providing information about the subsoil properties with greater precision. The maximum likelihood of the likelihood is related to the parameters of the convolution model **d** with the smallest difference between observed data **m** and the acquired data, while the maximum a priori probability is related to the slightest difference from the low-frequency model. Therefore, the maximum a posteriori (MAP) for the inverse problem is the optimal configuration of the model that manages to minimize the following function:

$$J(\mathbf{m}) = \left((\mathbf{d} - g(\mathbf{m})^T \boldsymbol{\Sigma}_{\mathbf{d}}{}^{-1}(\mathbf{d} - g(\mathbf{m}))\right) + \left((\mathbf{m} - \boldsymbol{\mu}_m)^T \boldsymbol{\Sigma}_{\mathbf{m}}{}^{-1}(\mathbf{m} - \boldsymbol{\mu}_m)\right), \tag{7}$$

where the covariances $\boldsymbol{\Sigma}_{\mathbf{d}}$ and $\boldsymbol{\Sigma}_{\mathbf{m}}$ act as the weights in the optimization process. The function $g(\mathbf{m})$ is a weighted sum of mixture components, where $g(\mathbf{m}) = \sum_{i=1}^{k} w_i g_i(\mathbf{m})$. Here, each $g_i(\mathbf{m})$ represents a specific mixture component, and the coefficient w_i indicates the weight associated with the ith component. These weights w_i control the relative contribution of each component to the overall mixture, adhering to the conditions $0 \leq w_i \leq 1$ for all i, and the sum of all weights equals one, $\sum_{i=1}^{k} w_i = 1$ [31].

2.2.4. Seismic Model

Iterative geostatistical seismic inversion and Bayesian linearized seismic inversion are the two most frequently employed seismic inversion techniques in the oil and gas sector [27]. The model requires seismic data with varying acquisition angles θ since it relies on the linearized approximation for weak elastic contrasts. It is possible to estimate the sequence of reflection coefficients with the models described in [25,32,33]. In terms of math, it is stated as:

$$d(t,\theta) = \int_{-\infty}^{\infty} s(\tau,\theta)c(t-\tau,\theta)\mathrm{d}\tau + e_d(t,\theta), \tag{8}$$

where $d(t,\theta)$ represents the seismic amplitude at time t, $s(\tau,\theta)$ is the wavelet, $e_d(t,\theta)$ denotes the observation error, and $c(t,\theta)$ represents the reflectivity at the interface, as modeled by:

$$c(t,\theta) = a_\alpha(t,\theta)\frac{\partial}{\partial t}\ln\alpha(t) + a_\beta(t,\theta)\frac{\partial}{\partial t}\ln\beta(t) + a_\rho(t,\theta)\frac{\partial}{\partial t}\ln\rho(t), \tag{9}$$

where α and β are the P-wave and S-wave velocities, respectively. The ρ parameter is the density, and the coefficients $a_\alpha(t,\theta)$, $a_\beta(t,\theta)$, and $a_\rho(t,\theta)$ are given by:

$$a_\alpha(t,\theta) = \frac{1}{2}(1 + \tan^2\theta), \tag{10}$$

$$a_\beta(t,\theta) = -4\left(\frac{\beta(t)^2}{\alpha(t)^2}\right)\sin^2\theta, \tag{11}$$

$$a_\rho(t,\theta) = \frac{1}{2}(1 + a_\beta(t,\theta)). \tag{12}$$

2.3. *Prior Knowledge*

Since the data of the prior distribution $p(\mathbf{m})$, used as input data for the model, are the combination of several normal distributions, the Gaussian mixture model [34] can be used. This process can be represented as:

$$P(m) = \sum_{k=1}^{k} \sigma_k N_k\left(s; \mu_s^k, \Sigma_s^k\right), \tag{13}$$

where μ_r is the mean and Σ_r the covariance, both from the Gaussian distribution $\mathcal{N}_i$, which is corresponding to the i-th factor. The value of σ_i describes the weight that is associated with the Gaussian distribution, where the sum of all these values must be equal to unity, that is, $\sum_{i=1}^{i} \sigma_i = 1$, where i is the initial total number of Gaussian distributions to consider for the model. Estimating Gaussian Mixture Model parameters during the inversion process entails calculating means and covariances for each Gaussian component. Additionally, subsurface parameters obtained from well logs (e.g., V_p, V_s, ρ, ϕ, V_{sh}) are incorporated into the model during this phase. Cross-validation determines the value of k, allowing for its iterative modification during the modeling procedure in response to performance metric evaluations.

2.4. Posterior Distribution

In the BLI inversion process, the resulting distribution (a posteriori distribution) is the probability of the model parameters that are conditioned, which is based on the observed data and a prior model. The multivariate analysis, developed in the work of [35], together with a posterior distribution, results in a measure of uncertainty of the parameters based on the observed data and the a priori model. This distribution shows that the probability distribution, which represents the elastic parameters **m** conditioned on the observed data **d**, results in a multivariate Gaussian distribution representing different values of model parameters:

$$P(\mathbf{m}|\mathbf{d}) = \mathcal{N}\left(\mu_{m|d}, \Sigma_{m|d}\right), \tag{14}$$

where the covariance matrix is $\boldsymbol{\Sigma}_{\boldsymbol{m}}|\boldsymbol{d}$ and the mean is $\boldsymbol{\mu}_{\boldsymbol{m}}|\boldsymbol{d}$, developed by:

$$\mu_{(m|d)} = \mu_m + \boldsymbol{\Sigma}_m\mathbf{G}^T\left(\mathbf{G}\boldsymbol{\Sigma}_m\mathbf{G}^T + \boldsymbol{\Sigma}_d\right)^{-1}\mathbf{d} - \mathbf{G}\mu_m, \tag{15}$$

and

$$\boldsymbol{\Sigma}_{(m|d)} = \boldsymbol{\Sigma}_m - \boldsymbol{\Sigma}_m\mathbf{G}^T\left(\mathbf{G}\boldsymbol{\Sigma}_m\mathbf{G}^T + \boldsymbol{\Sigma}_d\right)^{-1}\mathbf{G}\boldsymbol{\Sigma}_m. \tag{16}$$

Moreover, the direct linear operator, denoted by **G** and derived from the convolutional model, is given by **G = SAD**. As the posterior distribution of elastic properties follows a Gaussian distribution, the maximum a posteriori (MAP) solution involves maximizing the logarithm of the posterior probability density function, which is proportional to the product of the prior probability and the likelihood function. Mathematically, this solution is expressed as:

$$MAP\{(\alpha, \beta, \rho)^T\} = \exp\left(\mu_{m|d} - \sigma^2_{m|d}\right), \tag{17}$$

where $\sigma_{m|d}$ is the vector with the posterior variances, obtained from the diagonal of the covariance matrix $\boldsymbol{\Sigma}_{m|d}$. The computation of the solution to Equation (17) is efficient and stable due to the availability of the analytical solution for the inverse problem.

3. Methodology

Figure 1 depicts the steps in predicting porosity data using several tasks. The first part of the workflow explains the stages taken throughout the Bayesian linearized inversion (BLI) process, which include data preparation, model change or modulation, and data inversion finalization. The initial phase includes the Bayesian Linearized Inversion (BLI) procedure, which requires the discovery, selection, and processing of source data. In the following stage, wavelet extraction was performed using sparse spike deconvolution techniques, while the inversion model fit was optimized by selecting optimal parameter values. This was followed by training and evaluating the inversion model using machine learning algorithms. During this step, the low-frequency model and the objective function were developed. The final stage of the inversion procedure comprises inverting data (seismic and well) using previous knowledge and the probability function. In the last stage of the workflow, the inversion model results served as the starting point for two prediction models designed to calculate porosity values.

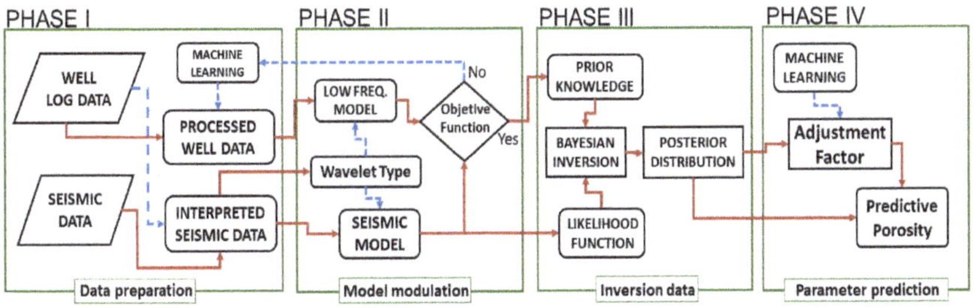

Figure 1. The workflow illustrates four distinct phases. In the initial part (data preparation, model modulation, data inversion, and parameter prediction), the focus is on applying the Bayesian Linearized Inversion (BLI). The selected and preprocessed data are modeled, and the BLI inversion is subsequently applied. In the latter part (Porosity Prediction), the results of the BLI inversion serve as the initial dataset for predictive models targeting porosity data. This second stage involves the utilization of predictive models to derive porosity information based on the results of the BLI inversion.

3.1. Data Preparation

The Norne oil field is characterized by a combination of petrophysical and seismic data providing comprehensive insight into its structure and production potential. Petrophysical information obtained from logs of two main wells reveal properties such as compressional velocity (V_P), shear velocity (V_S), density (ρ), effective porosity (ϕ), and clay content (Vsh). Despite sections with missing data and the presence of noise and outliers, these logs offer a detailed understanding of the geological formations present in the field. On the other hand, seismic data, depicted in Figure 2b, show a pronounced anticlinal geological structure, as well as the presence of olistric faults derived from the anticlinal. The presence of anomalies in the range of 2000 to 3200 milliseconds suggests the possible existence of gas or oil accumulations in these areas. The combination of petrophysical and seismic data provides a solid foundation for the assessment and efficient exploitation of resources in the Norne field. More details about the geology of the Norne field dataset can be found in Yan [36], Correia and Schiozer [37], and Maleki et al. [38]. At this point, identifying, assessing, choosing, and filtering source data constitute the main preprocessing tasks (see phase I, Figure 1). The research includes well logs and three-dimensional seismic data from seventeen wells gathered from different points in the study region. Essential characteristics, including density, porosity, clay content, and P- and S-wave velocities, are included in these well logs. Both seismic data and well logs are subjected to a comprehensive review process. In addition, a synthetic seismic model is selected for investigation and testing, acting as a standard for different inversion processes.

Processing Flow

Estimating a wavelet requires careful consideration of many elements, including Earth filtering, data processing, and seismic source characteristics. This is especially true when dealing with unknown quantities, such as the noise component. Therefore, to guarantee correctness in analysis and interpretation, choosing and estimating a wavelet calls for a methodical and exacting procedure. The inversion and prediction processes are made more complex by these unknown characteristics, necessitating the development of strong estimating and application techniques. In contrast to the amplitude, the phase spectrum is more difficult [39] to determine in the seismic section because it contains traces of reflection coefficients that must be convolutioned with an estimated wavelet [40].

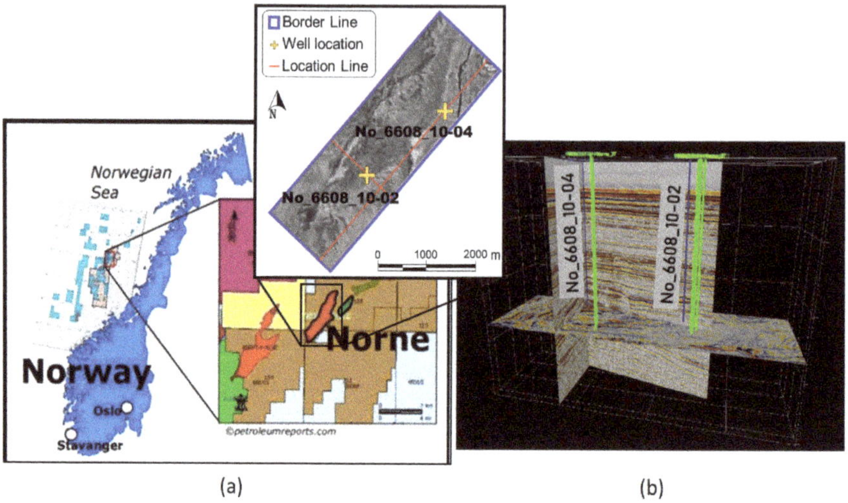

Figure 2. (**a**) Map delineating the regions of the Norne field into seismic sections and target well-logs locations. (**b**) Three-dimensional cross-sectional seismic profiles displaying the marked well-log locations.

Hence, ensuring the correct phase spectrum is essential for the results of various inversion and prediction processes. In this study, the Ricker wavelet was initially used but was later replaced with a wavelet determined by sparse spike deconvolution, developed in [29,41]. The resulting wave exhibits high-frequency behavior, which allows the modeling of key features of the data that differ from the Ricker wavelet, which has a wide range of frequencies and facilitates the preservation of most of the geological features in the seismic data. This change ensures the precision required for the BLI processes and the predictive models employed for porosity estimation.

To enhance the resolution and quality of the results, in this phase, apart from the Ricker wavelet, a wavelet sparse spike, extracted directly from the seismic data, was employed using the sparse spike deconvolution process developed by [29] (see Figure 3). At this stage, a meticulous cleaning was carried out on various well logs to identify and eliminate extreme and outlier data. Machine learning (ML) algorithms, coupled with the Savitzky–Golay filter algorithm [42], were employed for this purpose. Two distinct algorithms were assessed: the Isolation Forest (IF) algorithm and the Local Outlier Factor (LOF) [43]. Following thorough analysis utilizing a confusion matrix, the IF algorithm exhibited superior performance and was, thus, chosen for outlier detection and removal.

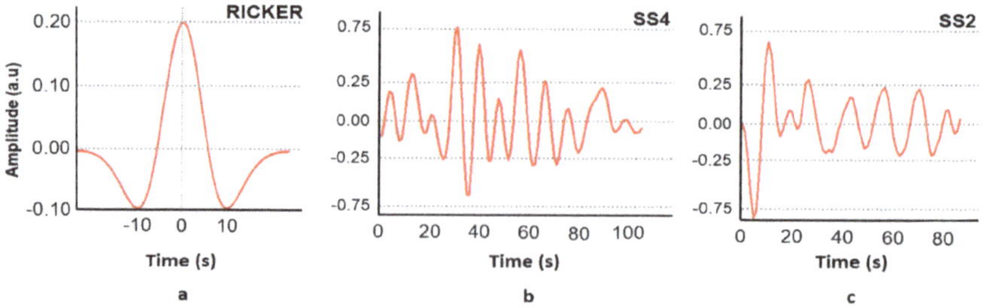

Figure 3. The wavelets used in the study consist of: (**a**) Ricker wavelet, (**b**) sparse spike wavelets derived from well No-6608-10-04, and (**c**) from well No-6608-10-02.

Even so, we made some very important changes that will be incorporated into the models used to detect outliers. These changes were made directly to the model's hyperparameters or configuration, which cannot be changed or optimized while the model is being trained. The ML models are subjected to two analyses: sensitivity analysis and cross-validation (CV). Finding the most relevant hyperparameters to improve the results is the goal. Two distinct search algorithms, GridSearchCV and RandomizedSearchCV, were investigated using cross-validation analysis, primarily focusing on two important dimensions: computational time and accuracy [43]. With the best evaluation results, the RandomizedSearchCV algorithm performed better than the others.

3.2. Model Modulation

During this phase (which we call Phase II, see Figure 1), we meticulously crafted crucial models, most notably the seismic model, which played a pivotal role in calibrating and refining the overarching model responsible for estimating fundamental parameters. The seismic model was meticulously constructed through the interpretation of the target horizon interlinked with fault planes identified in the seismic section (see Figure 4a,b). This approach aimed to enhance the fidelity of seismic data, ensuring a more accurate representation of actual conditions and subsurface structures. The development of the low-frequency model paralleled this process. This model encapsulated the low-frequency components, serving as a detailed approximation of the subsurface model. These parts are very important because they show important subsurface properties and make it easier to describe large structures and low-contrast features below the surface. To augment the precision of this model, additional insights from well logs were incorporated. Well logs, offering detailed and highly precise information on subsurface properties at well locations, played a significant role in fine-tuning and validating the low-frequency model derived from seismic data. This integrated approach ensured a more comprehensive and accurate representation of the subsurface, blending the strengths of seismic data and well-log information.

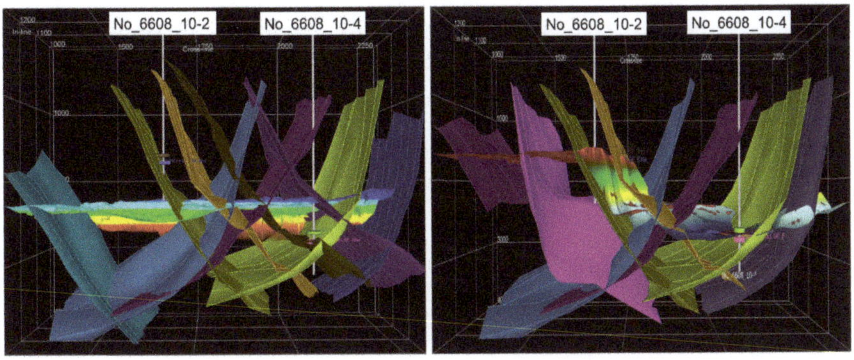

(a) (b)

Figure 4. The interpretation of the target stratigraphic horizon was executed utilizing the seismic cube through two distinct workflows: (**a**) employing the automatic tracking algorithm integrated into the Opendtect software, devoid of the fault planes; and (**b**) utilizing a manual approach that integrates fault planes and geological reference markers to conduct the interpretation of the seismic horizon.

Again, machine learning (ML) algorithms are used at this stage. The RandomizedSearchCV algorithm was used to tune the hyperparameters of the models. Because the well logs in the vicinity of the site presented gaps (between 3300 and 3700 m), it was necessary to develop regression models to extrapolate the missing data, and thus, reduce the possible errors present. For this, four different types of models were developed and tested: linear regression (LR), random forest (RF), K-nearest neighbors (KNN), and support vector regression (SVR); the initial settings of the models were determined by using the RandomizedSearchCV algorithm. The sensitivity of the initial data was then tested through a sensitivity analysis concerning the amount of data needed (see Figure 5). Likewise, an evaluation of the well logs was initially carried out to identify the type of underlying distribution of the data and optimize the results' quality. In particular, regression algorithms tend to exhibit improved performance when applied to datasets that exhibit a normal or Gaussian distribution pattern in the initial data (see Figure 6).

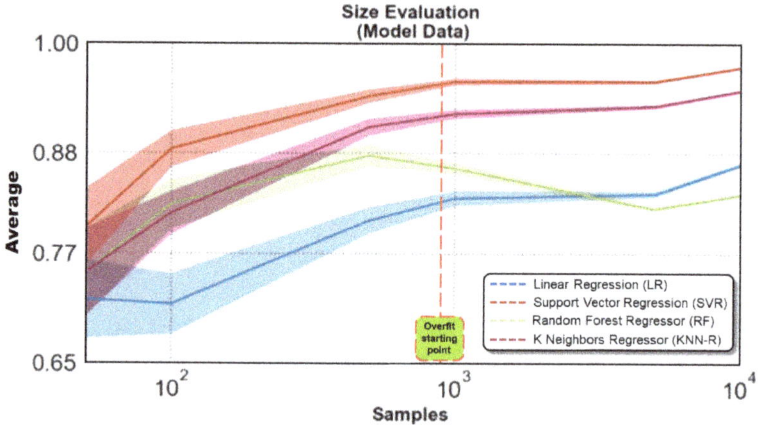

Figure 5. The assessment of model performance was conducted on the regression models, and the displayed confidence intervals were calculated at a 95% confidence level. The figure depicts the behavior observed during the evaluation of the initial data size of the four selected algorithms (Linear Regression, Support Vector Machine, Random Forest, and K-neighbors regression). The error and uncertainty exhibit an inverse relationship with the number of elements or data evaluated, as the area of the confidence interval tends to decrease with an increase in the evaluated data.

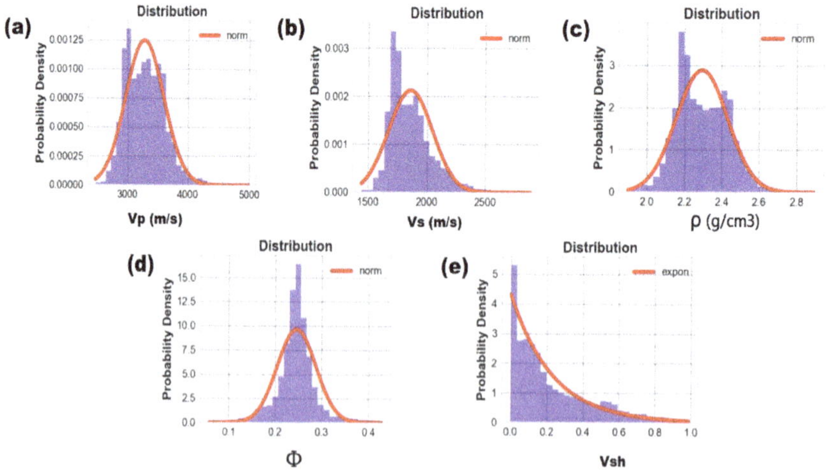

Figure 6. The graphs depict the evaluation of the well-log distribution type that served as the starting point for the inversion procedure. All of the used well logs (**a**–**d**) have a Gaussian or normal distribution, except one well log (**e**) that has an exponential distribution.

Within this stage, the BLI inversion process was carried out. A priori information plays a primary role, which is used to introduce essential knowledge to the model to identify possible geological alterations in the area, which can potentially affect the final results. This was enhanced to improve the accuracy of the data significantly, by introducing area information and providing subsurface details, before fitting the model with the observed data, which, in turn, mitigates ambiguity in the interpretation of the data. This helped establish a framework for comparing various stages of the inversion process. It also helped create various initial models, which were used to find the best adjustment of values in the hyperparameters. To prepare these models, various machine learning (ML) algorithms were used. Once the properly adjusted inversion model was obtained, it was applied to real data (seismic data and well logs) from the Norne field in the North Sea.

To select the best regression model, the four developed models were analyzed, with precision and standard deviation mainly evaluated to evaluate the fit and understand the predictive capacity. With the information on the precision of the regression model, we ensure that we obtain the best value fitted to the training data; this allows us to generalize the process to data not used by the model (blind data). The results of the standard deviation will be applied to be able to evaluate how far (dispersed) the data resulting from the model are concerning the initial observed data. Figure 7 shows us the analysis results of the different models, using precision and standard deviation as indicators. The SVR model is shown to have the highest precision and one of the lowest standard deviation values. Once the model with the highest precision and lowest dispersion was determined, the model was trained with the available data using the logs from 10 wells (P and S wave speed logs, density logs, and clay quantity logs) using the classic 70-30 split. This method divides the dataset into two parts: 70% of the data was used to train the model, allowing it to learn and adjust its parameters. At the same time, the remaining 30% was reserved for testing and validating its performance. Figure 8 shows the result of this process.

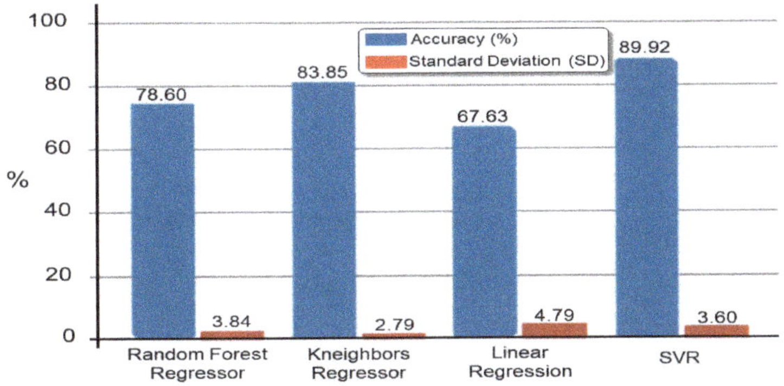

Figure 7. The results of the analysis of the precision and standard deviation carried out on the regression models are shown, this represents the level of precision in the estimation and the degree of dispersion or variability that the data has. Among the tested models, support vector regression (SVR) demonstrated the best performance, exhibiting the highest precision and one of the lowest standard deviation values.

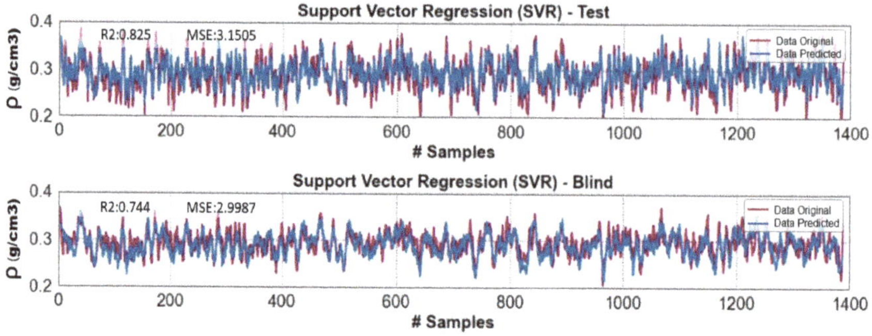

Figure 8. Shown here are the results of the training stage of the selected regression model (SVR model) used to predict missing data in well logs. The model was trained using data from 10 well logs in the area, split into 70% for training data (test) and 30% for test data (blind).

3.3. Paremter Estimation

Phase III is described in Section 2 under the heading Theoretical Background. Phase III output is utilized for the porosity models that will be elaborated upon in the subsequent sections. Porosity is a fundamental attribute with a significant impact on a reservoir's capacity for storing and producing hydrocarbons. To tackle this intricate challenge, researchers have explored various techniques that utilize seismic and well-log data to infer high-resolution porosity properties. The framework employed in this study is rooted in geophysical theories and empirical relationships that establish connections between wave speeds and rock characteristics, including porosity. More sophisticated techniques with enhanced precision were additionally introduced by machine learning algorithms. The model employed in this study was the SVR model derived by [44]. It was selected for its adaptability [45–47] as displayed in the analysis shown in Figures 6 and 8. The empirical observations, coupled with the results of the SVR model, corroborate the notion that porosity estimates can yield reasonably accurate results across diverse geological contexts.

3.3.1. Empirical Model

The empirical model used is described below. This model is adapted from the experimental work of [26]. Mathematically, this model is given by:

$$\phi = \alpha[\ln \mathbf{V_P} - \ln(\beta \mathbf{V_S})], \quad (18)$$

where ϕ represents the porosity to be predicted and V_P and V_S denote the P- and S-wave velocities. The parameters α and β work as control factors to adjust the relationship between velocities and porosity. The values of these factors are closely related to the geological environment, which is a function of the specific characteristics of the reservoir. In this study, the values of α and β were selected from the study developed by [26] related to the clean sandstone geological environment ($\alpha \approx 1/0.326$ and $\beta \approx 1.494$). This equation presents a valuable approach for estimating porosity based on P- and S-wave velocities, taking advantage of differences in wave propagation through subsurface formations (the derivation of the equation is detailed in Appendix A).

3.3.2. SVR Model

The main goal of support vector regression (SVR) is predicated on the concept of dimensionality transformation, where it projects the original dataset into a higher-dimensional feature space. This process is achieved by utilizing a kernel function that implicitly performs this mapping. In this elevated feature space, SVR aims to construct a linear regression model that embodies the simplest possible form to mitigate overfitting and ensure robust generalization capabilities. The formulation of the SVR model is centered on identifying a linear equation that encapsulates the relationship between the input vectors $\mathbf{x} \in \mathbb{R}^n$ and the output variable $y \in \mathbb{R}$ (petrophysical parameters like porosity). The linear model is given by:

$$y = f(\mathbf{x}) = \mathbf{w}^T\mathbf{x} + b, \tag{19}$$

where b signifies the bias term and $\mathbf{w}$ represents the weight vector. Together, they define the slope and position of the regression hyperplane within the feature space. The optimization of the SVR model involves adjusting the values of b and $\mathbf{w}$ to minimize the cost function R, which is described as:

$$R = \frac{1}{2}\|\mathbf{w}\|^2 + C \cdot \frac{1}{l}\sum_{i=1}^{l} \mathcal{L}_\epsilon(y_i, f(\mathbf{x}_i)), \tag{20}$$

where $\mathcal{L}_\epsilon$ represents the ϵ-insensitive loss function. This loss function only penalizes predictions that deviate from the actual values by more than a margin of ϵ, thus allowing for some errors within this margin, which contributes to the model's ability to ignore outliers and prevent overfitting [48–50].

The kernel choice, particularly the Radial Basis Function (RBF) denoted by $K(\mathbf{x}_i, \mathbf{x}_j) = \exp\left(-\frac{\|\mathbf{x}_i - \mathbf{x}_j\|^2}{2\sigma^2}\right)$, is instrumental in defining the SVR's performance [51]. The RBF kernel is favored due to its proficiency in dealing with non-linear relationships and its requirement for only one tuning parameter, σ [52]. This parameter influences the flexibility of the decision boundary: a small σ results in a highly flexible boundary, whereas a large σ leads to a smoother decision function that can approximate a linear boundary.

The RBF kernel has the unique advantage of potentially transforming the training data into an infinite-dimensional space, facilitating the SVR to handle complex non-linear relationships, which is often the case in predicting petrophysical properties such as porosity. The efficacy of SVR, therefore, is dependent on the optimal configuration of its learning parameters—primarily the RBF kernel parameter σ, the error margin ϵ, and the regularization parameter C. These parameters are typically tuned through a systematic search, often employing a logarithmic grid that spans a broad range of potential values: σ usually varies within 2^{-5} to 2^5, ϵ may be set between 0.01 and 0.6 to capture the permissible deviation in the predictions, and C could range from 2^{-5} to 2^{15}, balancing the trade-off between the model's complexity and the degree to which deviations beyond ϵ are penalized. This meticulous tuning process is important for developing an SVR model that can accurately predict porosity from well-log inputs, encompassing the necessary generalization to yield reliable predictions for unseen data.

In this study, initially, the testing and training of the model were conducted using well-log inputs. Subsequently, the developed model was applied to a 2D scenario. The petrophysical inputs logs utilized included compressional velocity (V_P), shear-wave velocity (V_S), and bulk density (ρ). These logs were specifically selected as they represent the outputs of Linearized Inversion (LI) and Amplitude Versus Offset (AVO) inversions.

3.4. Model Metrics Evaluation

Additionally, the mean square error (MSE) and Pearson correlation coefficient (CC) were chosen as accuracy measures in this study to check how well the models (BLI inversion, empirical, and SVR models) worked. Mean Squared Error (MSE) is a widely used metric for assessing the accuracy of a regression model. It is computed as the mean of the squares of the differences between the predicted values and the actual values. The mathematical formula is as follows:

$$MSE = \frac{1}{n}\sum_{k=1}^{n}(y_k - \hat{y}_k)^2, \tag{21}$$

where n is the total number of samples in the dataset, y_k is the dependent variable's actual value for sample k, and $\hat{y}_k$ is the value that the model predicted for sample k. The Pearson correlation coefficient, denoted as CC, is a statistical measure quantifying the strength and direction of the linear relationship between two variables. The mathematical formula for calculating the Pearson correlation coefficient is as follows:

$$CC = \frac{\sum_{i=1}^{n}(x_i - \bar{x})(y_i - \bar{y})}{\sqrt{\sum_{i=1}^{n}(x_i - \bar{x})^2 \sum_{i=1}^{n}(y_i - \bar{y})^2}}, \tag{22}$$

where n is the number of paired observations, x_i and y_i are the individual data points for variables X and Y, and $\bar{x}$ and $\bar{y}$ are the means of variables x and y, respectively.

4. Results and Discussion

This section begins by showing the use of AVO and BLI inversions applied to the dataset (seismic data, well logs, etc.) obtained from the Norne field in the North Sea. Initially, we worked with data obtained in 1D and then expanded to a synthetic case of 2D dimensionality, to evaluate the method's effectiveness and reduce the uncertainty inherent in changing dimensionality from 1D to 2D. Finally, we demonstrate how porosity inversion and estimation can be applied to a real dataset.

4.1. Elastic Inversion Results

The first iteration of the BLI process was carried out using log data from the main wells (6608-10-02 and 6608-10-04) (see Figure 1, phases I and II). The result of this first iteration indicated that the signal-to-noise ratio (SNR) value was in the range of 0 to 10 dB, which is considerably low. To improve this value, it was decided to implement machine learning algorithms using the Pylops library. Additionally, it was decided to perform an iterative adjustment of the model hyperparameters to reach the optimal adjustment value. Next, the inversion process was applied to the velocity log data corresponding to both wells. To establish a comparison framework, two different inversion processes were applied to the data: the BLI model and the AVO model. The results are presented in Figure 9. The results of the one-dimensional inversion, that is, applied to well logs, were extrapolated to cover the entire seismic section, resulting in an inverted seismic section. This process was carried out using the BLI inversion module of the OpendTect software v6.4.5. This technique was carried out using the OpendTect program, which includes a BLI inversion module.

Although Figure 9 shows an example of the inversion only for V_P, this procedure was also applied to the other parameters (V_S and ρ); however, it is not displayed here. The black line on the graph represents the original data, while the red line for the AVO inversion and the blue line for the BLI inversion represent the settings received from the inversion processes. Pearson's correlation coefficient (CC) values analyzed between the original values and the values of the inversion models for primary wells No-6608-10-02 and No-6608-10-04 are also presented. For well No-6608-10-02, a coefficient of 0.8680 (86.8%) is observed for the original values with respect to the BLI inversion values and 0.8332 (83.32%) for the original values regarding the AVO inversion values. On the other hand, in well No-6608-10-04, the values of the Pearson coefficient are 0.8758 (87.58%) for the original values about the BLI inversion values and 0.8021 (80.21%) for the values about AVO inversion values. These values indicate a strong correlation between the predicted values and the original values since they are in the range of 0.9 (or 90%). This implies that the model used to make porosity predictions is quite accurate, as the predicted values align well with the original values. To achieve these values, an iterative refinement was carried out, achieving the best possible data fit in approximately seven rounds.

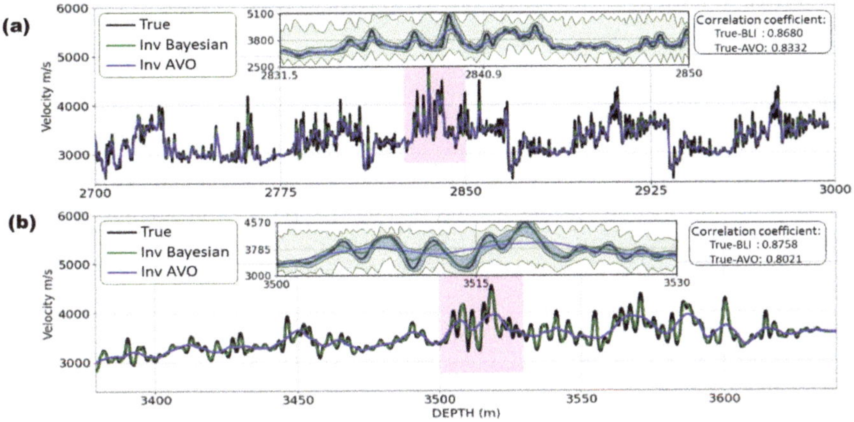

Figure 9. A comparative analysis is presented between the results of the AVO and BLI inversion processes concerning the V_p well log. The black line represents the original data, the blue line represents the BLI inversion results, and the red line represents the AVO inversion results. The results highlight the superior accuracy achieved through Bayesian inversion. Additionally, the results illustrate the shift between the initial and final states of the 95% confidence interval (CI). Image (**a**) corresponds to data from well No-6608-10-02, while image (**b**) displays data from well No-6608-10-04.

Table 1 provides a record of the various values obtained during each iterative process for V_P, V_S, and ρ parameters. In Figure 9, the target area highlighted in pink, which has more uncertainty and could be home to the reservoir, shows a big difference in how well the AVO inversion fits the original data. Specifically, the AVO inversion demonstrates inferior fit, while the BLI inversion demonstrates superior fit in this region. Figure 9 provides a comparative analysis that reveals that the BLI inversion process shows a superior fit to the original data compared to the AVO inversion process. More accurate results were obtained, as seen in Figure 9.

This is likely due in part to the changes that were made to the machine learning (ML) models and the models themselves when the well logs were first cleaned. These strategic adjustments improved the performance and accuracy of the BLI investment process, resulting in a more favorable match between the obtained results and the original data. Figure 9 shows a difference in the target area that is highlighted in red. This shows that the Bayesian Linear Inversion (BLI) method works better when used with the wavelet sparse spike to obtain the geological features in the target area accurately. This improvement is likely attributed to incorporating data uncertainty and the strategic introduction of a different wavelet into the investment process.

To assess the model's performance and ensure control, we utilized metrics such as mean square error (MSE) and the coefficient (R^2) in each step of the process. The aim was to evaluate the results and enhance the quality of each iteration. These metrics helped us identify and choose the most favorable results of the Bayesian linearized inversion (BLI) process. The R^2 metric is commonly used in machine learning to measure how well the model fits the observed data in regression analysis results. These control measures were applied to attain optimal hyperparameter values by iteratively refining the models used. These metrics together imply that, within this hypothetical scenario, Bayesian inversion is the option that provides a better definition of the inversion compared to the AVO method. Figure 10 we can see that, specifically, both the amplitude versus offset inversion (AVO) method and the Bayesian linear inversion (BLI) method exhibit an exceptional ability to recover values closely aligned with those of the original velocity model. However, the BLI inversion process outperforms the AVO method regarding recovery accuracy, as demonstrated by the results presented in Figure 10. The comparison is especially notable at higher values where the shape of the horizon is distorted, showing that the BLI inversion accurately captures the main features of the model.

In this study, we incorporate prior knowledge by using prior information from the low-frequency components of the low-frequency model to establish constraints for the model. To improve quality, we use a likelihood function to evaluate how well the model describes the observed data, considering assumptions and limitations based on what we already know. The objective was to maximize this likelihood function to determine the values of the model parameters that maximize the probability of obtaining the observed data. During the data inversion stage, the study used two different inversion processes: AVO and BLI inversion. To facilitate this, in the initial phase, a synthetic model was selected to apply the inversion processes.

Table 1. The values of the correlation coefficients for V_P, V_S, and ρ are shown for each well and each iteration of the AVO and BLI inversion procedures. The final results are depicted in Figure 11.

P-Wave Velocity	**WELL No-6608-10-02**					
	INITIAL	**ITER. 1**	**ITER. 2**	**ITER. 3**	**ITER. 4**	**FINAL**
AVO Inversion	0.7050	0.6508	0.7065	0.7505	0.7999	0.8332
BLI Inversion	0.7865	0.6878	0.7154	0.8004	0.8289	0.8680
P-Wave Velocity	**WELL No-6608-10-04**					
	INITIAL	**ITER. 1**	**ITER. 2**	**ITER. 3**	**ITER. 4**	**FINAL**
AVO Inversion	0.6989	0.6604	0.7150	0.7515	0.8001	0.8021
BLI Inversion	0.7059	0.6941	0.7180	0.8105	0.8320	0.8758
S-Wave Velocity	**WELL No-6608-10-02**					
	INITIAL	**ITER. 1**	**ITER. 2**	**ITER. 3**	**ITER. 4**	**FINAL**
AVO Inversion	0.6698	0.6183	0.6712	0.7130	0.7599	0.7915
BLI Inversion	0.7472	0.6534	0.6796	0.7604	0.7875	0.8246
S-Wave Velocity	**WELL No-6608-10-04**					
	INITIAL	**ITER. 1**	**ITER. 2**	**ITER. 3**	**ITER. 4**	**FINAL**
AVO Inversion	0.6640	0.6274	0.6793	0.7139	0.7601	0.7620
BLI Inversion	0.6706	0.6594	0.6821	0.7700	0.7904	0.8320
ρ	**WELL No-6608-10-02**					
	INITIAL	**ITER. 1**	**ITER. 2**	**ITER. 3**	**ITER. 4**	**FINAL**
AVO Inversion	0.6557	0.6052	0.6570	0.6980	0.7439	0.7749
BLI Inversion	0.7314	0.6397	0.6653	0.7444	0.7709	0.8072
ρ	**WELL No-6608-10-04**					
	INITIAL	**ITER. 1**	**ITER. 2**	**ITER. 3**	**ITER. 4**	**FINAL**
AVO Inversion	0.6500	0.6142	0.6649	0.6989	0.7441	0.7459
BLI Inversion	0.6565	0.6455	0.6677	0.7538	0.7738	0.8145

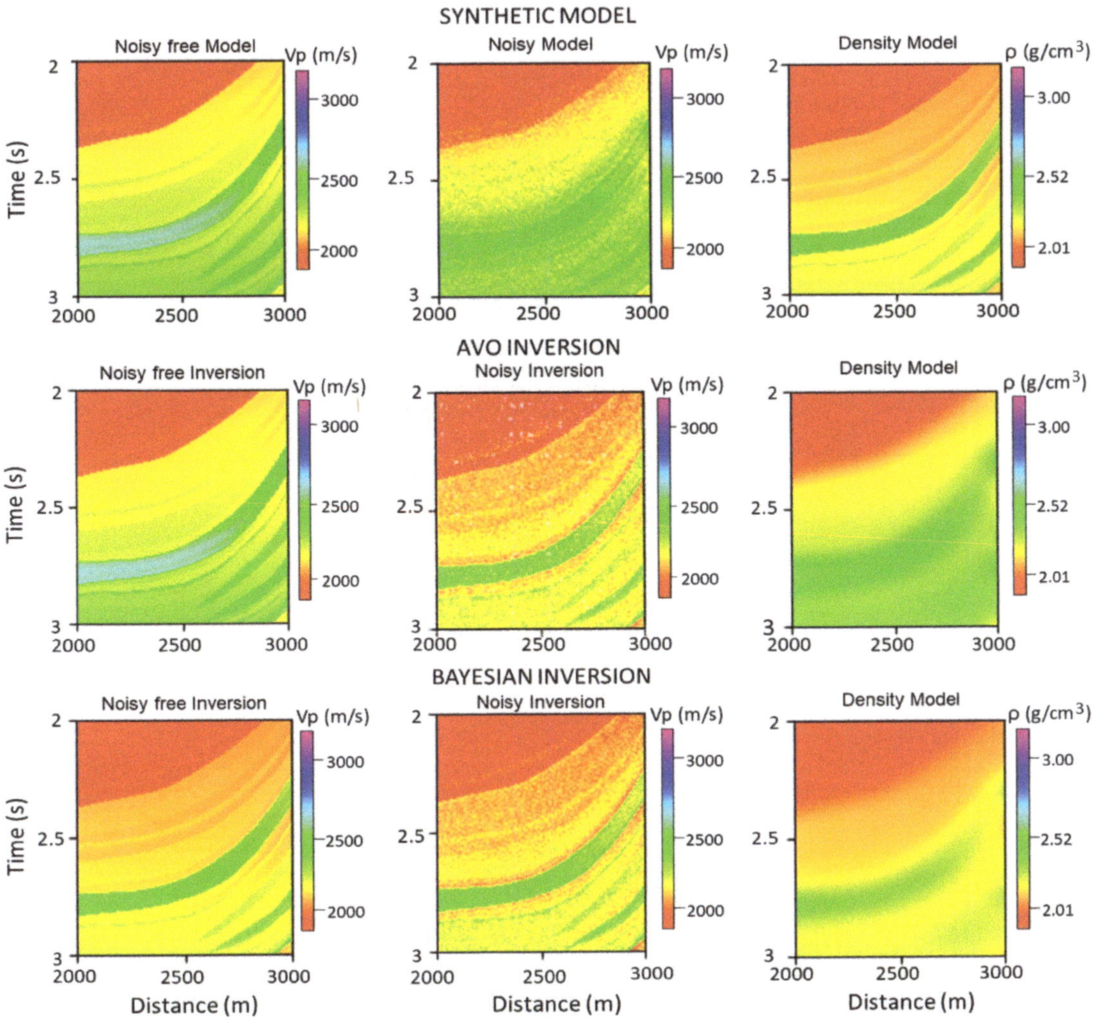

Figure 10. The results of the inversion performed using Bayesian Inversion (BLI) and Amplitude Versus Offset (AVO) methodologies are displayed. The inversion procedures were implemented on a synthetic model, which was used in two distinct scenarios: in the first, the model was utilized without any introduced noise; in the second, noise was introduced into the initial model. The purpose of deliberately introducing noise was to examine the inversion process's behavior and enable a comparative analysis of the two. The inversion was executed on the V_P and density (ρ) models in the current case.

In Figure 10, we can see the results of the AVO and BLI inversion processes applied to two synthetic models: one without noise and the other with noise included. The AVO inversion process produces slightly higher velocities than the original model, while the BLI inversion process produces lower velocities very close to the original model. A comprehensive comparison between the AVO and BLI inversion processes was performed using various metrics, and the results are included in Table 2.

These metrics together imply that, within this hypothetical scenario, Bayesian inversion is the option that provides a better definition of the investment compared to the AVO method. Figure 10 we can see that, specifically, both the amplitude versus offset inversion

(AVO) method and the Bayesian linear inversion (BLI) method exhibit an exceptional ability to recover values closely aligned with those of the original velocity model. However, the BLI inversion process outperforms the AVO method regarding recovery accuracy, as demonstrated by the results presented in Figure 10.

This difference is especially noticeable at higher values where the horizon's shape is distorted, demonstrating the BLI inversion's ability to capture the fundamental features of the model. When comparing the final model performance parameters to the AVO investment method, Table 2 indicates that Bayesian inversion (BLI) performs marginally better. This is because BLI exhibits a higher R-squared value (0.78 vs. 0.68), a lower mean square error (MSE) (0.014 versus 0.020), and a lower mean root square error (RMSE) (0.118 vs. 0.141). These changes in light are probably caused by the low-frequency model's inclusion of past knowledge and data uncertainty. Particularly in cases where the data are noisy, the BLI inversion procedure is frequently more dependable and effective. This sets it apart from other techniques for seismic inversion, such as AVO inversion.

Table 2. Performance metrics of ρ estimation and P- and S-wave velocities using AVO and BLI inversion methods. This table presents the R^2, MSE, and RMSE values for each parameter estimated, corresponding to the results depicted in Figure 10.

Method		Metrics		
		R^2	MSE	RMSE
AVO Inversion	V_P	0.677	0.020	0.141
	V_S	0.542	0.018	0.131
	ρ	0.607	0.022	0.153
BLI inversion	V_P	0.784	0.014	0.118
	V_S	0.646	0.013	0.108
	ρ	0.711	0.016	0.125

Figure 11 displays a segment of the inverted seismic section, including P- and S-velocities and the density section, over 500 ms $\times$ 100 traces using three wavelets: the Ricker wavelet, the sparse spike wavelet from well 2 (SS2), and the sparse spike wavelet from well 4 (SS4). Notable features in these images are areas of extremely low and high velocities. The figure also illustrates the results of BLI inversion with each wavelet, highlighting the SS4 wavelet us superior vertical resolution, followed by SS2. The improved model results are due to the wavelets being generated through sparse spike deconvolution from seismic data, enhancing resolution by retaining high-frequency waveforms and amplitudes during inversion.

4.2. Porosity Prediction Results

For porosity prediction, the Support Vector Regression (SVR) machine learning model was utilized, which was trained on well-labeled data divided into training and test sets. In total, 70% of the data was used for model training, with the remaining 30% used as blind data to evaluate model performance. The prediction model employed a radial basis function (RBF) kernel with parameter C set to one. Setting ϵ to 10^{-3} lowered the model's margin of error, improving precision. In the case of the porosity empirical equation, we used the fit equation parameters for clean sandstone to estimate porosity based on Han's equation, as previously described.

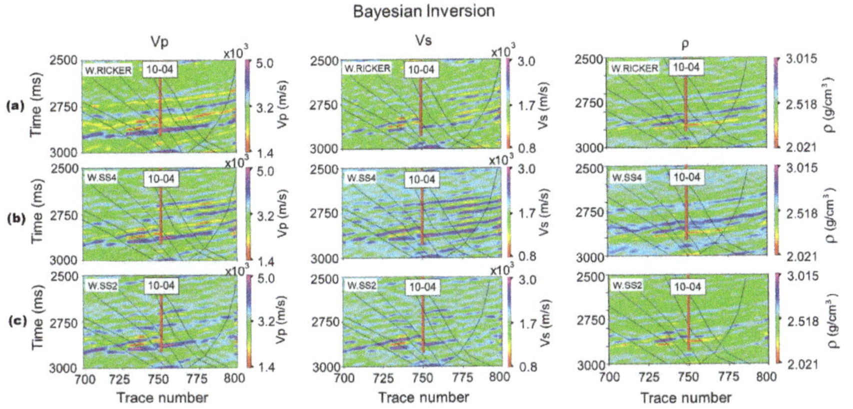

Figure 11. The 2D seismic section shows the inverted findings for P- and S-velocities, as well as density inversion achieved using the proposed approach. These maps represent inversions performed with (**a**) the Ricker wavelet, the sparse spike wavelets from well No-6608-10-04 (**b**), and (**c**) the sparse spike wavelets from well No-6608-10-02. It is worth noting that the best prediction occurs when we utilize the wavelet associated with the region surrounding the well.

Although the models closely reflect the well-log data, Figure 12 illustrates variations in the prediction models' outputs for porosity values. Especially when compared to the model based on Han's empirical equation, the SVR algorithm-based model closely matches the actual porosity records. This performance difference could be explained by the SVR model's hyperparameters' correctness about the empirical model's control and adjustment constants. The optimal hyperparameter configuration is most likely derived from effective pre-processing of the original data during the BLI inversion, as well as continuing adjustments performed throughout each iteration of the process.

Figure 12 illustrates the differences in resolution capacities and influence between empirical and SVR models. Even when using the wavelet in the exact position, there are discrepancies in resolution across the outputs. The empirical model's higher vertical resolution allows for detailed observation of the surrounding geological features and structures near the target horizon. On the other hand, the SVR model exhibits better horizontal continuity, making it easier to understand the connectivity inside the objective horizon, and thus, supporting more correct interpretation attempts.

Figure 13 presents the well logs for P- and S-wave velocities, density (ρ), and porosity (ϕ). Each record displays the original data (black line), the early results with the SS4 wavelet (green line), the final results with the SS4 wavelet (blue line), and the Ricker wavelet (red line). This picture allows for a direct comparison of the results of the BLI inversion and the SVR predictive and empirical models. Mean square error (MSE) and Pearson correlation coefficient (CC) were chosen to evaluate capacity for prediction. These criteria were selected to offer a thorough evaluation of the model's accuracy and precision. To improve computational efficiency, analyses were performed on individual traces for each 2D model result. The comparative analysis was performed on a single trace of each 2D model output, yielding three evaluations for each wavelet (see Figure 14).

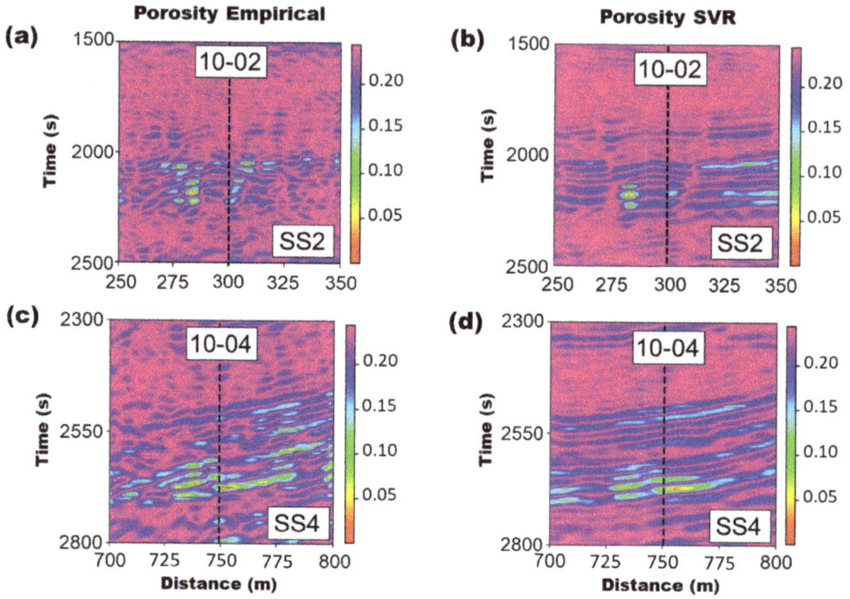

Figure 12. The 2D seismic section depicts the four results of the provided predictive models applied to regions locating wells No-6608-10-02 and No-6608-10-04. (**a**,**c**) The porosity estimations produced from the model using Han's empirical equation. Similarly, (**b**,**d**) show the porosity calculated using the Support Vector Regression (SVR) model. These techniques used the SS2 and SS4 wavelets, which were applied in their predetermined locations.

Furthermore, Figure 13 shows that the initial Bayesian Linear Inversion (BLI) employing the SS4 wavelet (green line) has a low signal-to-noise ratio, which is most likely due to a poor fit of the model's hyperparameters. Furthermore, the image depicts the findings after the iterative adjustment method was performed on the inversion model, demonstrating an increased signal-to-noise ratio (blue line) and a better fit when compared to the original results. The inversion results using the Ricker wavelet are also provided, with a strong signal-to-noise ratio despite a rather low fit to the original data.

A comparison is made between the initial results (represented by the black line) and the predicted porosity values (represented by the blue line) in Figure 14. This comparison is based on two distinct models: the SVR machine learning model (represented by the red line) and the empirical model (represented by the blue line), estimated using Han's equation. The Pearson correlation coefficient and MSE metrics are presented herein as iterative techniques for quantifying the extent of correction. The results were derived through an iterative process of contrasting the initial data with the results generated by the SVR and empirical models. Figure 14 additionally compares the initial porosity data and the results obtained from our predictive models (empirical and SVR) while modifying the initial wavelet.

Furthermore, these results suggest that the SVR model fits better than the empirical predictive model; this is most likely due to the empirical model's constant values, which are approximations of ideal geological conditions (rock type and geological environment). It should also be noted that the prediction models, both empirical and SVR, perform better when using the SS4 wavelet, which is most likely due to its similarity to a high-frequency wavelet. The procedure was iterative, thus the Pearson's Correlation Coefficient (CC) and MSE were computed in each iteration to determine the values that provide the best fit to the models (see Table 3).

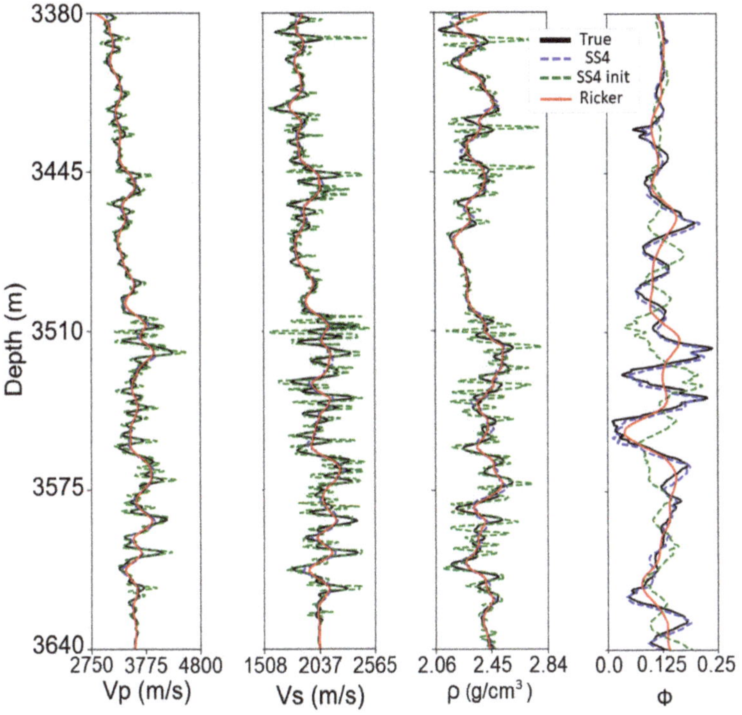

Figure 13. The well-log curves from No-6608-10-04 show the results obtained by using both the BLI inversion and prediction models for V_P,V_S, ρ, and ϕ. The logs show the following inversion using data inverted (with the wavelet SS4) in the first iteration (green line) and the final iteration (blue line) with the same wavelet. The data are inverted with the Ricker wavelet (red line). It should be noted that inversion with the Ricker wavelet tends to smooth the estimated parameter.

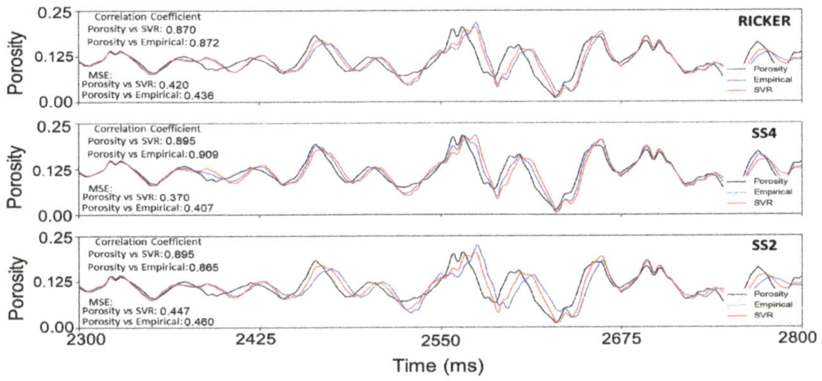

Figure 14. A comparison between the porosity prediction findings and the original values (black line) is shown. Two different models were applied, to achieve the predictions utilizing different wavelets (Ricker, SS4, and SS2). The models are the empirical model (blue line), which is estimated from Han's equation, and the Support Vector Regression (red line). These comparisons are related to well No-6608-10-04. The correlation coefficient (CC) and mean squared error (MSE) values are also reported, demonstrating that the Support Vector Regression (SVR) model exhibits superior performance compared to the Han models in terms of predictive accuracy.

Table 3. The prediction model's fit was evaluated using the Pearson correlation coefficient (C.C.) and Mean Squared Error (MSE) values from three iterations, as shown in Figure 14 (referring to well No-6608-10-04).

C.C.	ϕ—SVR INITIAL	ITER. 1	FINAL	ϕ—EMP INITIAL	ITER. 1	FINAL
Ricker	0.698	0.782	0.870	0.605	0.755	0.870
SS4	0.731	0.805	0.895	0.619	0.795	0.909
SS2	0.668	0.795	0.895	0.589	0.706	0.865
M.S.E	**INITIAL**	**ITER. 1**	**FINAL**	**INITIAL**	**ITER. 1**	**FINAL**
Ricker	0.650	0.590	0.420	0.630	0.600	0.436
SS4	0.680	0.560	0.370	0.585	0.567	0.407
SS2	0.700	0.570	0.477	0.620	0.595	0.460

5. Conclusions

This study presents a machine learning (ML) approach using Bayesian linear inversion (BLI) to optimize initial data for porosity prediction using seismic and well-log data along with a support vector regression (SVR) model and an empirical model. The BLI inversion functions as an optimal model prediction in our methodology, mitigating the intrinsic uncertainty of the data before implementing the empirical and SVR formulations. We were able to verify this by using the Pearson coefficient, which in both results showed that the value was in the range of 0.9 (90%), which demonstrates that by effectively integrating the BLI model, the performance of the empirical models and SVR improves significantly. Thus, also the use of a procedure such as BLI inversion, which takes into account the uncertainty in the initial data, can be a valuable asset in the evaluation and interpretation of reservoirs within the context of predictive modeling, as This study demonstrates.

Seismic and well-log data from the Norne field in the North Sea will be used for the workflow application. By using iterative analysis, the best fit for the accuracy and hyperparameter values of the predictive investment models was achieved. This is verified thanks to the implementation of different evaluation metrics. In particular, the Pearson correlation coefficient (CC) of the BLI model experienced a substantial improvement, increasing from 0.7865 to 0.8680. Likewise, the coefficient R^2 reached a value of 0.784, indicating a greater proportion of variance explained by the model. The root mean square error (MSE) was reduced to 0.020, suggesting a decrease in the discrepancy between the predicted values and the observed values, while the root mean square error (RMSE) was set at 0.141. These results show a notable improvement in the performance of the BLI model, which highlights the effectiveness of the hyperparameter tuning process. The incorporation of machine learning algorithms (in several steps of the work) was mainly responsible for improving the accuracy and quality of the predictions produced by the models. Because of this, our ability to identify complicated patterns and complex correlations in the dataset improved significantly, and as a result, selecting optimal hyperparameter values for inversion and prediction models became easier. However, it is worth noting that although this approach was carried out in the Norne field, the methodology described in a previous workflow is still relevant for other domains and environments that depend on data accessibility.

Also, it was shown that the prediction of seismic porosity (ϕ) using Support Vector Regression (SVR) outperformed the forecast made using Han's empirical equation, which used a variety of wavelets (Ricker, SS4, and SS2). To acquire the ϕ attribute, it was also necessary to use the well-log curves that showed the results of Bayesian Linearized Inversion (BLI) and prediction models for V_P, V_S, and ρ. Notably, the inversion prediction based on the SS2 and SS4 wavelets appeared to be the better one, whereas the data inversion using the Ricker wavelet seemed to smooth the estimated parameters. Evaluation metrics, such as Mean Squared Error (MSE) values and Pearson correlation coefficient (C.C.), were used to evaluate the model's performance over several iterations. The results indicate that, in most cases, the inversion using SS4 concerning the well location is essential for precise porosity prediction. These measures offered insightful information about the models' reliability and precision.

It is important to emphasize that the methodology can be applied in areas other than the one used because values in the models and equations serve as adjustment parameters, allowing the effect of discrepancies between the values of the variables, in this case, V_P and V_S, to be increased or decreased. In addition, values are used to change the scale of the variables, notably the velocities V_P and V_S. These factors are influenced by geological conditions, allowing the methodology to be used in a range of places as long as the impact of the site's geological context is well-valued.

Author Contributions: J.A.T.M.: Conceptualization, Methodology, Writing—review and editing, Writing—original draft, Investigation, Validation, and editing; J.J.S.d.F.: Conceptualization, Methodology, Supervision, Writing—review and editing, Writing—original draft, Investigation, Validation, Writing—review and editing. C.E.S.A.: Editing, Validation and Software. All authors have read and agreed to the published version of the manuscript.

Funding: The authors extend their sincere appreciation to CAPES, CNPq, and PROPESP-UFPA for their generous financial support, which facilitated the completion of this research.

Institutional Review Board Statement: Not applicable.

Informed Consent Statement: Not applicable.

Data Availability Statement: The modified codes used for Bayesian Linearized Inversion (BLI) and Amplitude Versus Offset (AVO) analyses are accessible at https://pylops.readthedocs.io/en/stable/, accessed on 9 August 2021. For 3D data visualization, we utilized the Opendetect software v6.4.5, which can be accessed at https://www.dgbes.com/software/opendtect, accessed on 9 August 2021. The Norne field dataset is available on request at http://www.ipt.ntnu.no/~norne/wiki1/doku.php?id=english:nornebenchmarkcase2, accessed on 9 August 2021. The supporting datasets and codes are available upon request from J.J.S. Figueiredo at UFPA (email: jadsom@ufpa.br).

Conflicts of Interest: The authors declare no conflicts of interest.

Appendix A. Derivation of an Empirical Porosity Model

We present the empirical porosity equation based on experimental measurements conducted by [26] in clean sandstones, which relates the V_P/V_S ratio to ϕ as follows:

$$\frac{V_P}{V_S} = 1.494\exp(0.326\phi), \tag{A1}$$

or equivalently:

$$\frac{V_P}{1.494V_S} = \exp(0.326\phi). \tag{A2}$$

To isolate ϕ, we take the natural logarithm (ln) of both sides of Equation (A2). Since $\ln(\exp(x)) = x$ and $\ln(a/b) = \ln(a) - \ln(b)$, we obtain:

$$\phi = \frac{1}{0.326}(\ln(V_P) - \ln(1.494V_S)). \tag{A3}$$

For the general case, Equation (A3) can be expressed as:

$$\phi = \alpha(\ln(V_P) - \ln(\beta V_S)). \quad \text{(A4)}$$

It is important to note that this model was developed for predictive seismic analysis, with the parameters α and β serving as control and adjustment factors to enhance the accuracy of porosity prediction. Specifically, the parameter α acts as a general calibration parameter, while β adjusts the S-wave velocity values corresponding to the porosity predictions.

References

1. Zare, A. ; Bagheri, M.; Ebadi, M. Reservoir facies and porosity modeling using seismic data and well logs by geostatistical simulation in an oil field. *Carbonates Evaporites* **2020**, *35*, 65. [CrossRef]
2. Serra, O. Fundamentals of Well-log Interpretation: The acquisition of logging data. In *The Acquisition of Logging Data*; Developments in Sedimentology; Elsevier: Amsterdam, The Netherlands, 1984; pp. 1–4.
3. Goodyear, G.; Parker, T.J.; Kok, K.H.; Kyi, K.K.; Mansur, H.; Zaman, S.K. Unexpected behavior of logging-while-drilling neutron porosity logs in water-wet, sand-shale sequences. *Interpretation* **2017**, *5*, T265–T277. [CrossRef]
4. Selley, R. *Elements of Petroleum Geology*; Elsevier Science: Amsterdam, The Netherlands, 1998.
5. Zong, Z.; Yin, X.; Wu, G. Geofluid discrimination incorporating poroelasticity and seismic reflection inversion. *Surv. Geophys.* **2015**, *36*, 659–681. [CrossRef]
6. Aleardi, M.; Ciabarri, F.; Peruzzo, F.; Garcea, B.; Mazzotti, A. Bayesian Estimation of Reservoir Properties by Means of Wide-angle AVA Inversion and a Petrophysical Zoeppritz Equation. In Proceedings of the 78th EAGE Conference and Exhibition 2016, Vienna, Austria, 30 May–2 June 2016; pp. 1–4. [CrossRef]
7. Aleardi, M.; Ciabarri, F.; Gukov, T.; Giussani, M.; Mazzotti, A. A Single-step Bayesian Petrophysical Inversion Algorithm Based on a Petrophysical Reformulation of the P-wave Reflection Coefficients. In Proceedings of the 79th EAGE Conference and Exhibition 2017, Paris, France, 12–15 June 2017; pp. 1–4. [CrossRef]
8. Cyz, M.; Azevedo, L. Direct geostatistical seismic amplitude versus angle inversion for shale rock properties. *IEEE Trans. Geosci. Remote Sens.* **2021**, *59*, 5335–5344. [CrossRef]
9. Fjeldstad, T.; Avseth, P.; Omre, H. A one-step Bayesian inversion framework for 3D reservoir characterization based on a Gaussian mixture model—A Norwegian Sea demonstration. *Geophysics* **2021**, *86*, R221–R236. [CrossRef]
10. Su, Z.; Cao, J.; Xiang, T.; Fu, J.; Shi, S. Seismic prediction of porosity in tight reservoirs based on transformer. *Front. Earth Sci.* **2023**, *11*, 11376453. [CrossRef]
11. Boateng, C.D.; Fu, L.Y.; Yu, W.; Xizhu, G. Porosity inversion by Caianiello neural networks with Levenberg–Marquardt optimization. *Interpretation* **2017**, *5*, SL33–SL42. [CrossRef]
12. Teillet, T.; Fournier, F.; Zhao, L.; Borgomano, J.; Hong, F. Geophysical pore type inversion in carbonate reservoir: Integration of cores, well logs, and seismic data (Yadana field, offshore Myanmar). *Geophysics* **2021**, *86*, B149–B164. [CrossRef]
13. Guo, Q.; Ba, J.; Luo, C. Nonlinear petrophysical amplitude variation with offset inversion with spatially variable pore aspect ratio. *Geophysics* **2022**, *87*, M111–M125. [CrossRef]
14. Feng, R. Estimation of reservoir porosity based on seismic inversion results using deep learning methods. *J. Nat. Gas Sci. Eng.* **2020**, *77*, 103270. [CrossRef]
15. Kolbjørnsen, O.; Buland, A.; Hauge, R.; Røe, P.; Ndingwan, A.O.; Aker, E. Bayesian seismic inversion for stratigraphic horizon, lithology, and fluid prediction. *Geophysics* **2020**, *85*, R207–R221. [CrossRef]
16. Tarantola, A.; Valette, B. Generalized nonlinear inverse problems solved using the least squares criterion. *Rev. Geophys. Space Phys.* **1982**, *20*, 219–232. [CrossRef]
17. Duijndama, J.W. Bayesian estimation in seismic inversion. Part I: Principles. *Geophys. Prospect.* **1988**, *36*, 878–898. [CrossRef]
18. Scales, J.A. Prior information and uncertainty in inverse problems. *Geophysics* **2001**, *66*, 389–397. [CrossRef]
19. Grana, D. Bayesian linearized rock-physics inversion. *Geophysics* **2016**, *81*, D625–D641. [CrossRef]
20. de Figueiredo, L.P.; Grana, D.; Bordignon, F.L.; Santos, M.; Roisenberg, M.; Rodrigues, B.B. Joint Bayesian inversion based on rock-physics prior modeling for the estimation of spatially correlated reservoir properties. *Geophysics* **2018**, *83*, M49–M61. [CrossRef]
21. Guo, Q.; Luo, C.; Grana, D. Bayesian linearized rock-physics amplitude-variation-with-offset inversion for petrophysical and pore-geometry parameters in carbonate reservoirs. *Geophysics* **2023**, *88*, MR273–MR287. [CrossRef]
22. Teruya Monroe, J.A.; de Figueiredo, J.J.S.; de J. S. Souza, I. Enhancement of Bayesian Seismic Inversion Using Machine Learning and Sparse Spike Wavelet: Case Study Norne Field Dataset. *IEEE Trans. Geosci. Remote Sens.* **2024**, *62*, 1–12. [CrossRef]
23. Mukerji, T.; Avseth, P.; Mavko, G.; Takahashi, I.; González, E.F. Statistical rock physics: Combining rock physics, information theory, and geostatistics to reduce uncertainty in seismic reservoir characterization. *Lead. Edge* **2001**, *20*, 313–319. [CrossRef]

24. Avseth, P.; Veggeland, T.; Horn, F. Seismic screening for hydrocarbon prospects using rock-physics attributes. *Lead. Edge* **2014**, *33*, 266–274. [CrossRef]
25. Buland, A.; Omre, H. Bayesian linearized AVO inversion. *Geophysics* **2003**, *68*, 185–198. [CrossRef]
26. Han, D. Effects of Porosity and Clay Content on Acoustic Properties of Sandstones and Unconsolidated Sediments. Ph.D. Thesis, Stanford University, Stanford, CA, USA, 1986; 219p.
27. Bordignon, F.L.; Figueiredo, L.P.; Azevedo, L.; Soares, A.; Roisenberg, M.; Neto, G.S. Hybrid Global Stochastic and Bayesian Linearized Acoustic Seismic Inversion Methodology. *IEEE Trans. Geosci. Remote Sens.* **2017**, *55*, 4457–4464. [CrossRef]
28. Joyce, J. Bayes' Theorem. In *The Stanford Encyclopedia of Philosophy*, Fall 2021 ed.; Zalta, E.N., Ed.; Metaphysics Research Lab, Stanford University: Stanford, CA, USA, 2021.
29. Macedo, I.A.S.; Figueiredo, J.J.S. On the seismic wavelet estimative and reflectivity recovering based on linear inversion: Well-to-seismic tie on a real data set from Viking Graben, North Sea. *Geophysics* **2020**, *85*, D157–D165. [CrossRef]
30. Rimstad, K.; Avseth, P.; Omre, H. Hierarchical Bayesian lithology/fluid prediction: A North Sea case study. *Geophysics* **2012**, *77*, B69–B85. [CrossRef]
31. Grana, D.; Fjeldstad, T.; Omre, H. Bayesian Gaussian Mixture Linear Inversion for Geophysical Inverse Problems. *Math. Geosci.* **2017**, *49*, 494–515. [CrossRef]
32. Stolt, R.H.; Weglein, A.B. Migration and inversion of seismic data. *Geophysics* **1985**, *50*, 2458–2472. [CrossRef]
33. Aki, K.; Richards, P.G. *Quantitative Seismology*, 2nd ed.; University Science Books: Herndon, VA, USA, 2002; 700p.
34. Grana, D.; Della Rossa, E. Probabilistic petrophysical-properties estimation integrating statistical rock physics with seismic inversion. *Geophysics* **2010**, *75*, O21–O37. [CrossRef]
35. Andreson, T. *An Introduction to Multivariate Statistical Analysis*; A. John & Sons Inc.: New York, NY, USA, 1984; 721p.
36. Yan, J. Reservoir parameters estimation from well log and core data: A case study from the North Sea. *Pet. Geosci.* **2002**, *8*, 63–69. [CrossRef]
37. Correia, G.G.; Schiozer, D.J. Reservoir characterization using electrofacies analysis in the sandstone reservoir of the Norne Field (offshore Norway). *Pet. Geosci.* **2016**, *22*, 165–176. [CrossRef]
38. Maleki, M.; Davolio, A.; Schiozer, D.J. Qualitative time-lapse seismic interpretation of Norne Field to assess challenges of 4D seismic attributes. *Lead. Edge* **2018**, *37*, 754–762. [CrossRef]
39. Umar, A.; Aqsa, A.; Wanzhong, S.; Ren, W.; Muhammad, A.; Ren, J.; Hung, V.T.; Ibrar, I.; Zhang, X.; Zhang, H. Estimation of porosity and facies distribution through seismic inversion in an unconventional tight sandstone reservoir of Hangjinqi area, Ordos basin. *Front. Earth Sci.* **2022**, *10*, 1014052.
40. Wang, Q.; Guo, K. Constrained sparse spike inversion applied in the reservoir prediction. *Well Logging Technol.* **2008**, *32*, 35–37.
41. Velis, D.R. Stochastic sparse-spike deconvolution. *Geophysics* **2008**, *73*, R1–R9. [CrossRef]
42. Savtzky, A.; Golay, M. Smoothing and Differentiation of Data by Simplified Least Squares Procedures. *Anal. Chem.* **1964**, *36*, 1627–1639. [CrossRef]
43. Pedregosa, F. Scikit-learn: Machine Learning in Python. *J. Mach. Learn. Res.* **2011**, *12*, 2825–2830.
44. Vapnik, V. *The Nature of Statistical Learning Theory*; Springer: New York, NY, USA, 1999; 314p.
45. Al-Anazi, A.; Gates, I. Support vector regression for porosity prediction in a heterogeneous reservoir a comparative study. *Comput. Geosci.* **2010**, *36*, 1494–1503. [CrossRef]
46. Jiang, B.; Zhao, F. Combination of support vector regression and artificial neural networks for prediction of critical heat flux. *Int. J. Heat Mass Transf.* **2013**, *62*, 481–494. [CrossRef]
47. Wu, Q.; Law, B. Fuzzy support vector regression machine with penalizing Gaussian noises on triangular fuzzy number space. *Expert. Syst. Appl.* **2010**, *37*, 7788–7795. [CrossRef]
48. Huang, C.; Dun, J. A distributed PSO–SVM hybrid system with feature selection and parameter optimization. *Appl. Soft. Comput.* **2008**, *4*, 1381–1391. [CrossRef]
49. Huang, C.; Wang, C. A GA-based feature selection and parameters optimizationfor support vector machines. *Expert. Syst. Appl.* **2006**, *2*, 231–240. [CrossRef]
50. Sarafrazi, S.; Nezamabadi-pour, H. Facing the classification of binary problems with a GSA-SVM hybrid system. *Math. Comput. Model.* **2013**, *57*, 270–278. [CrossRef]
51. Ranaee, V.; Ebrahimzadeh, A.; Ghadseri, R. Application of the PSO–SVM model for recognition of control chart patterns. *ISA Trans.* **2010**, *49*, 577–586. [CrossRef] [PubMed]
52. Friedrichs, F.; Igel, C. Evolutionary tuning of multiple SVM parameters. *Neurocomputing* **2005**, *64*, 107–117. [CrossRef]

Disclaimer/Publisher's Note: The statements, opinions and data contained in all publications are solely those of the individual author(s) and contributor(s) and not of MDPI and/or the editor(s). MDPI and/or the editor(s) disclaim responsibility for any injury to people or property resulting from any ideas, methods, instructions or products referred to in the content.

applied sciences

MDPI

Article

Three-Dimensional Forward Modeling and Inversion Study of Transient Electromagnetic Method Considering Inhomogeneous Magnetic Permeability

Chenyu Wang [1], Yan Dong [1], Jingyu Gao [1,*], Handong Tan [1], Yingge Wang [2] and Weiyu Dong [2]

[1] School of Geophysics and Information Technology, China University of Geosciences (Beijing), Beijing 100083, China; 2010220050@cugb.edu.cn (C.W.); dongyan_cugb@163.com (Y.D.); thd@cugb.edu.cn (H.T.)

[2] Aerospace Information Research Institute, Chinese Academy of Sciences, Beijing 100094, China; wangyingge@aircas.ac.cn (Y.W.); dongwy@aircas.ac.cn (W.D.)

* Correspondence: gaojingyu@cugb.edu.cn

Abstract: Traditional studies on Transient Electromagnetic Method (TEM) typically assume that the subsurface medium is non-magnetic. However, in regions with igneous rock accumulations or where the subsurface is rich in ferromagnetic minerals, neglecting the magnetic properties of the underground medium may lead to erroneous interpretations for TEM data. This paper conducts a 3-D TEM forward modeling and inversion study considering the non-uniformity cases of magnetic permeability. 3-D TEM forward modeling employs an edge-based finite element method using unstructured grids and a second-order implicit backward Euler method, achieving a modeling algorithm that simultaneously considers non-uniform models of magnetic permeability and resistivity. The accuracy of the modeling algorithm is verified by comparing it with the analytical solution of a homogeneous half-space model and the solution of a 1-D TEM forward modeling algorithm. 3-D TEM inversion employs the L-BFGS algorithm and synthetic examples considering non-uniform magnetic permeability are presented. The inversion results show good recovery for the resistivity and magnetic permeability models. Comparisons with the inversion results that neglect the non-uniformity of magnetic permeability validate the importance of considering the variation of permeability in 3-D TEM forward modeling and inversion.

Keywords: transient electromagnetic method; non-uniform magnetic permeability model; 3-D forward modeling; 3-D inversion

1. Introduction

Transient electromagnetic method (TEM) is a classic geophysical electromagnetic (EM) exploration technique and has been widely used in mineral and petroleum exploration, engineering and environmental monitoring, groundwater and geothermal detection, as well as geological disaster prediction, etc.

TEM utilizes artificial sources and normally adopts a periodic step-off source waveform to emit EM signals. During the switch-off period, the underground medium modifies the induced EM fields that decay over time. By measuring and analyzing the transient response, the information of the underground anomalous bodies can be obtained, which helps to infer the spatial distribution of the target body. TEM is characterized by lightweight instruments, high sensitivity, and strong resolution. Additionally, TEM observes the pure secondary EM fields with fewer requirements on the transmitter-receiver offset compared to the frequency-domain EM methods, and therefore, TEM has attracted more attention in recent years.

Ward, Wait, Spies, and Singh were among the first to use simple spherical models to study the impact of magnetic susceptibility on electromagnetic response [1–4]. Dey and Ward included susceptibility effects in a study using 1-D layered models [5]. Ryu et al. investigated the effect of the magnetic susceptibility of underground media on the EM field induced by a ring current source [6]. Ward and Fraser demonstrated that the effect of magnetic susceptibility is observable for geological models with a high-magnetic-susceptibility thin layer [7,8]. Zhiyi Zhang and D.W. Oldenburg, using a 1-D layered model, with known resistivities, obtained the 1-D vertical distribution of magnetic susceptibility from the inversion of frequency-domain EM data [9]. Beard and Nyquist experimentally demonstrated that, under certain conditions, assuming constant magnetic susceptibility in the inversion could not determine the location of the anomalous body [10]. However, when the magnetic susceptibility was treated as the inversion parameters, the location, resistivity, and magnetic susceptibility values of the anomalous body could be accurately determined. Zhiyi Zhang et al. developed a 1-D inversion algorithm for magnetotelluric (MT) data that simultaneously inverts both conductivity and magnetic susceptibility, which was later applied to the interpretation of 3-D data [11]. Farquharson et al. published a 1-D inversion algorithm for TEM using a loop source, which simultaneously inverts both conductivity and magnetic susceptibility [12].

Currently, research on EM forward modeling and inversion considering the non-homogeneity of magnetic susceptibility is mainly focused on the MT method. In TEM forward and inverse modeling, it is typically assumed that the magnetic permeability or the magnetic susceptibility of the entire medium is homogeneous. The magnetic permeability of a vacuum is normally adopted, with calculations and analyses focusing on the non-homogeneity of resistivity in the subsurface medium. However, in areas with high magnetic susceptibility, such as regions with igneous rocks of high magnetic content or areas rich in ferromagnetic minerals [13], neglecting the effect of magnetic susceptibility may lead to incorrect data-interpretation results. Therefore, it is important to develop 3-D forward modeling and inversion algorithms for TEM data that take into account the influence of magnetic susceptibility.

Based on the current state of research, this paper develops a 3-D forward and inverse modeling algorithm for ground-based loop-source TEM data, which can simultaneously handle the inhomogeneous condition for both resistivity and magnetic susceptibility models. Several synthetic examples are presented to demonstrate the accuracy of the forward modeling algorithm and the inversion results for the resistivity and magnetic susceptibility models.

Additionally, we present a comparison of the 3-D inversion with and without considering the non-homogeneity of the underground magnetic susceptibility model, and the results verify the importance of incorporating magnetic susceptibility as the inversion parameters in 3-D TEM inversion.

2. 3-D TEM Forward Modeling Based on the Unstructured Finite Element Method

In TEM modeling, the effect of displacement current is generally negligible, which leads to the time-domain diffusion equation for the electric field [14]:

$$\nabla \times \frac{1}{\mu(\mathbf{r})} \nabla \times \mathbf{e}(\mathbf{r}, t) + \sigma(\mathbf{r}) \frac{\partial \mathbf{e}(\mathbf{r}, t)}{\partial t} + \frac{\partial \mathbf{j_s}(\mathbf{r}, t)}{\partial t} = 0 \tag{1}$$

where $\mathbf{r}$ denotes the spatial coordinate; t is time; $\mathbf{e}$ represents the electric field; σ and μ represent the conductivity and magnetic permeability, respectively; and $\mathbf{j_s}$ denotes the source current density. As the artificial source is fixed during the measurement, we can separate the source current density term as $\mathbf{j_s}(\mathbf{r}, t) = I(t)\bar{\mathbf{J}}_\mathbf{s}(\mathbf{r})$, where $I(t)$ is the source current intensity that is employed for simulating the source waveform and $\bar{\mathbf{J}}_\mathbf{s}$ denotes the spatial pattern of $\mathbf{j_s}$ using unit current intensity.

In this study, forward modeling uses unstructured tetrahedral meshes for the 3-D model discretization and applies the edge-based finite element method for the spatial discretization of Equation (1) that discretize the electric fields on the grid edges. The electric field at any point within the k-th tetrahedral element of the mesh can be expressed as [15]:

$$\mathbf{e}(\mathbf{r},t) = \sum_{i=1}^{6} e_i^k(t)\mathbf{n}_i^k(\mathbf{r}) \tag{2}$$

where $e_i^k(t)$ represents the magnitude of the electric field component along the i-th edge direction in the k-th element, and $\mathbf{n}_i^k(\mathbf{r})$ is the first-order edge-based interpolation function. By applying the Galerkin method, a semi-discrete system can be established [16]:

$$\mathbf{M}\frac{\mathrm{d}\mathbf{e}(t)}{\mathrm{d}t} + \mathbf{S}\mathbf{e}(t) + \frac{\mathrm{d}I(t)}{\mathrm{d}t}\bar{\mathbf{J}} = 0 \tag{3}$$

where $\mathbf{M}$ denotes the mass matrix; $\mathbf{S}$ is the stiffness matrix; $\mathbf{e}(t)$ represents the solution vector of the discretized electric fields at time t; and $\bar{\mathbf{J}}$ denotes the source term vector [17].

$$\begin{gathered} \mathbf{M}_{i,j}^e = \sigma^e \iiint_{v^e} \mathbf{n}_i^e(\mathbf{r})\cdot\mathbf{n}_j^e(\mathbf{r})\mathrm{d}v \\ \mathbf{S}_{i,j}^e = \frac{1}{\mu_0\mu_r^e}\iiint_{v^e}\left[\nabla\times\mathbf{n}_i^e(\mathbf{r})\right]\cdot\left[\nabla\times\mathbf{n}_j^e(\mathbf{r})\right]\mathrm{d}v \\ \bar{\mathbf{J}}_i^e = \iiint_{v^e}\mathbf{n}_i^e(\mathbf{r})\cdot\bar{\mathbf{J}}_\mathbf{s}(\mathbf{r})\mathrm{d}v \\ i = 1,\ldots,6 \\ j = 1,\ldots,6 \end{gathered} \tag{4}$$

Equation (4) gives the expansion forms of the local matrix elements for a specific tetrahedral element (e), where μ_r^e denotes the relative magnetic permeability of the element, and different elements can be assigned with different values.

Using the second-order implicit backward Euler method to discretize Equation (3) in the time domain [18], we can derive the unconditional stable time-domain iterative formulation:

$$\begin{gathered} (3\mathbf{M} + 2\Delta t_k\mathbf{S})\mathbf{e}_k = \mathbf{M}(4\mathbf{e}_{k-1} - \mathbf{e}_{k-2}) - (3I_k - 4I_{k-1} + I_{k-2})\bar{\mathbf{J}} \\ k = 1,\ldots,n_{\mathrm{step}} \end{gathered} \tag{5}$$

where $\mathbf{e}_k$ and I_k represent the discrete electric field solution vector and current intensity at the m-th time point, and Δt_m is the time step. The time iteration calculation requires a total of n_{step} steps to complete. In each time step, we solve the above system of linear equations to derive the $\mathbf{e}_k$ at the new time point.

In this work, the loop transmitters are adopted as the artificial sources, and the time-stepping process starts from a steady state (e.g., the moment before switching off the source), so the initial condition of the electric field (i.e., $\mathbf{e}_0$) is set to 0. The Dirichlet boundary conditions [19] are used, which assume that the tangential components of the electric field are always 0 at the distant boundaries, which are far from the study region.

3. Accuracy Validation for 3-D Forward Modeling Algorithm

To validate the accuracy of the forward modeling algorithm, we compared our 3-D forward modeling response with the analytical solution of a uniform half-space model and the layered model response calculated by an open-source 1-D forward program [20]. A 400 m × 400 m loop transmitter was designed, and a unit step-off source current waveform was used. A measurement site was located at the center of the loop, with both source and receiver deployed on the surface. In addition, a 1-D layered model was designed for testing, as shown in Figure 1. The resistivity values from the top to bottom layers are 10^6 Ω·m, 10 Ω·m, 100 Ω·m, and the relative magnetic permeabilities are 1, 2, 1, respectively. The interfaces between the three layers are located at 0 m and 50 m. Figure 2a shows the comparison between our 3-D modeling response and the analytical solution of the uniform half-space model. Figure 2b shows a comparison between our 3-D forward modeling response and the 1-D numerical solution.

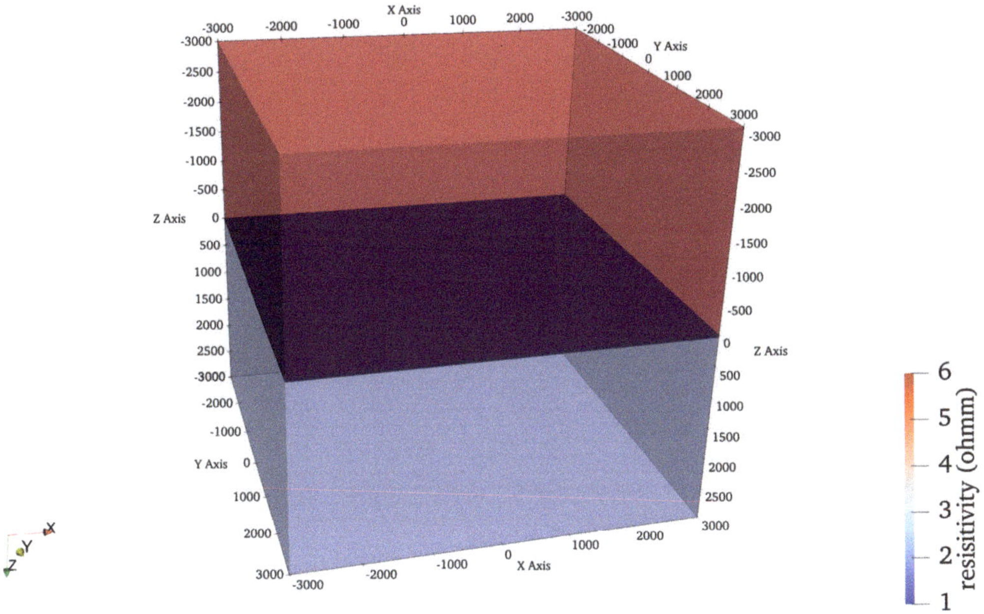

Figure 1. Horizontal view of the 1-D layered model.

It can be observed from Figure 2a that for the uniform half-space model, the relative differences between our 3-D modeling results and the 1-D analytical solution were less than 3%. In Figure 2b, when both resistivity and magnetic permeability anomalies existed, the relative differences between our 3-D response and the 1-D numerical solution were generally below 5%. Therefore, this example validated the computational accuracy of our 3-D forward algorithm presented in this work.

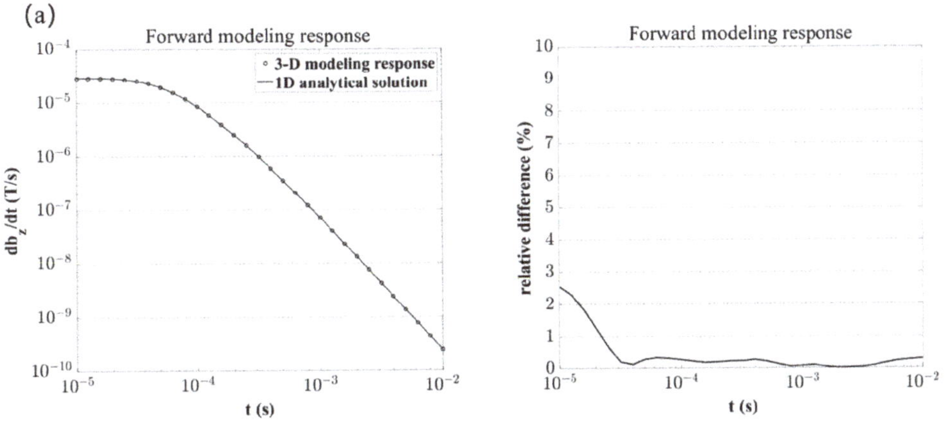

Figure 2. *Cont.*

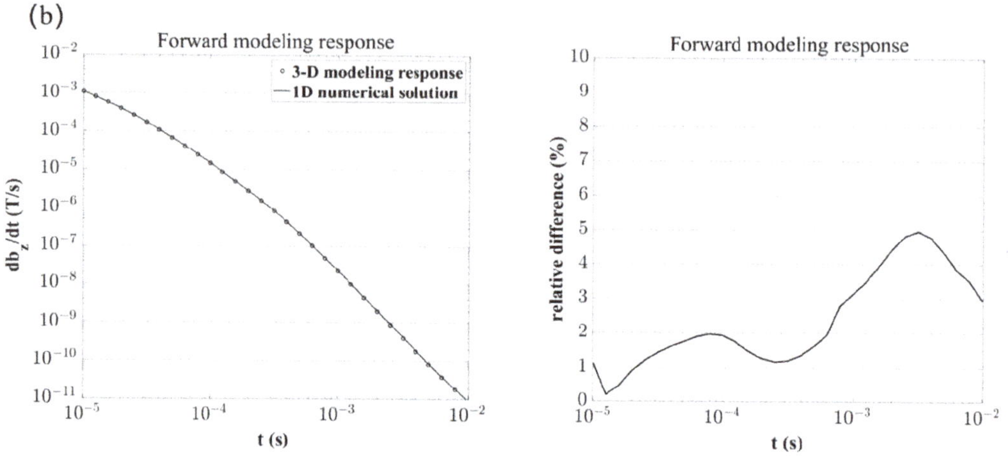

Figure 2. Modeling accuracy validation: (**a**) Comparison with the 1-D analytical solution; (**b**) Comparison with the 1-D numerical solution.

4. Calculation and Characteristic Analysis of 3-D Forward Modeling Response Considering Magnetic Permeability

In this example, a 3-D model was designed to investigate the impact of magnetic permeability on the 3-D TEM response. The measurement configuration in this example was the same as that in the 1-D example described above. The uniform half-space model contained an anomaly block, located at $x, y \in [-50, 50]$ m and $z \in [50, 100]$ m. The background resistivity was 100 Ω·m, and the anomaly was located below the loop center. Figure 3 shows the modeling responses for the anomaly with (ρ = 100 Ω·m, μ_r = 1), (ρ = 10 Ω·m, μ_r = 1), (ρ = 1000 Ω·m, μ_r = 1), (ρ = 100 Ω·m, μ_r = 1.5), and (ρ = 10 Ω·m, μ_r = 1.5), respectively, along with the relative differences of these responses compared to the uniform half-space model response.

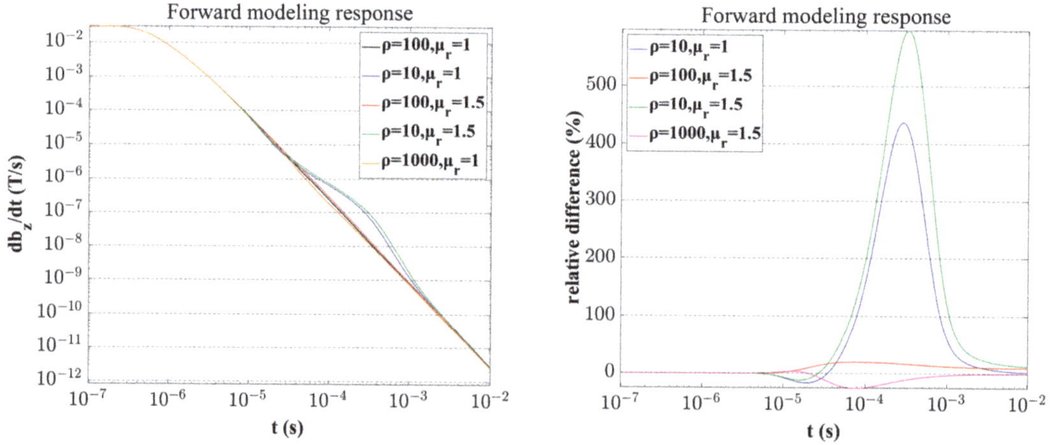

Figure 3. Comparison of 3-D modeling responses for the anomaly with different parameters.

As shown in Figure 3, the conductive block caused a distinct anomaly response; when the block included both resistivity and magnetic permeability anomalies, its modeling response had an obvious difference compared to the model response affected only by the resistivity. Additionally, for the response at the middle time stage, the differences rapidly reached a maximum value (596%) and then faded out.

Next, a 3-D model was designed to study the influence of a highly magnetizable anomaly (e.g., manhole cover) on the modeling response when it was present in shallow layers. The measurement configuration was also the same as in the previous 1-D example. A dodecahedron prism-shaped anomaly was placed below the center of the loop in the uniform half-space with a background resistivity of 100 Ω·m. The dodecahedron prism had an edge length of 10 m, a height of 1 m, and a burial depth of 10 m. Three cases were set with the anomaly holding different parameters: (ρ = 10 Ω·m, μ_r = 1), (ρ = 100 Ω·m, μ_r = 5), and (ρ = 10 Ω·m, μ_r = 5).

Figure 4 shows a comparison of the forward modeling responses for the three cases. As can be seen, due to the anomaly being at a shallow depth near the surface, the maximum difference compared with the uniform half-space model response occurs at the early times, rapidly reaching its peak value, which was 34%, and then decreasing. Simply the presence of a magnetic object (μ_r = 5) caused more than 5% differences in the response.

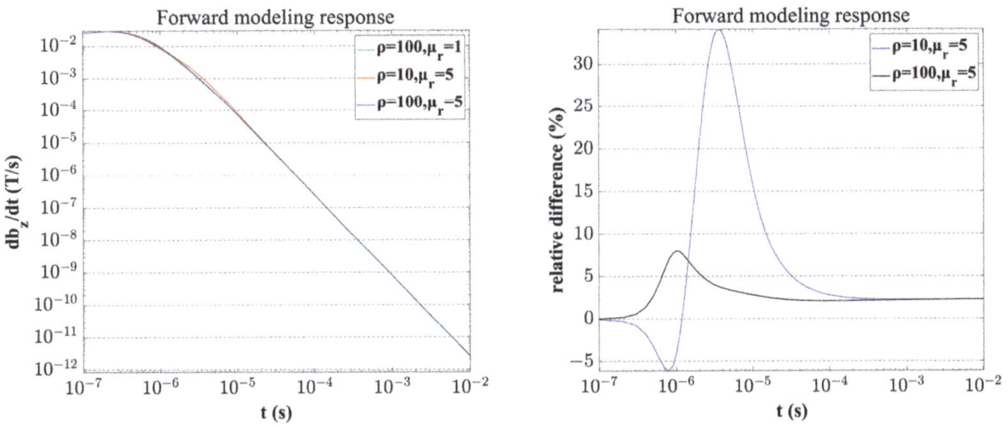

Figure 4. Comparison of forward responses for the anomaly with different parameters.

The above examples demonstrate the influence of the magnetic permeability on the TEM responses. Therefore, when there is a significant magnetic permeability anomaly in the subsurface, the 3-D TEM forward modeling algorithm must be capable of simulating the models with heterogeneous magnetic permeability to accurately compute such anomalous responses.

5. 3-D TEM Inversion Algorithm Considering Magnetic Permeability

The objective function for the 3-D TEM Inversion is [21]:

$$\Phi(\mathbf{m}) = \frac{1}{2}\|\mathbf{W}_{\mathrm{d}}[\mathbf{d} - \mathbf{d}_{\mathrm{obs}}]\|_2^2 + \frac{\lambda}{2}\|\mathbf{W}_{\mathbf{m}}(\mathbf{m} - \mathbf{m}_{\mathrm{ref}})\|_2^2 \quad (6)$$

where $\mathbf{m}$ denotes the model vector, which includes both resistivity and magnetic permeability; $\mathbf{d}_{\mathrm{obs}}$ is the observation data vector, representing the vertical component of the induced electromotive force ($\mathrm{d}\mathbf{b}_{\mathbf{z}}/\mathrm{d}t$) response corresponding to multiple time gates for each measurement site; $\mathbf{W}_{\mathrm{d}}$ represents a diagonal matrix of the inverse of the data uncertainties, and $\mathbf{W}_{\mathrm{m}}$ is the model regularization matrix; $\mathbf{d}$ denotes the forward modeling response of $\mathbf{m}$; and $\mathbf{m}_{\mathrm{ref}}$ represents the reference model.

The forward problem involves a time-stepping process corresponding to Equation (5) that can be expressed as solving a large-scale system of linear equations:

$$\begin{pmatrix} \ddots & & & & & \\ & -4\mathbf{M} & 3\mathbf{M}+2\Delta t_{k-2}\mathbf{S} & & & \\ & \mathbf{M} & -4\mathbf{M} & 3\mathbf{M}+2\Delta t_{k-1}\mathbf{S} & & \\ & & \mathbf{M} & -4\mathbf{M} & 3\mathbf{M}+2\Delta t_{k}\mathbf{S} & \\ & & & & & \ddots \end{pmatrix} \begin{bmatrix} \vdots \\ \mathbf{e}_{k-2} \\ \mathbf{e}_{k-1} \\ \mathbf{e}_{k} \\ \vdots \end{bmatrix} = \begin{bmatrix} \vdots \\ (4I_{k-3}-3I_{k-2}-4I_{k-4})\bar{\mathbf{J}} \\ (4I_{k-2}-3I_{k-1}-4I_{k-3})\bar{\mathbf{J}} \\ (4I_{k-1}-3I_{k}-4I_{k-2})\bar{\mathbf{J}} \\ \vdots \end{bmatrix} \quad \mathbf{Ae}=\mathbf{B} \tag{7}$$

where $\mathbf{A}$ is the coefficient matrix of the forward equation, and $\mathbf{B}$ is the right-hand side of the forward equation.

Since we observe the $\mathrm{d}\mathbf{b_z}/\mathrm{d}t$ response, we have:

$$\frac{\mathrm{d}\mathbf{b_z}}{\mathrm{d}t} = \mathbf{Qe} = \mathbf{QA}^{-1}\mathbf{B} \tag{8}$$

where $\mathbf{Q}$ denotes the spatial interpolation matrix. The sensitivity matrix can be calculated as:

$$\begin{aligned} \mathbf{J} &= \frac{\partial\left(\frac{\mathrm{d}\mathbf{b_z}}{\mathrm{d}t}\right)}{\partial\mathbf{m}} = \mathbf{Q}\frac{\partial\mathbf{e}}{\partial\mathbf{m}} = \mathbf{Q}\frac{\partial\mathbf{A}^{-1}\mathbf{B}}{\partial\mathbf{m}} = \mathbf{Q}\left(\mathbf{A}^{-1}\frac{\partial\mathbf{B}}{\partial\mathbf{m}} + \frac{\partial\mathbf{A}^{-1}}{\partial\mathbf{m}}\mathbf{B}\right) \\ &= \mathbf{Q}\frac{\partial\mathbf{A}^{-1}}{\partial\mathbf{m}}\mathbf{B} = -\mathbf{QA}^{-1}\frac{\partial\mathbf{A}}{\partial\mathbf{m}}\mathbf{A}^{-1}\mathbf{B} = -\mathbf{QA}^{-1}\frac{\partial\mathbf{A}}{\partial\mathbf{m}}\mathbf{e} \end{aligned} \tag{9}$$

where $\partial\mathbf{B}/\partial\mathbf{m} = 0$.

This paper uses the Limited-memory Broyden–Fletcher–Goldfarb–Shanno (L-BFGS) method to solve the inversion problem, which only requires the gradient vector of the objective function. The core of the calculation is to obtain: $\mathbf{J}^{\mathrm{T}}\mathbf{r} = -\mathbf{e}^{\mathrm{T}}\left(\frac{\partial\mathbf{A}}{\partial\mathbf{m}}\right)^{\mathrm{T}}\mathbf{A}^{-\mathrm{T}}\mathbf{Q}^{\mathrm{T}}\mathbf{r}$.

The main computational cost concentrate in solving $\mathbf{A}^{-\mathrm{T}}\mathbf{Q}^{\mathrm{T}}\mathbf{r}$. For such a problem, the virtual electric field $\tilde{\mathbf{e}}$ can be defined to satisfy:

$$\mathbf{A}^{\mathrm{T}}\tilde{\mathbf{e}} = \mathbf{Q}^{\mathrm{T}}\mathbf{r} \tag{10}$$

Solving the above linear system is similar to the forward process, and it is typically referred as the adjoint modeling process.

Through the adjoint modeling, the expression can be obtained as: $\mathbf{J}^{\mathrm{T}}\mathbf{r} = -\mathbf{e}^{\mathrm{T}}\left(\frac{\partial\mathbf{A}}{\partial\mathbf{m}}\right)^{\mathrm{T}}\tilde{\mathbf{e}}$.

When the inversion model parameters include both magnetic permeability and conductivity, the coefficient matrix $\mathbf{A}$ is differentiated with respect to the conductivity m_σ and the magnetic permeability m_μ, yielding:

$$\frac{\partial\mathbf{A}}{\partial m_\sigma} = \begin{pmatrix} \frac{\partial(3\mathbf{M}+2\Delta t\mathbf{S})}{\partial m_\sigma} & & & & \\ \frac{\partial(-4\mathbf{M})}{\partial m_\sigma} & \frac{\partial(3\mathbf{M}+2\Delta t\mathbf{S})}{\partial m_\sigma} & & \cdots & \\ \frac{\partial\mathbf{M}}{\partial m_\sigma} & \frac{-\partial(4\mathbf{M})}{\partial m_\sigma} & \frac{\partial(3\mathbf{M}+2\Delta t\mathbf{S})}{\partial m_\sigma} & & \\ & \vdots & & \ddots & \vdots \\ & & & \cdots & \frac{\partial(3\mathbf{M}+2\Delta t\mathbf{S})}{\partial m_\sigma} \end{pmatrix}, \tag{11}$$

$$\frac{\partial \mathbf{A}}{\partial m_\mu} = \begin{pmatrix} \frac{\partial(3\mathbf{M}+2\Delta t\mathbf{S})}{\partial m_\mu} & & & & \\ \frac{\partial(-4\mathbf{M})}{\partial m_\mu} & \frac{\partial(3\mathbf{M}+2\Delta t\mathbf{S})}{\partial m_\mu} & & \cdots & \\ \frac{\partial \mathbf{M}}{\partial m_\mu} & \frac{\partial(-4\mathbf{M})}{\partial m_\mu} & \frac{\partial(3\mathbf{M}+2\Delta t\mathbf{S})}{\partial m_\mu} & & \\ & \vdots & & \ddots & \vdots \\ & & & \cdots & \frac{\partial(3\mathbf{M}+2\Delta t\mathbf{S})}{\partial m_\mu} \end{pmatrix}, \quad (12)$$

where **M** is only related to the conductivity, and **S** is only related to the magnetic permeability.

Therefore, we have:

$$\frac{\partial \mathbf{A}}{\partial m_\sigma} = \begin{pmatrix} \frac{\partial(3\mathbf{M})}{\partial m_\sigma} & & & & \\ \frac{\partial(-4\mathbf{M})}{\partial m_\sigma} & \frac{\partial(3\mathbf{M})}{\partial m_\sigma} & & \cdots & \\ \frac{\partial \mathbf{M}}{\partial m_\sigma} & \frac{\partial(-4\mathbf{M})}{\partial m_\sigma} & \frac{\partial(3\mathbf{M})}{\partial m_\sigma} & & \\ & \vdots & & \ddots & \vdots \\ & & & \cdots & \frac{\partial(3\mathbf{M})}{\partial m_\sigma} \end{pmatrix}, \quad (13)$$

$$\frac{\partial \mathbf{A}}{\partial m_\mu} = \begin{pmatrix} \frac{\partial(2\Delta t\mathbf{S})}{\partial m_\mu} & & & & \\ & \frac{\partial(2\Delta t\mathbf{S})}{\partial m_\mu} & & \cdots & \\ & & \frac{\partial(2\Delta t\mathbf{S})}{\partial m_\mu} & & \\ & \vdots & & \ddots & \vdots \\ & & & \cdots & \frac{\partial(2\Delta t\mathbf{S})}{\partial m_\mu} \end{pmatrix}, \quad (14)$$

Based on the above formulations, the two matrix-vector products $\mathbf{J}_\sigma^{\mathrm{T}}\mathbf{r}$ and $\mathbf{J}_\mu^{\mathrm{T}}\mathbf{r}$ can be computed separately. This allows the gradient vectors of the objective function with respect to the magnetic permeability and conductivity models to be obtained. Subsequently, based on the L-BFGS algorithm, the iterative updates for the model parameters are progressively completed.

The regularization matrix $\mathbf{W}_m$ is normally required to stabilize the inversion. For using the traditional structured grids, $\mathbf{W}_m$ can be chosen as a discrete gradient operator [22]. However, for the unstructured tetrahedral grids, the model roughness matrix is influenced by factors such as the irregular shape, distance, and volume of the tetrahedron, etc. Therefore, additional constraints need to be imposed to make the regularization constraint work in this case. Key [23] presented the definition of $\mathbf{W}_m$ for the 2-D grids using irregular triangular elements. Hongzhu Cai [24] further modified the work and made $\mathbf{W}_m$ adapted for the 3-D tetrahedral grids, as proposed by:

$$\mathbf{W}_m(i,j) = \frac{V_i}{r_{ij}}\left(\sqrt{\frac{V_{ij}}{\sum_{k=1}^{N} V_{ik}}}\right) \quad (15)$$

where N is the number of neighboring cells for the i-th cell; V_i is the volume of the i-th cell; V_{ij} is the volume of the j-th neighboring cell; V_{ik} is the volume of the k-th neighboring cell; and r_{ij} is the distance between the centroids of the i-th and j-th cells:

$$r_{ij} = \sqrt{(x_i - x_j)^2 + (y_i - y_j)^2 + (z_i - z_j)^2} \quad (16)$$

6. Synthetic Model Inversion Results and Analysis

To verify the accuracy and effectiveness of the inversion algorithm, we designed a 3-D geoelectric model as shown in Figure 5. Figure 6 shows the horizontal view of the source and the subsurface anomaly. The resistivity of the background medium was 100 Ω·m, and a block-shaped anomaly was placed in the underground at $x \in [100, 350]$ m, $y \in [-100, 100]$ m, and $z \in [50, 180]$ m. This example employed a large-loop source, with the shape being a square and the side length was 420 m. The unit step-off waveform was adopted for transmitting the primary field. The survey consisted of 21 profiles, each with 21 sites, totaling 441 measurement points. The intervals between the profiles and sites were 50 m, and the survey area spanned $x, y \in [-500, 500]$ m, as indicated by the yellow points in Figure 5. The $\mathrm{d}\mathbf{b}_z/\mathrm{d}t$ response for each measurement point was computed within the time range from 10^{-5}s to 10^{-2}s, with a total of 31 logarithmically spaced time points. The inversion area is $x \in [-2000, 2000]$ m, $y \in [-2000, 2000]$ m, $z \in [0, 500]$ m. Gaussian random noise with a 5% standard deviation was added to the true model responses to generate synthetic data for the inversion test.

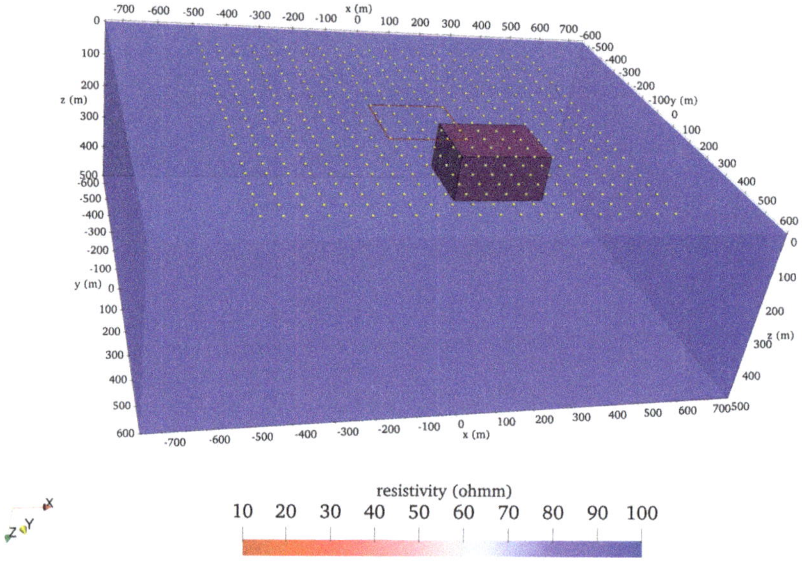

Figure 5. Horizontal view of the source and the subsurface anomaly.

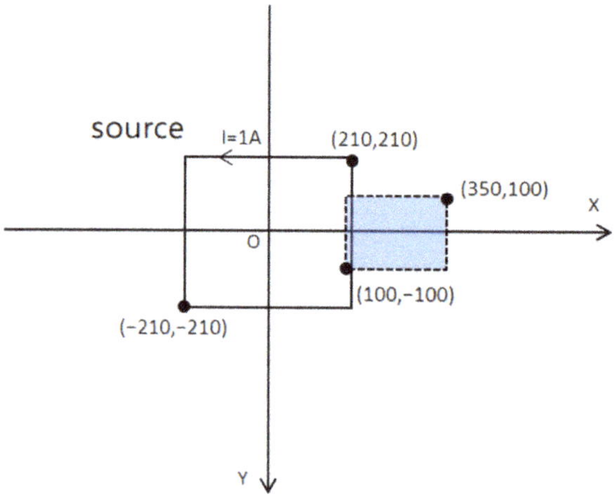

Figure 6. Top view of the surface source and subsurface anomalies.

This example consisted of several inversion tests to validate the reliability of the algorithm and to demonstrate the necessity of considering the impact of magnetic permeability in the 3-D TEM inversion. These tests include: (1) Resistivity model inversion, (2) Magnetic permeability model inversion, (3) 3-D TEM inversion considering the impact of magnetic permeability, and (4) 3-D TEM inversion ignoring the impact of magnetic permeability.

6.1. Resistivity Model Inversion

The first test focused on the inversion effect considering only the resistivity anomalies in the underground target body (using a vacuum magnetic permeability model for the entire space). The resistivity values of the anomaly were set as 10 Ω·m for a low-resistivity block and 1000 Ω·m for a high-resistivity block.

As shown in Figure 7a, for the low-resistivity anomaly, the inversion algorithm accurately recovered the position and shape of the anomaly, and the resistivity value of the anomaly was close to the true value of 10 Ω·m. As shown in Figure 7b, the 3-D TEM inversion also accurately characterized the position of the high-resistivity anomaly. However, because TEM has a relatively lower sensitivity for the resistive target, the anomaly's resistivity value was only recovered to around 400 Ω·m, which still showed a certain discrepancy with the true value.

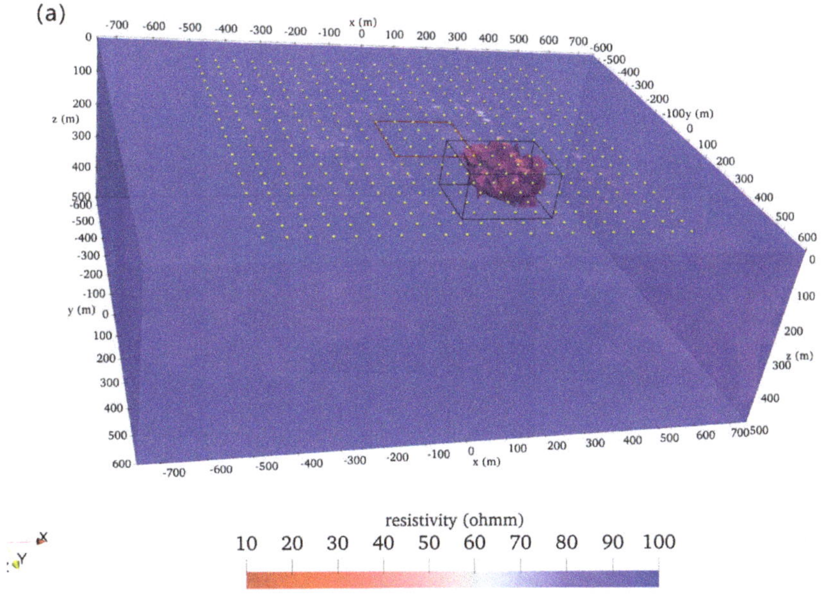

Figure 7. *Cont.*

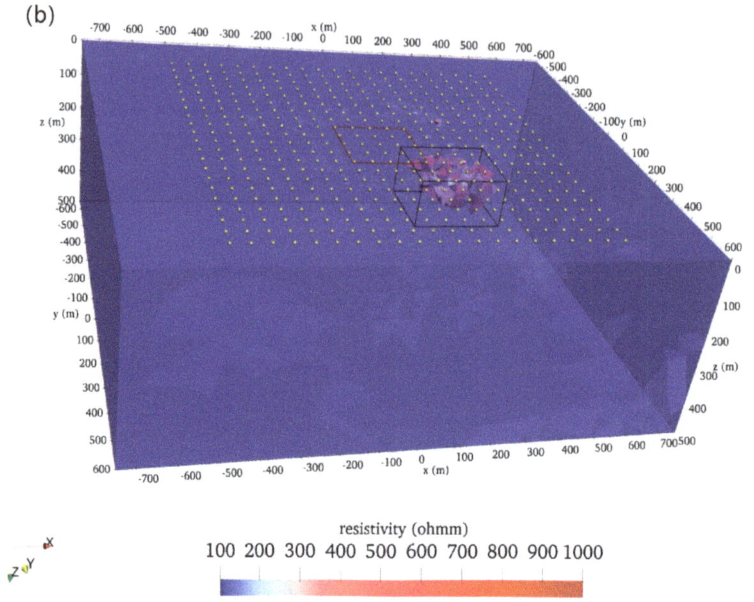

Figure 7. Data inversion results: (**a**) Inversion result for anomaly with ρ = 10 Ω·m; (**b**) Inversion result for anomaly with ρ = 1000 Ω·m.

6.2. Magnetic Permeability Model Inversion

Next, the inversion case considering only the magnetic permeability anomalies in the underground target body was tested (using a uniform half-space resistivity model with 100 Ω·m). The inversion was carried out by considering only the change in the relative magnetic permeability of the anomaly. The relative magnetic permeability of the anomaly was set to 2 and 5 for the two rectangular blocks.

As shown in Figure 8a,b, the inversion results clearly reflect the position and shape of the anomaly. When the relative magnetic permeability of the anomaly was 2, the relative magnetic permeability within the anomaly region was accurately recovered to approximately 1.8–2.0. However, when the relative magnetic permeability of the anomaly was 5, only part of the anomaly area had the relative magnetic permeability recovered to 5, and most of the area was recovered to around 4.

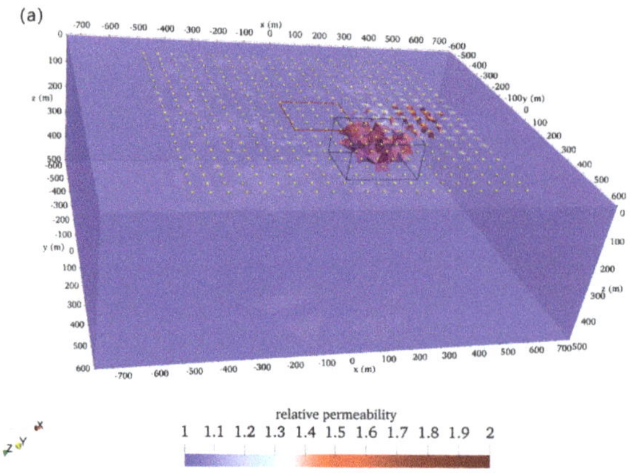

Figure 8. *Cont.*

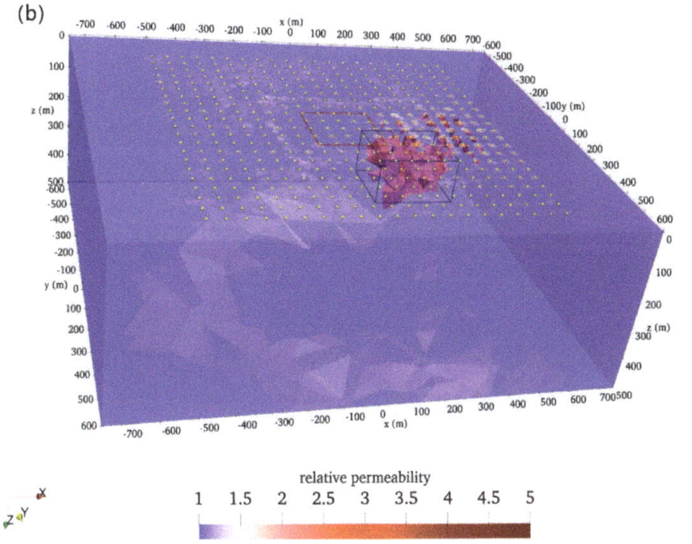

Figure 8. Data inversion results: (**a**) Inversion result for anomaly with $\mu_r = 2$; (**b**) Inversion result for anomaly with $\mu_r = 5$.

6.3. 3-D TEM Inversion Considering the Impact of Magnetic Permeability

To analyze the effect of the magnetic permeability parameter on the 3-D TEM inversion, synthetic data was generated using a model that included both resistivity and magnetic permeability anomalies. This study tested the 3-D TEM inversions with and without considering magnetic permeability impact. The inversion results for the resistivity and magnetic permeability parameters are presented, as shown in Figure 9.

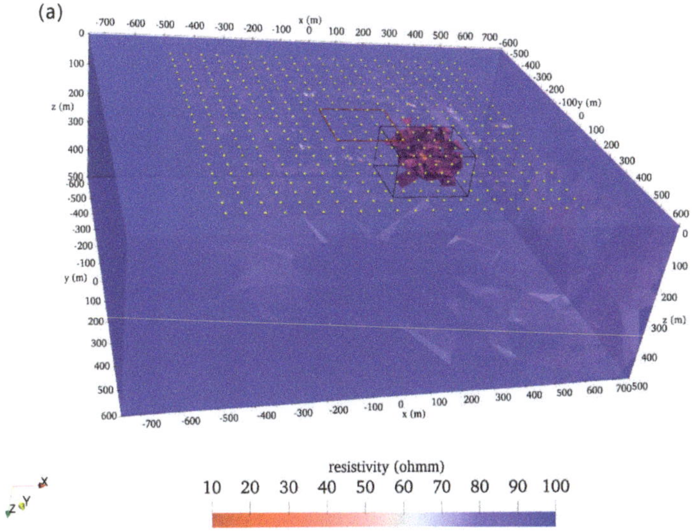

Figure 9. *Cont.*

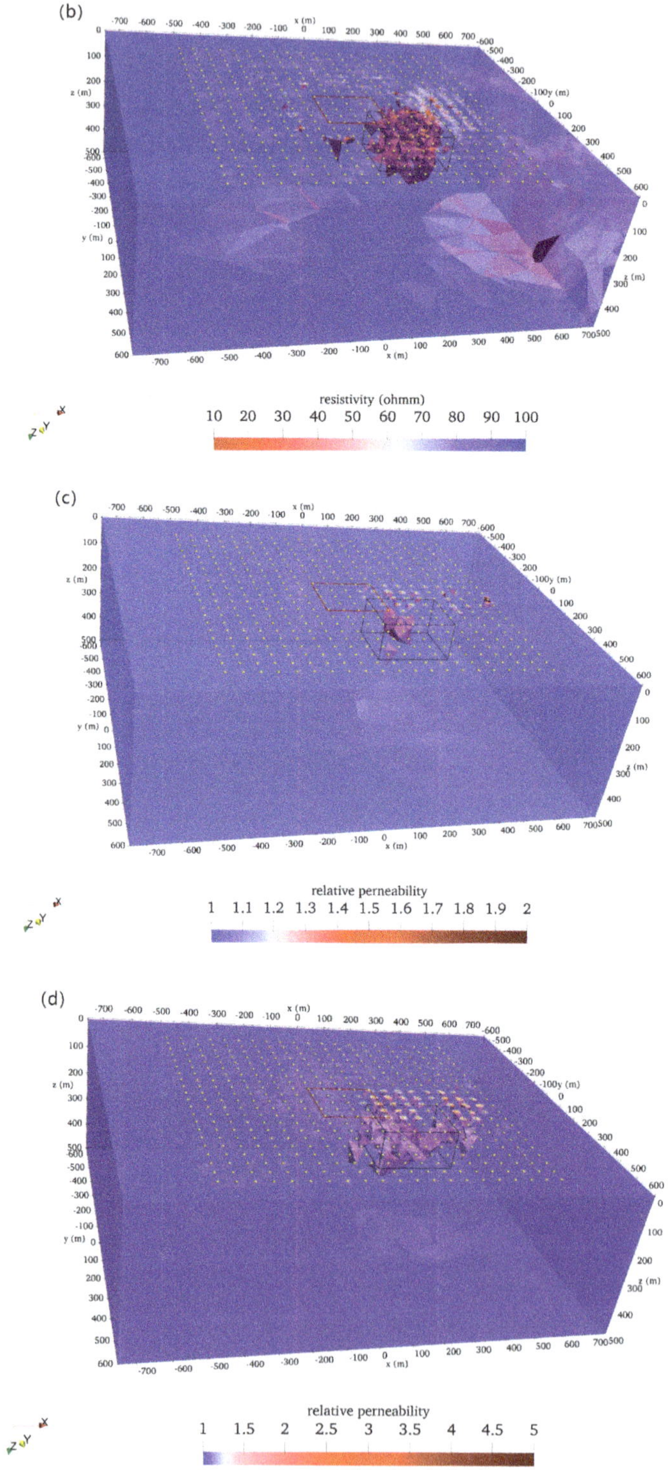

Figure 9. *Cont.*

(e)

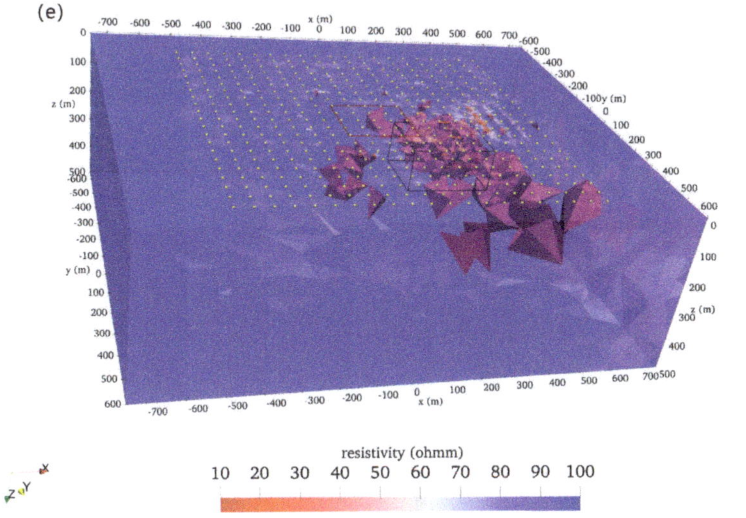

(f)

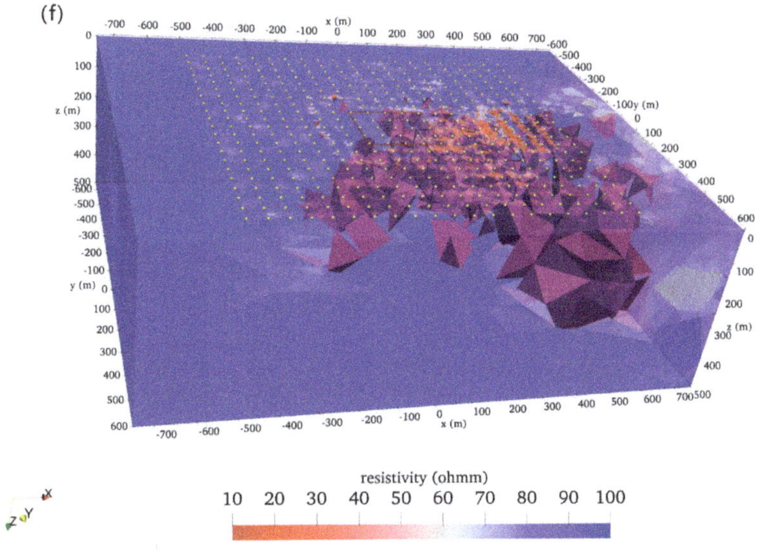

Figure 9. Data inversion results: (**a**) Resistivity inversion result for anomaly with (ρ = 10 Ω·m, μ_r = 2), considering the impact of magnetic permeability; (**b**) Resistivity inversion result for anomaly with (ρ = 10 Ω·m, μ_r = 5), considering the impact of magnetic permeability. (**c**) Relative magnetic permeability inversion result for anomaly with (ρ = 10 Ω·m, μ_r = 2); (**d**) Relative magnetic permeability inversion result for anomaly with (ρ = 10 Ω·m, μ_r = 5). (**e**) Resistivity inversion result for anomaly with (ρ = 10 Ω·m, μ_r = 2), ignoring the impact of magnetic permeability; (**f**) Resistivity inversion result for anomaly with (ρ = 10 Ω·m, μ_r = 5), ignoring the impact of magnetic permeability.

Figure 9a,b display the resistivity inversion results for the anomalies with (ρ = 10 Ω·m, μ_r = 2) and (ρ = 10 Ω·m, μ_r = 5), considering the impact of magnetic permeability. Figure 9c,d show the inversion results for the relative magnetic permeability. As it can be seen, the resistivity inversion results matched the true model (Figure 5) quite well, clearly reflecting the size, location, and resistivity of the subsurface target body. However, when the magnetic permeability value of the anomaly was large, a few spurious anomalies appeared in the inversion results. When the anomaly's relative magnetic permeability was set to 2, the inversion of the relative magnetic permeability parameter was not able to properly recover the underground target. In contrast, when the anomaly's relative magnetic permeability increased to 5, the position and size of the underground target were well-recovered, but the relative magnetic permeability value was only recovered to approximately 2, and a few spurious anomalies appeared near the anomaly.

Comparing Figure 9a,b with Figure 9e,f, it can be observed that if the algorithm ignored the magnetic permeability and tried to invert the data that was influenced by magnetic permeability, although the general distribution of the underground target can be roughly obtained, large spurious anomalies would appear in the inversion results. These false anomalies were mostly concentrated around the anomaly.

This example clearly demonstrates that when there is a significant magnetic permeability contrast in the underground target, ignoring magnetic permeability in the inversion algorithm (by assuming vacuum magnetic permeability) will result in poor inversion results, with large spurious anomalies near the target. This will complicate the subsequent interpretation for the inversion result.

Figure 9c,d shows the inversion results for the magnetic permeability model when both resistivity and magnetic permeability were considered as inversion parameters. Because both physical parameters were inverted simultaneously, the solution space was larger compared to the inversion case only considering the resistivity variation. Moreover, since TEM has a higher sensitivity to resistivity compared to magnetic permeability, the inversion did not recover the magnetic permeability model well. Additionally, the 3-D TEM inversion algorithm considering magnetic permeability took 2.18 h, while ignoring magnetic permeability caused the difficulties for fitting the data, resulted in the computational task that took 4.42 h.

7. Conclusions

We employed the edge-based finite element method using unstructured grids, combined with the second-order backward Euler method for spatial and time discretization, to develop a 3-D time-domain forward modeling algorithm for TEM that simultaneously considers the non-uniformity of resistivity and magnetic permeability. The accuracy of the forward modeling algorithm was verified through the comparisons with a uniform half-space analytical solution and a 1-D numerical solutions, and the influence of the magnetic permeability variations on the 3-D TEM forward response was demonstrated.

Based on the forward modeling work, a 3-D TEM inversion algorithm was further implemented, which simultaneously inverts both resistivity and magnetic permeability parameters. Through tests on three types of model cases: resistivity model inversion, magnetic permeability model inversion, and simultaneous inversion for both resistivity and magnetic permeability, the positions of the anomalies were accurately recovered, validating the effectiveness of the inversion algorithm. Additionally, tests were conducted using synthetic data affected by magnetic permeability, where an inversion algorithm ignoring the magnetic permeability parameter was applied. The results showed that the inversion considering only the resistivity model led to significant spurious anomalies around the target body, whereas considering magnetic permeability greatly improved the recovery for the target body, highlighting the importance of accounting for magnetic permeability in 3-D TEM inversion.

The 3D forward and inversion algorithm developed in this paper has certain limitations in its applicability. First, the algorithm only considers scenarios with a large-loop magnetic source and does not yet accommodate cases involving an electric field source. Second, the algorithm is designed specifically for isotropic resistivity models and does not yet account for the more complex anisotropic resistivity conditions typically encountered in real-world scenarios. Finally, the effects of induced polarization are not considered in the example models, suggesting that there is still potential for further enhancement of the approach presented in this paper.

Our work significantly extends the flexibility and applicability of the 3-D TEM inversion algorithm, which is expected to aid in the quantitative interpretation of the TEM field data.

Author Contributions: C.W.: methodology; software; validation; writing—original draft preparation; visualization. Y.D.: software; writing—review and editing. J.G.: supervision; funding acquisition; writing—review and editing. H.T.: supervision; conceptualization; writing—review and editing. Y.W.: funding acquisition; writing—review and editing. W.D.: funding acquisition; writing—review and editing. All authors have read and agreed to the published version of the manuscript.

Funding: This work was funded by the National Natural Science Foundation of China (Grant No. 42204081), National Key Research and Development Program of China (Grant No. 2023YFB3907703), and Key Laboratory of Target Cognition and Application Technology (Grant No. 2023-CXPT-LC-005).

Data Availability Statement: The original contributions presented in the study are included in the article, further inquiries can be directed to the corresponding author.

Conflicts of Interest: The authors declare no conflict of interest.

References

1. Ward, S.H. Unique determination of conductivity, susceptibility, size, and depth in multifrequency electromagnetic exploration. *Geophysics* **1959**, *24*, 531–546. [CrossRef]
2. Wait, J.R. Electromagnetic induction in a solid conducting sphere enclosed by a thin conducting spherical shell. *Geophysics* **1969**, *34*, 753–759. [CrossRef]
3. Wait, J.R.; Spies, K.P. Electromagnetic induction in a conducting sphere with a concentric shell. *Radio Sci.* **1969**, *4*, 557–560. [CrossRef]
4. Verma, S.K.; Singh, R.N. Transient electromagnetic response of an inhomogeneous conducting sphere. *Geophysics* **1970**, *35*, 331–336. [CrossRef]
5. Dey, A.; Ward, S.H. Inductive sounding of a layered earth with a horizontal magnetic dipole. *Geophysics* **1970**, *35*, 660–703. [CrossRef]
6. Ryu, J.; Morrison, H.F.; Ward, S.H. Electromagnetic fields about a loop source of current. *Geophysics* **1970**, *35*, 862–896. [CrossRef]
7. Ward, S.H. The electromagnetic response of a magnetic iron ore deposit. *Geophys. Prospect.* **1961**, *9*, 191–202. [CrossRef]
8. Fraser, D.C. Magnetite ore tonnage estimates from an aerial electromagnetic survey. *Geoexploration* **1973**, *11*, 97–105. [CrossRef]
9. Zhang, Z.; Oldenburg, D.W. Recovering magnetic susceptibility from electromagnetic data over a one-dimensional earth. *Geophys. J. Int.* **1997**, *130*, 422–434. [CrossRef]
10. Beard, L.P.; Nyquist, J.E. Simultaneous inversion of airborne electromagnetic data for resistivity and magnetic permeability. *Geophysics* **1998**, *63*, 1556–1564. [CrossRef]
11. Zhang, Z.; Oldenburg, D.W. Simultaneous reconstruction of 1-D susceptibility and conductivity from electromagnetic data. *Geophysics* **1999**, *64*, 33–47. [CrossRef]
12. Farquharson, C.G.; Oldenburg, D.W.; Routh, P.S. Simultaneous 1D inversion of loop–loop electromagnetic data for magnetic susceptibility and electrical conductivity. *Geophysics* **2003**, *68*, 1857–1869. [CrossRef]
13. Telford, W.M.; Geldart, L.P.; Sheriff, R.E. *Applied Geophysics*; Cambridge University Press: Cambridge, UK, 1990.
14. Yin, C.; Hodges, G. Influence of displacement currents on the response of helicopter electromagnetic systems. *Geophysics* **2005**, *70*, G95–G100. [CrossRef]
15. Jin, J.M. *The Finite Element Method in Electromagnetics*; John Wiley & Sons: New York, NY, USA, 2015.
16. Yin, C.; Qi, Y.; Liu, Y. 3D time-domain airborne EM modeling for an arbitrarily anisotropic earth. *J. Appl. Geophys.* **2016**, *131*, 163–178. [CrossRef]
17. Yin, C.; Qi, Y.; Liu, Y.; Cai, J. 3D time-domain airborne EM forward modeling with topography. *J. Appl. Geophys.* **2016**, *134*, 11–22. [CrossRef]
18. Oldenburg, D.W.; Haber, E.; Shekhtman, R. Three dimensional inversion of multisource time domain electromagnetic data. *Geophysics* **2013**, *78*, E47–E57. [CrossRef]
19. Um, E.S.; Harris, J.M.; Alumbaugh, D.L. 3D time-domain simulation of electromagnetic diffusion phenomena: A finite-element electric-field approach. *Geophysics* **2010**, *75*, F115–F126. [CrossRef]
20. Werthmüller, D. An open-source full 3D electromagnetic modeler for 1D VTI media in Python: Empymod. *Geophysics* **2017**, *82*, WB9–WB19. [CrossRef]
21. Tikhonov, A.N. *Solutions of Ill-Posed Problems*; VH Winston and Sons: New York, NY, USA, 1977.
22. Li, Y.; Oldenburg, D.W. 3-D inversion of magnetic data. *Geophysics* **1996**, *61*, 394–408. [CrossRef]

23. Key, K. MARE2DEM: A 2-D inversion code for controlled-source electromagnetic and magnetotelluric data. *Geophys. J. Int.* **2016**, *207*, 571–588. [CrossRef]
24. Cai, H.; Long, Z.; Lin, W.; Li, J.; Lin, P.; Hu, X. 3D multinary inversion of controlled-source electromagnetic data based on the finite-element method with unstructured mesh. *Geophysics* **2021**, *86*, E77–E92. [CrossRef]

Disclaimer/Publisher's Note: The statements, opinions and data contained in all publications are solely those of the individual author(s) and contributor(s) and not of MDPI and/or the editor(s). MDPI and/or the editor(s) disclaim responsibility for any injury to people or property resulting from any ideas, methods, instructions or products referred to in the content.

applied sciences

Article

Modeling of Characteristics of Complex Microstructure and Heterogeneity at the Core Scale

Yiwei Chen [1,2] and Pingchuan Dong [1,2,*]

[1] State Key Laboratory of Petroleum Resources and Engineering, China University of Petroleum (Beijing), Beijing 102249, China; yiwei.chen_@outlook.com
[2] College of Petroleum Engineering, China University of Petroleum (Beijing), Beijing 102249, China
* Correspondence: dpcfem@163.com

Abstract: Complex pore structures and strong matrix heterogeneity distinguish carbonate rocks, but there is a lack of comprehensive methods to describe these characteristics. In this study, a integrated approach is proposed to improve the accuracy and adaptability of velocity prediction methods, using a modified squirt flow model based on microcrack structures to characterize complicated pore structures, and a mixed random medium model to represent significant heterogeneity. In addition, the microcrack structure is obtained by inversion, but different from the D-Z method, each group of microcracks corresponds to a different equivalent medium model, so as to improve the accuracy of the inversion results. And the modified squirt flow model takes into account the attenuation caused by local flow between microcracks. The random medium model simulates the inhomogeneous body in the core by adjusting the autocorrelation length a and b, the rounding coefficient n, and the angle θ. A comparative study of the measured data of five limestone and dolomite samples reveals that the P-wave prediction error of the new model is less than 5%, whereas the Biot model is less than 10%, implying that the prediction accuracy of the new model is better.

Keywords: petrophysics; velocity analysis; squirt flow model; random medium model; multiple microfracture

1. Introduction

Carbonate rocks have major economic significance; sixty percent of the world's oil reserves are in carbonate reservoirs, and the potential for additional gas reserves is considerable [1–4]. As a result, studies focusing on the pore structure and petrophysical properties of carbonate are attracting widespread interest. Carbonate rocks are characterized by complex pore structures and strong heterogeneity [5]. For samples with the same porosity, their different pore structures greatly influence the ultrasonic wave velocity. Since the rock pore distribution is very complex and the pore structure is difficult to characterize quantitatively, many studies have used ideal geometries for approximation, with the most commonly used geometry being the ellipsoid, which employs the ratio of the short and long axes of the ellipsoid as its structural parameter [6–8].

Many studies have confirmed that the pore structure plays a key role in affecting the elastic wave velocity and elastic moduli [6,9–16], particularly in carbonate rocks with complex pore structures. Consequently, it is necessary to consider the effect of the pore structure in petrophysical modeling. Many models have considered the effect of local flow within pores arranged in different pore structures when elastic waves pass and have focused on changes in petrophysical properties. Among them, a traditional local flow mechanism model called the squirt model was developed to simulate the dispersion and attenuation of seismic wave velocities at the microscale (in nonhomogeneous spaces smaller than the wavelength) [17–24]. The squirt flow mechanism assumes that when seismic waves pass through rock, the small "compliant" pores in the rock close, causing the fluid inside to be ejected into the large "stiff" pores, and the local flow phenomenon tends to result in a large energy loss, which is the main cause of the induced large-scale seismic wave velocity [25]. This phenomenon of fluid flow between pores of different shapes and orientations occurs due to the pressure relaxation of seismic waves [26]. Squirt models are applicable to uniformly saturated rocks with low permeability (preventing the free flow of fluids within the pores) and low pressure (ensuring that the compliant pores are still open) [27]. With the development of computer hardware, simulations can provide more details of the pore structure [28]. In recent years, the development of multicore devices has significantly advanced current modelling capabilities, making numerical modelling based on the actual pore structure of rocks a popular research topic [21,28–32]. Squirt flow models consider the squirt flow action of a compliant pore with a single aspect. For complex pore structures, numerous scholars have proposed improvement schemes, including the inclusion of medium stiffness pores [26], iterative inclusion of compliant pores with different aspect ratios [33], and squirt flow between stiff pores and compliant pores with different aspect ratios [34].

It is generally accepted that strong heterogeneity is a vital characteristic of carbonate rocks. The random medium model can be used to appropriately characterize this heterogeneity. This method allows us to determine the size and distribution of mineral particles, microstructures, and pores in samples, as well as to quantitatively evaluate the heterogeneity of the rock. At present, three kinds of random medium models are widely used, namely, Gaussian, exponential and von Kármán autocorrelation function models [35–37], which are used to simulate the random distribution of rock particles with different geometric shapes. In addition, to meet various research needs, variable-scale reconstruction methods have been proposed. These include the Gaussian window function, simulated annealing, sequence classification, multipoint statistics, and their combinations [38–43].

In summary, although existing squirt flow models fully consider the local fluid flow between the stiff and compliant pores, they overlook the fluid pressure gradients present between microcracks with different aspect ratios. In addition, simulations of carbonate rocks often set the matrix as uniform, which is far from the real situation. Therefore, this paper combines an improved squirt flow model that considers local flow between microcracks and a mixed random medium model that describes matrix heterogeneity, thereby enhancing the accuracy and adaptability of carbonate rock core simulations. The pore structure in the rock is inverted from dry ultrasonic experimental data. Unlike the D-Z method, in the equivalent medium model of each group of microcracks, the background elastic modulus needs to be set to the elastic modulus after the closure of this group of microcracks, rather than to the elastic modulus after the closure of all microcracks. This aligns better with the assumption of gradual closure of microcracks. The modified squirt model, which takes into account the local flow between microcracks, is saturated by the Gassmann model. Finally, the predictions of the saturation velocity from the new model in this study are compared with experimental measurements and the predictions from the Biot model to validate their accuracy.

2. Methods

In previous studies [34,44], a relationship between microcracks and effective pressure was obtained. The high-pressure modulus of elasticity from the experiments is usually used to represent the elasticity modulus of rock in which the microcracks are completely closed [45], but in some rock samples, the microcracks are not completely closed until approximately 100 MPa. In the absence of high-pressure data, determining the bulk modulus and shear modulus of the rock with all cracks closed through experimental data is impossible. To address this situation, the modulus of elasticity after the microcracks completely close can be obtained by fitting the experimental data before using them.

In this paper, we inversely determine the density and porosity distributions of microfractures with different aspect ratios, which are used to describe the complex pore structure in carbonate rocks, based on the changes in elastic modulus during experimental pressurization. Based on the distribution characteristics of the microfractures, combined with the different bulk moduli of each group of microfractures, the suirt flow model [45] is modified, and the new model can describe the attenuation of ultrasonic waves in carbonate cores containing multiple groups of microfractures with different aspect ratios. The heterogeneity of the matrix is simulated by adjusting the length, width, and arrangement angle of the parameters of the mixed random medium model.

2.1. Pore Structure Analysis Method

By inverting the reduction in the density of microcracks with different aspect ratios during pressurization in the stress experiment, the microcrack density and porosity distribution of microcracks in the sample can be obtained, the flow chart shown in Figure 1. $\Gamma(\alpha)$ represents the cumulative fracture density of all microfractures whose aspect ratio is greater than α.

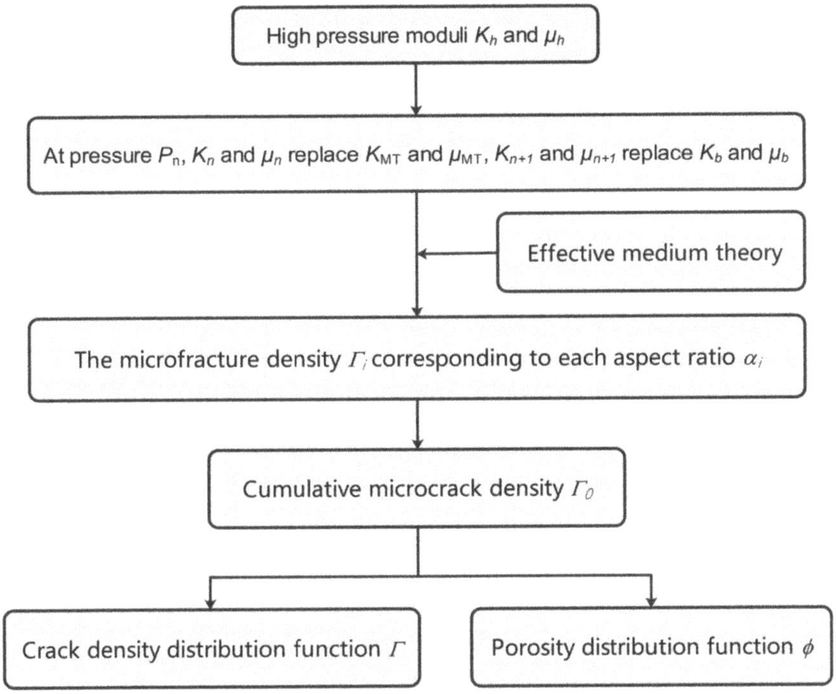

Figure 1. Inversion workflow of the aspect ratio distribution based on the pressurization process.

To obtain the initial aspect ratio of the microcracks, it is necessary to calculate the cumulative fracture density of microcracks. The equivalent medium theory is an important tool for linking the micropore structure with the elastic properties of rock. The equivalent medium model used in this paper is the Mori–Tanaka theory [15], whose geometric parameters use coin-shaped microcracks with small vertical and horizontal ratios [46]:

$$\frac{K_b}{K_{MT}} - 1 = \frac{16(1-\nu_b)(1+\nu_b)}{9(1-2\nu_b)}\Gamma, \tag{1}$$

$$\frac{G_b}{G_{MT}} - 1 = \frac{32(1-\nu_b)(5-\nu_b)}{45(2-\nu_b)}\Gamma, \tag{2}$$

where K_b, G_b and ν_b represent the bulk modulus, shear modulus and Poisson's ratio of the background matrix, respectively, and K_{MT} and G_{MT} represent the bulk modulus and shear modulus after the addition of pores, respectively.

In contrast to the D-Z inversion method [44], this research determines the fracture density corresponding to each group of microcracks with varying aspect ratios by using the appropriate equivalent medium model for each group of microcracks rather than a single equivalent medium model, the selection of the background bulk modulus and shear modulus are different.

During pressurization, the aspect ratios of all micropores in the rock decrease by the same amount [47]. In this study, it is assumed that at the beginning of the pressurization process P_1, the microcracks of the group with the smallest aspect ratio (α_1^0) begin to close and the aspect ratio of all microcracks decreases α_1^0. In this stage, compared to the rock in its original state without pressure $P(0)$, it is equivalent to taking the rock under pressure P_1 as the matrix and adding pores with aspect ratio α_1^0. (1) That is, under pressure P_n, in Equations (1) and (2), K_{n+1}, G_{n+1} and ν_{n+1} replace K_b, G_b and ν_b, respectively, K_n and G_n replace K_{MT} and G_{MT}, respectively, and (2) the microfracture density Γ_n corresponding to the group of microfractures with the smallest porosity ratio α_n^0 under P_n is obtained by the least squares method. (3) When P_{n+1} is high and all microcracks are closed, we obtain multiple groups of microcrack densities corresponding to different aspect ratios. (4) The cumulative microfracture density $\bar{\Gamma}_i$ under different pressures P_i is obtained by summing the densities of multiple groups of unclosed microfractures with closing pressures greater than P_i and minimum aspect ratios greater than α_i^0. When the sample is in the initial state without pressure, all microfractures are open and the cumulative microfracture density $\bar{\Gamma}_0$ is the sum of the microfracture density of all groups. (5) The crack density of the soft pores in dry samples decreases exponentially with increasing effective pressure [48]. By fitting the obtained results, the crack density under any pressure P can be obtained:

$$\bar{\Gamma}(P) = \bar{\Gamma}_0 e^{-P/\hat{P}}, \tag{3}$$

where $\bar{\Gamma}_0$ is the initial crack density at zero pressure and $\hat{P}$ is a constant on the order of magnitude of that of the pressure P. (6) For the MT model, the minimum aspect ratio of micropores and the cumulative microfracture density [33,44] are expressed as

$$\alpha_i = \frac{4\hat{P}(1-\nu_h^2)\ln\left(\frac{\bar{\Gamma}_0}{\bar{\Gamma}}\right)}{3\pi K_h(1-2\nu_h)}. \tag{4}$$

(7) Finally, the fracture density distribution function $\gamma(\alpha_0)$ and porosity distribution function $c(\alpha_0)$ of microfractures can be obtained by [49]

$$\gamma(\alpha_0) = -\frac{d\bar{\Gamma}}{d\alpha_0}, \tag{5}$$

$$c(\alpha_0) = \frac{4\pi\alpha_0}{3}\gamma(\alpha_0), \tag{6}$$

$$\phi(\alpha_0) = c(\alpha_0)d\alpha_0. \tag{7}$$

Under the effective pressure P, the fracture density distribution function $\gamma(\alpha)$ is the same as $\gamma(\alpha_0)$ in the initial state of zero pressure because the fracture density is only related to the major axis a of the microfracture. The ratio of the porosity to the aspect ratio is constant,

$$c(\alpha) = \frac{4\pi\alpha}{3}\gamma(\alpha), \tag{8}$$

$$\phi(\alpha) = \frac{4\pi\alpha}{3}\bar{\Gamma}(\alpha), \tag{9}$$

where $\bar{\Gamma}$ can be obtained by Equations (1) and (2) and $\bar{\Gamma}_0$ is replaced by the fitted $\bar{\Gamma}(P)$ in Equation (3), which represents the cumulative density of microfractures with a closing pressure greater than P as the original cumulative microfracture density at that pressure.

2.2. Squirt Flow Theory Considering Different Aspect Ratios

Assuming that a rock is composed of stiff pores and microfractures (compliant pores), the total porosity of the rock is represented as $\phi = \phi_s + \phi_c \approx \phi_s$, where ϕ_s represents the porosity of the stiff pores, ϕ_c represents the total porosity of the microfractures, and $\phi_c \ll \phi_s$. The bulk modulus and shear modulus of the rock mineral matrix are represented as K_g and μ_g, respectively, while the dry skeleton elastic moduli are represented as K_d and μ_d. The permeability of the rock is represented as κ. The bulk modulus of the modified frame $K(P)$ and $\mu(P)$ can be expressed as [45]

$$\frac{1}{K(P,\omega)} = \frac{1}{K_h} + \frac{1}{\left(\frac{1}{K_d(P)} - \frac{1}{K_h}\right)^{-1} + \frac{1}{\phi_c(P)}\left(\frac{1}{K_f^*(\omega)} - \frac{1}{K_g}\right)^{-1}}, \tag{10}$$

$$\frac{1}{\mu(P)} = \frac{1}{\mu_d(P)} - \frac{4}{15}\left(\frac{1}{K_d(P)} - \frac{1}{K(P)}\right), \tag{11}$$

where $\phi_c(P)$ represents the microcrack porosity under the current effective pressure P and η and $K_f^*(\omega)$ are the viscosity and bulk modulus of the pore fluid, respectively. K_h represents the high-pressure bulk modulus of the dry rock when all microcracks are closed.

In this study, the pores obtained from the thin sections present different aspect ratios, and considering only the stiff and compliant pores is insufficient. Therefore, the improved squirt-flow model adopted in this study considers not only the squirt flow effect of the closure of compliant pores on stiff pores but also the squirt effect between compliant pores with various aspect ratios.

According to the pressurization method introduced in the previous section, compared to when the pressure is zero, microcracks with an aspect ratio of α_1 close in the rock at time P_1, where $\alpha_i(P)$ is the minimum aspect ratio of microcracks under pressure P. When ultrasonic waves approach before the ultrasonic wave peak arrives, the soft cracks with the minimum aspect ratio close first, and as the ultrasonic wave arrives, the microcracks close in order of increasing aspect ratio. When the rock contains only one kind of pore with aspect ratio α_c, the volume compression coefficient C_p of the microfracture satisfies [47]

$$C_p(P) = \frac{1/K_d(P) - 1/K_h}{\phi_c(P)}. \tag{12}$$

The pressure gradient between the microcracks with the minimum aspect ratio α_1 and those with larger aspect ratio is generated under the action of waves, forming a local flow,

$$\frac{1}{K_1} = \frac{1}{K_2} + \frac{\phi_1(p_1)}{\frac{1}{C_1(p_1)} + \left(\frac{1}{K_f^*(\omega)} - \frac{1}{K_g}\right)^{-1}}, \tag{13}$$

$$C_1(p_1) = \frac{1/K_1(p_1) - 1/K_h}{\phi_1(p_1)}, \tag{14}$$

where $\phi_1(p_1)$, $\alpha_1(p_1)$, and $C_1(p_1)$ are the porosity, aspect ratio, and volume compressibility coefficient, respectively, of the first group of microcracks with aspect ratio α_1. p_1 is the closing pressure of this group of microcracks. There is also a pressure gradient between the microcracks with aspect ratio α_2 and other microcracks and stiff pores,

$$\frac{1}{K_2} = \frac{1}{K_3} + \frac{\phi_2(p_2)}{\frac{1}{C_2(p_2)} + \left(\frac{1}{K_f^*(\omega)} - \frac{1}{K_g}\right)^{-1}}, \tag{15}$$

in the microcracks with the largest aspect ratio α_N,

$$\frac{1}{K_N} = \frac{1}{K_h} + \frac{\phi_N(p_N)}{\frac{1}{C_N(p_N)} + \left(\frac{1}{K_f^*(\omega)} - \frac{1}{K_g}\right)^{-1}}. \tag{16}$$

Combining Equations (12)–(16) gives

$$\frac{1}{K_{mf}(P_N,\omega)} = \frac{1}{K_h} + \sum_{n=1}^{N} \frac{\phi_n(p_n)}{\frac{1}{C_n(p_n)} + \left(\frac{1}{K_f^*(\omega)} - \frac{1}{K_g}\right)^{-1}}. \tag{17}$$

The modified squirt-flow model considering all microcracks is obtained:

$$\frac{1}{K_{mf}(P_N,\omega)} = \frac{1}{K_h} + \sum_{n=1}^{N} \frac{1}{\frac{1}{\frac{1}{K_n(p_n)} - \frac{1}{K_h}} + \frac{1}{\left(\frac{1}{K_f^*(\omega)} - \frac{1}{K_g}\right)\phi_n(p_n)}}, \tag{18}$$

the bulk modulus of the fluid is frequency dependent [45],

$$K_f^*(\omega) = \left[1 - \frac{2J_1(ka)}{kaJ_0(ka)}\right]K_f, \tag{19}$$

where, J is a homogeneous Bessel equation, and

$$ka = \frac{1}{\alpha_n}\sqrt{-\frac{3i\omega\eta}{K_f}}. \tag{20}$$

The saturated unrelaxed moduli K_s and μ_s can be obtained using the Gassmann equation [50],

$$\frac{1}{K_s(P,\omega)} = \frac{1}{K_g} + \frac{\phi_s\left(\frac{1}{K_f} - \frac{1}{K_g}\right)}{1 + \phi_s\left(\frac{1}{K_f} - \frac{1}{K_g}\right)\Big/\left(\frac{1}{K_{mf}(P,\omega)} - \frac{1}{K_g}\right)}, \tag{21}$$

and

$$\mu_s = \mu_{mf}. \tag{22}$$

2.3. Random Medium Model

As shown in Figure 2, numerical simulations are important when proposing a new model because they can check whether the new model captures the key factors of physical phenomena. The method above is used to describe the complex pore structure, while the inhomogeneity of the matrix is described using the random medium model presented in this section.

For a generally uniform, isotropic elastic medium, the velocity and density at spatial coordinates (x, y) can be expressed as

$$v(x,z) = v_0 + \delta v(x,z), \tag{23}$$

$$\rho(x,z) = \rho_0 + \delta\rho(x,z), \tag{24}$$

where, the variables v_0 and ρ_0 represent the velocity and density of a macroscopic nonuniform medium, respectively, while perturbations δv and $\delta\rho$ represent the velocity and density of a microscopic nonuniform medium, respectively. According to Birch's theory [51], it can be postulated that the relative perturbations of longitudinal and shear wave velocities are equivalent, while the relative perturbations of density and velocity exhibit a linear relationship. To represent small-scale inhomogeneous bodies within a random medium, a perturbation parameter can be employed. For instance, the velocity perturbation can be mathematically expressed as follows:

$$\sigma(x,z) = \frac{\delta v}{v_0} = \frac{1}{\kappa}\frac{\delta\rho}{\rho_0}, \tag{25}$$

where v_0 is the background velocity, $\sigma(x,z)$ is the relative velocity perturbation in space, and κ is a scale factor ranging from 0.3 to 0.8. Then,

$$v(x,z) = v_0 + \delta v(x,z) = v_0(1+\sigma), \tag{26}$$

$$\rho(x,z) = \rho_0 + \delta\rho(x,z) = \rho_0(1+\sigma). \tag{27}$$

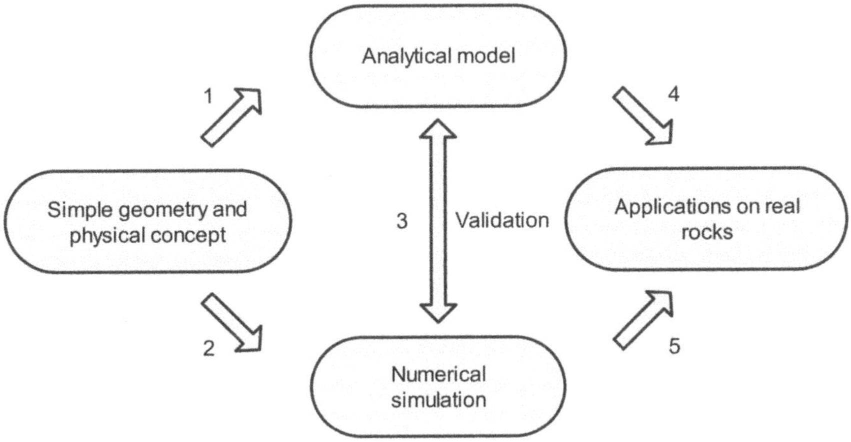

Figure 2. Workflow for benchmarking published analytical models and ensuring the quality of future models.

The process used to construct the random medium $\sigma(x,z)$ with autocorrelation function $\varphi(x,z)$ [52,53] is as follows.

1. The power spectral density of a stationary stochastic process σ can be obtained from the Fourier transform of its autocorrelation function $\varphi(x,z)$:

$$\Phi(k_x,k_z) = \int_{-\infty}^{+\infty}\int_{-\infty}^{+\infty} \varphi(x,z)e^{-i(k_x x+k_z z)}dxdz. \tag{28}$$

The autocorrelation function is

$$\varphi(x,z) = e^{-\left[\frac{(x\cos\theta+z\sin\theta)^2}{a^2}+\frac{(x\sin\theta-z\cos\theta)^2}{b^2}\right]^n}, \tag{29}$$

where, a and b are the autocorrelation length, and the angle factor θ describes the orientation of the reconstructed rock particles.

2. The initial random medium $\xi(k_x, k_y)$ is generated with a uniform distribution using a random number generator with the interval $[0, 2\pi]$.
3. A random medium power spectrum is generated:

$$\Psi(k_x, k_z) = \sqrt{\Phi(k_x, k_z) e^{i\xi(k_x, k_z)}}. \tag{30}$$

4. The random disturbance in the space domain is obtained with the inverse Fourier transform of the random power spectrum function $\Psi(k_x, k_z)$,

$$\psi(x, z) = \frac{1}{(2\pi)^2} \int_{-\infty}^{+\infty} \int_{-\infty}^{+\infty} \Psi(k_x, k_z) e^{i(k_x x + k_z z)} dk_x dk_z, \tag{31}$$

and its mean and variance are

$$E = \langle \psi(x, z) \rangle, \tag{32}$$

$$D = \left\langle (\psi(x, z) - E)^2 \right\rangle, \tag{33}$$

where $\langle \cdots \rangle$ indicates the mean.

5. Based on the above mean and variance, a spatial domain random medium distribution with $\varphi(x, z)$ as the autocorrelation function is generated:

$$\sigma(x, z) = \frac{\varepsilon}{\sqrt{D}} [\psi(x, z) - E]. \tag{34}$$

Its mean is zero, and its variance is ε^2.

2.4. Multiphysics Field Coupling and Workflow

According to the workflow in Figure 3, we can predict the velocities of elastic waves based on the modified Gurevich model. Figure 2 demonstrates that the analytical model needs to be validated through simulations and should also be able to describe key features when applied to real rocks [21]. In the ultrasonic pulse transmission method, two ultrasonic transducers (divided into two types according to the direction of the wave, compressional wave and shear wave) are installed on both sides of the rock sample so that the wave produced by the transmitter is transmitted through the measured material to the receiver. The first pulse waveform of each of the two transducers is recorded, and the travel time and amplitude are calculated [54–59]. The signal transmission and reception of the model are simulated by a set of piezoelectric transducers to simulate ultrasonic experiments more accurately.

Before presenting a detailed description of our model, we first list the key assumptions used in our model. (1) The pore network is simplified as elliptical pores and cracks, representing stiff pores and compliant pores. (2) The solid is isotropic and linearly elastic. (3) The pores are saturated by water. (4) The signal is transmitted and received by a piezoelectric transducer.

A two-dimensional geometrical model of width $W_c = 25$ mm and height $L_c = 50$ mm representing a sample of porous rock on the core scale is built. The porous material includes regular elliptical pores with a variety of major axis lengths and aspect ratios (α, the ratio of the minor axis to the major axis). The porosity $\phi_n(P)$ and aspect ratio $\alpha_n(P)$ of the microfractures are discussed in the next section. When the effective pressure $P = 0$, the porosity of the micropores $\phi_c(P) = \phi_{c0}$. When the effective pressure is P, the microcracks are gradually closed, the porosity of the stiff pores is unchanged, and the porosity of the microcracks is set to $\phi_n(P)$. The stiff pores and compliant pores are saturated by water. The moduli and density are perturbed by the random medium model; these coordinate-dependent data are used as the matrix modulus.

According to Newton's second law, the motion of the solid matrix and the pore fluid is represented as [60]

$$\rho \frac{\partial^2 \mathbf{u}}{\partial t^2} - \nabla \sigma = \mathbf{F}, \tag{35}$$

where ρ is the density of the medium, $\mathbf{u} = (u_x, u_y)$ is the two-dimensional displacement vector, $\sigma = [\sigma_{ij}]$ is the stress tensor, and $\mathbf{F}$ is the stress vector caused by an outer force. For a linear fluid, the stress–strain relation is

$$\begin{cases} \sigma = J^{-1}\mathbf{Z}\mathbf{S}\mathbf{Z}^T \\ \mathbf{Z} = \mathbf{I} + \nabla \mathbf{u}, J = \det(\mathbf{Z}) \\ \mathbf{S} = K_f trace(\boldsymbol{\varepsilon})\mathbf{I} + 2\mu_f\left(\boldsymbol{\varepsilon} - \frac{1}{3}trace(\boldsymbol{\varepsilon})\mathbf{I}\right) \end{cases}, \tag{36}$$

where K_f is the bulk modulus of the pore fluid; μ_f is the dynamic shear modulus of the fluid, $\mathbf{I}$ is the identity matrix, and $\mathbf{S}$ is the stiffness matrix. $trace(\boldsymbol{\varepsilon})$ is the trace of the strain tensor. For a linear elastic solid, the stress–strain relation is

$$\begin{cases} \boldsymbol{\varepsilon} = \frac{1}{2}(\nabla \mathbf{u} + \nabla \mathbf{u}^T) \\ \sigma_{ij} = (K_s - \frac{2}{3}\mu_s)trace(\boldsymbol{\varepsilon})\delta_{ij} + 2\mu_s\varepsilon_{ij} \end{cases}, \tag{37}$$

where $\boldsymbol{\varepsilon} = [\varepsilon_{ij}]$ is the strain tensor; K_s and μ_s are the bulk and shear moduli of the solid, respectively; δ_{ij} is the Kronecker delta.

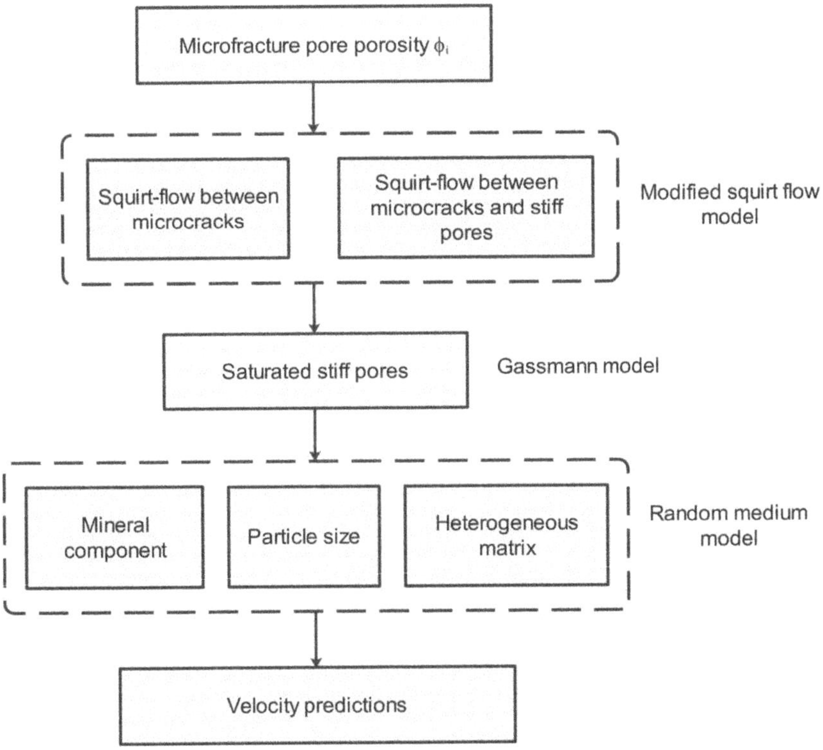

Figure 3. Workflow of the modified model considering multiple groups of microcracks and pores and a heterogeneous matrix.

The entire area is discretized using an unstructured mesh of tetrahedral elements, about totaling 334,737 elements. The smallest element in the thin cracks is 2.88×10^{-4} mm, while the largest element in the solid background matrix is 9.86×10^{-4} mm. The time step is 2.5×10^{-8} s. The stiff pores and fractures in the rock sample are represented by circles and ellipses, and the sizes and quantities of circles and ellipses are adjusted among different samples according to the descriptions in the dataset. The boundary between the solid domain and the fluid domain is naturally coupled, which means that displacement is described using the same variables. The model uses the finite element method to solve, and the direct PARDISO solver [61] is used to solve the linear equations.

The electrical circuit in the piezoelectric module in the transmitter is shown in Figure 4. During modelling, the input voltage signal ($V(t)$) is given by

$$V(t) = 100 \times \exp\left(-\left(\frac{t - T_0}{\frac{T_0}{2}}\right)^2\right) \times \sin(2\pi f_0 t), \tag{38}$$

where f_0 is the frequency in hertz and T_0 is the period in seconds. The P-wave velocity and attenuation are calculated by the voltage oscillogram:

$$v = L/(t_2 - t_1). \tag{39}$$

v is the P-wave velocity; t_1 and t_2 are time points at which the first pulse is received by the transmitter and the receiver, respectively (s); and L is the straight-line distance between the transmitter and the receiver (m).

$$\alpha = (\ln A_0 - \ln A)/L, \tag{40}$$

where A_0 is the first wave amplitude (V) of the incident waveform, A is the first wave amplitude (V) of the transmitted waveform, L is the straight-line distance between the transmitter and the receiver (m), and α is the attenuation, dB/m.

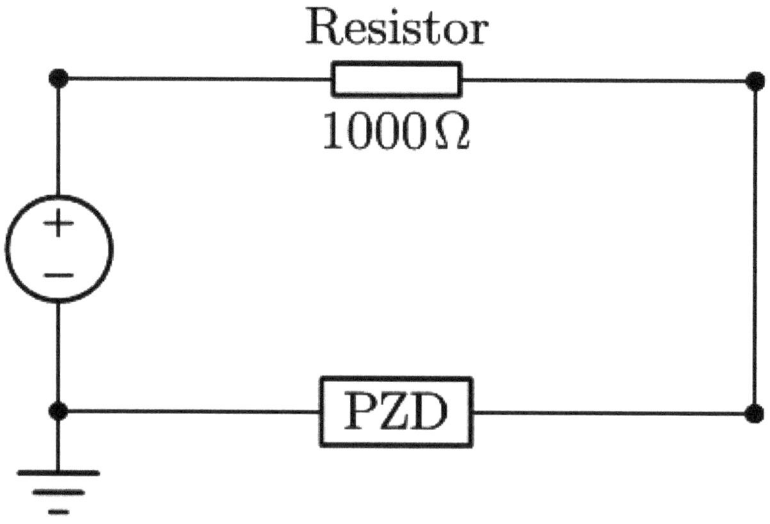

Figure 4. Electrical circuit connected to a piezoelectric device to drive and receive signals.

The electrical signals are converted into stress by the piezoelectric transducer via the following equations:

$$\sigma = c_E \varepsilon - e^T, \tag{41}$$

$$\mathbf{D} = e\varepsilon + \epsilon_S \mathbf{E}, \tag{42}$$

which express the relation between the stress σ, strain ε, electric field $\mathbf{E}$, and electric displacement field $\mathbf{D}$. For the stress-charge form, an elasticity matrix, Voigt notation (c_E), which is Young's modulus in all directions, is selected. e represents the coupling matrix, which is the piezoelectric constant and represents the capacity of electric energy conversion into displacement. ϵ_S is the dielectric matrix, the dielectric constant, represents the size of the capacitance. The transmitter of the piezoelectric transducer is applied to the top surface of the sample, and the receiver of the piezoelectric transducer is on the bottom surface.

3. Results

The experimental data of five carbonate rock samples reported by Long [62] (2020) are considered to illustrate the proposed model. The samples are from the Dengying Formation dolomite in the Sichuan Basin and the Ordovician limestone in the Tarim Basin, Xinjiang. According to thin section and scanning electron microscopy results and physics data for five rock samples obtained by Long [62], intergranular pores, intercrystalline pores, dissolved pores, and dissolved cracks are the main pore types.

As detailed in Figures 5–8, the thin section and scanning electron microscopy results for the core sample reveal different kinds of pores [62].

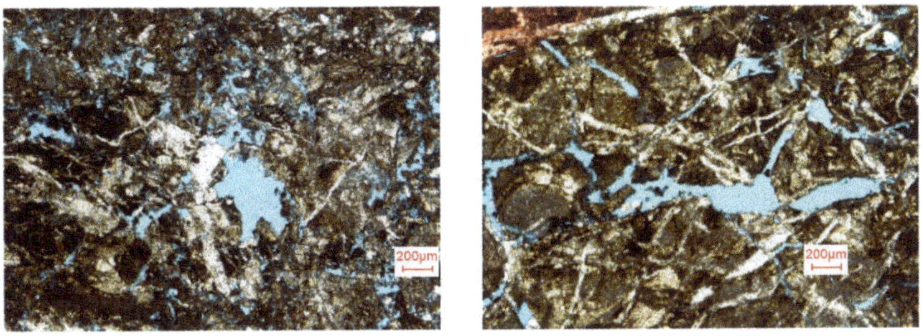

Figure 5. Thin section of limestone sample 24: dissolution pores and dissolution fissures.

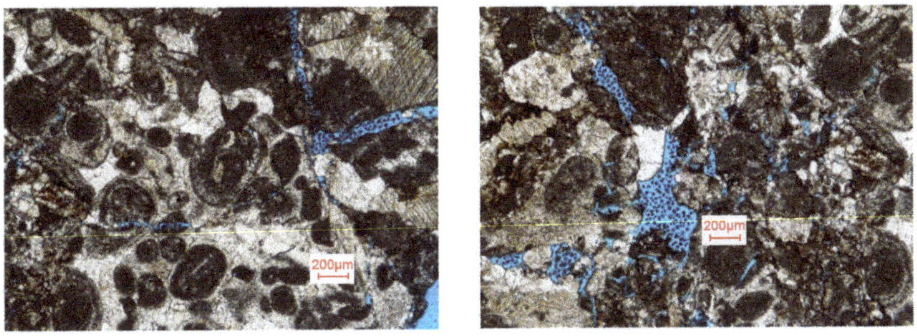

Figure 6. Thin section of limestone sample 25: dissolution fissures and dissolution pores.

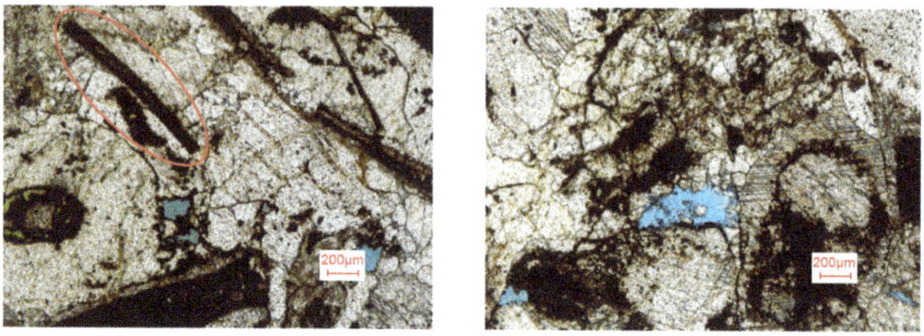

Figure 7. Thin section of limestone sample 26: dissolution fissures and dissolution pores.

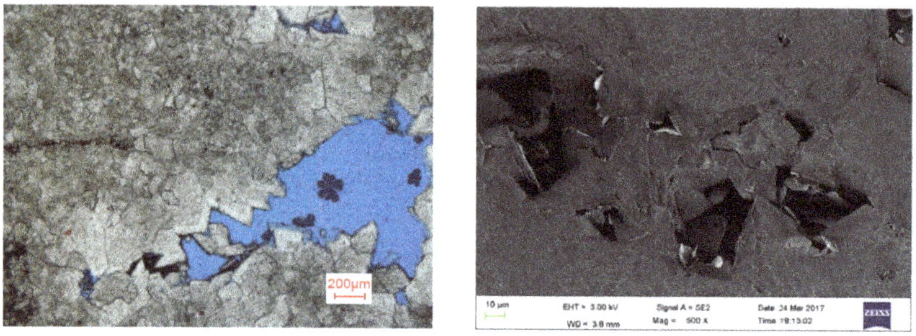

Figure 8. Thin section and scanning electron microscopy image of dolomite sample 132: intercrystalline pores.

Choquette and Pray [63] (1970) classified carbonate pores into 15 basic types, providing detailed geological characteristics of almost all types of porosity in sedimentary carbonates. The flat and thin pores deform easily, so the seismic velocity of rocks containing such holes is low. In contrast, circular pores and holes are difficult to deform, so the seismic velocities are correspondingly high, even when the porosity may be high. Experiments on 52 carbonate rock samples have shown that the stiffness of samples containing large pores is high, while the stiffness of samples containing micropores or cracks is relatively low [64]. The porosity-velocity relationship varies for rocks containing different pore types, such as rocks with micropores, rocks with moldic pores, rocks with interparticle pores, rocks with intraparticle porosity, and densely cemented rocks [65].

In this study, to simulate the easily deformable, closed cracks and the less deformable large pores in the assumptions of the squirt flow model, circular pores are selected to represent stiff pores, and long and narrow elliptical pores are selected to represent compliant microcracks. The sizes, numbers, and number ratios of the two types of pores and the aspect ratios of the elliptical pores of the model were obtained from the description of the cores and the pore morphology of several thin sections in the literature. The model of sample 26 is shown in Figure 9, based on the pore morphology shown in the description and thin sections of the core and the heterogeneity of the matrix. The green circles in Figure 9 represent saturated water pores and microfractures, while the blue stripes in the matrix are tilted 45 degrees to the left, simulating the many heterogeneous black, left-leaning striped bodies in the matrix in Figure 7. The length of the stripes is about five times the size of the circles, reflecting the size ratio of the heterogeneous body to the pores in Figure 7. The specific parameter settings are as follows: The radius of the circular pore is 0.4 mm, the long axis length of the crack is 0.3 mm, the aspect ratio is 0.01, the autocorrelation lengths are a = 2 mm and b = 0.5 mm, the rounding coefficient n = 1, and the angle θ = 135°.

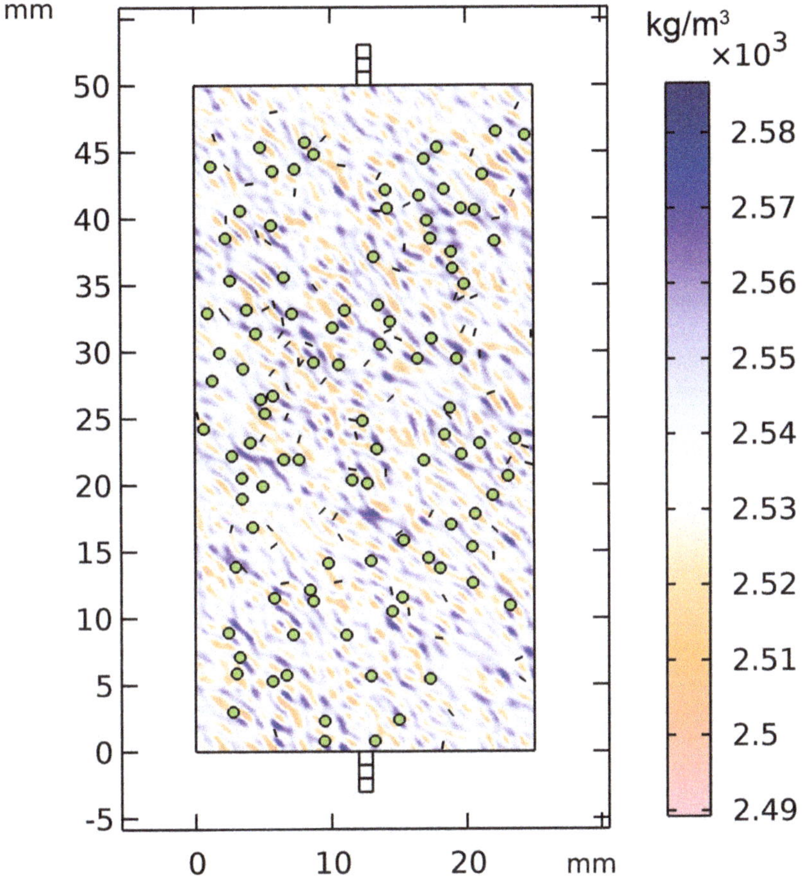

Figure 9. Local density distribution of sample No. 26.

The petrophysical parameters of the cores used in this study were obtained from ultrasonic tests performed on limestone cores and dolomite cores, as shown in Table 1. To verify the accuracy and practicality of the modified model in this study, we analysed and discussed the model parameters using the bulk and shear moduli measured for the dry carbonate core samples and information from thin section characterisation. Finally, the results were compared with the velocities obtained from ultrasonic saturated carbonate rock core experiments and predictions calculated by other models.

Table 1. Rock physics data of samples.

Number	No. 24	No. 25	No. 26	No. 30	No. 132
Lithology	limestone	limestone	limestone	limestone	dolomite
Dry density (g/m^3)	2.53	2.535	2.539	2.674	2.75
Porosity (%)	12.298	5.904	4.092	4.391	2.671
Permeability (mD)	3.293	8.518	0.176	1.641	0.0029
Grain bulk modulus (GPa)	75.47	76.61	76.8	76.8	87.83
Grain shear modulus (GPa)	32.22	32.03	32	32	44.92

3.1. Pore Structure Parameter Analysis

Shapiro [48] proposed the following empirical formula for the elastic modulus and pressure:

$$\frac{1}{K_d(P)} = \left(\frac{1}{K_d(0)} - \frac{1}{K_h}\right)e^{-\frac{P}{\hat{P}}} + \frac{1}{K_h}, \tag{43}$$

$$\frac{1}{G_d(P)} = \left(\frac{1}{G_d(0)} - \frac{1}{G_h}\right)e^{-\frac{P}{\hat{P}}} + \frac{1}{G_h}, \tag{44}$$

where K_d and G_d represent the elastic modulus of the dry core; $K_d(0)$ and $G_d(0)$ are the initial moduli of the dry rock at zero pressure; and K_h and G_h are obtained by fitting the measured results of the dry core and represent the elastic modulus when all the micropores are closed under high pressure (shown by the black arrow), as shown in Figure 10.

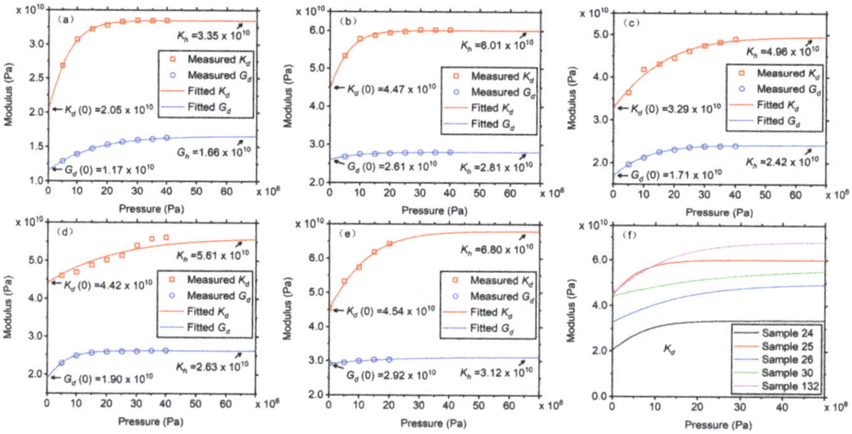

Figure 10. Measured (hollow point) and fitted (curve) bulk moduli K_d and shear moduli G_d of samples (**a**) 24, (**b**) 25, (**c**) 26, (**d**) 30, and (**e**) 132 and (**f**) comprehensive comparison of K_d with respect to pressure.

In samples 24, 25 and 132, the modulus increases greatly when the pressure is below 15 MPa, indicating that there are many microcracks with small aspect ratios and many microcracks with large aspect ratios in these two samples. When the pressure is slightly increased, many microcracks are closed, and the modulus increases rapidly. After reaching 15 MPa, most of the microcracks are closed, and the modulus decreases with increasing pressure. With the closure of a few unclosed microcracks with large aspect ratios, the high-pressure modulus K_h is finally reached. In contrast, for samples 26 and 30, the modulus increases more gradually with pressure, which means that the sample has less variation in the number of microcracks with different aspect ratios. Such cores are generally denser than cores containing microfractures with large differences in aspect ratios. Notably, in Figure 10d, the curve fitting error is much larger, and the modulus before 30 MPa is nearly linear with increasing pressure. Because there are large vertical cracks in sample 30, the pressure increases until the cracks close, and the microcracks are not the main cause of the modulus change. Figure 10f shows a comprehensive comparison of the bulk modulus change at 0 MPa for several samples, where samples 24, 25 and 132 show the same pattern of change, i.e., a sharp increase in the modulus in the low-pressure range, which tends to flatten at high pressures. The rates of change increase in the order 25, 24, and 132. In contrast, the modulus changes of samples 26 and 30 are more moderate.

By using the inversion process of the pore aspect ratio described in the Methods section, we obtained the cumulative microfracture density distribution $\bar{\Gamma}$ of five cores under effective pressures of 0 MPa, 5 MPa, 10 MPa, 15 MPa, 20 MPa, 25 MPa, 30 MPa, 35 MPa, and 400 MPa and performed a comprehensive comparison for all the samples at 0 MPa, as shown in Figure 11.

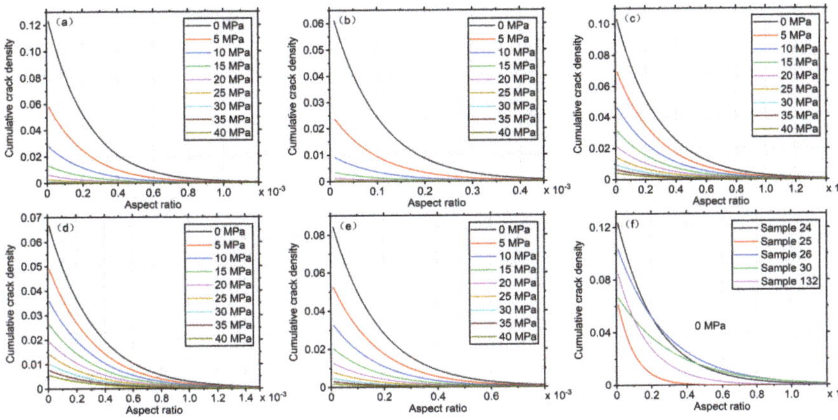

Figure 11. Cumulative crack density distribution as a function of the aspect ratio in samples (**a**) 24, (**b**) 25, (**c**) 26, (**d**) 30, and (**e**) 132 and (**f**) comprehensive comparison at 0 MPa

As shown in Figure 11, the cumulative crack density decreases with increasing pore aspect ratio. α represents the aspect ratio of the smallest microcrack in the sample, and the cumulative crack density $\bar{\tau}$ is the sum of the densities of microcracks with aspect ratios greater than or equal to α. Therefore, the cumulative densities of microcracks $\bar{\tau}$ that are larger than α decrease as the smallest aspect ratio α increases. With increasing pressure, the $\bar{\tau}$ curve moves toward the direction of a smaller aspect ratio. As mentioned previously, under pressure P, the reduction in the aspect ratio of all microcracks is the same, indicating that as microcracks with smaller aspect ratios close, microcracks with larger aspect ratios are also compressed, leading to a decrease in aspect ratio. Under a high effective pressure, the microcracks with initially larger aspect ratios become microcracks with smaller aspect ratios until $P \to P_h$, at which point all the microcracks close. The range of the horizontal axis shows that the aspect ratio ranges of the microcracks in each sample are different, as shown in Figure 11f. Notably, in samples 24, 25 and 132, the cumulative microcrack density varies significantly when $P = 0$ and $P = 5$ (approximately 0.06, 0.04 and 0.03), indicating that with an increase in pressure to 5 MPa, many microcracks close, and the proportion of microcracks with a small aspect ratio is the largest. As the pressure increases to 10 MPa, the decrease in the cumulative microcrack density gradually decreases, indicating that in these two samples, microcracks with smaller aspect ratios account for a larger proportion, and microcracks with smaller aspect ratios have an absolute advantage. Figure 11f shows that for samples 24, 132 and 25, the cumulative crack density decreases rapidly in the small aspect ratio interval with a similar trend, suggesting that the microcracks with small aspect ratios account for a similar percentage of the total density, but the density decreases sequentially due to the difference in the total density of the microcracks. On the other hand, for samples 30 and 132, the curves are flatter, indicating a more even distribution of microcracks that have a similar range of aspect ratios but slightly different densities.

Figure 12 shows the curves of porosity ϕ versus aspect ratio α for the microfractures in the five samples at the different effective pressures. As shown in Figure 12, the distribution interval of the aspect ratio becomes narrower and moves towards a small aspect ratio as the pressure increases. A comparison of the aspect ratio, porosity and range of the dominant microcracks in each sample is shown in Figure 12f. As shown in Figure 12, samples 24,

132, and 25 have a rapid decrease in porosity at low pressures, and at high pressures, the microporosity is almost completely closed. Figure 12f shows that not only does the porosity of these three samples decrease gradually but also the range of the aspect ratio distribution becomes narrower and the centre moves in the direction of a small aspect ratio. The decrease in porosity of samples 26 and 30 is more uniform with decreasing pressure, and the microcracks are not completely closed at 40 MPa. Figure 12f shows that the distributions of the aspect ratios of these two samples are almost the same, and the porosity in the centre decreases.

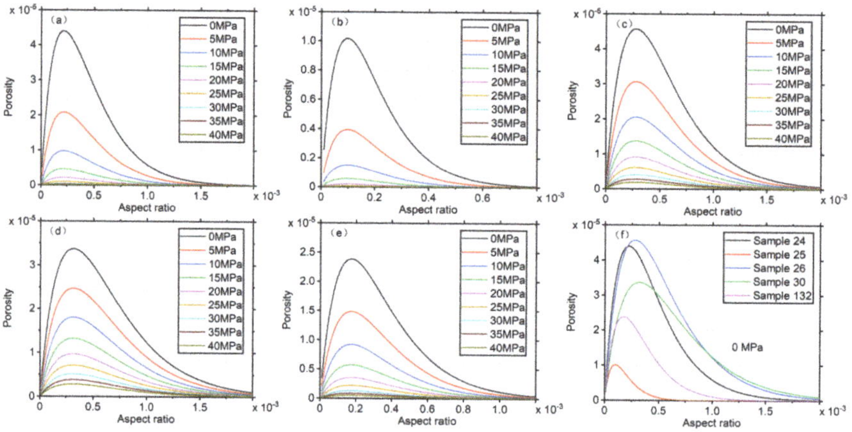

Figure 12. Porosity as a function of aspect ratio for samples (**a**) 24, (**b**) 25, (**c**) 26, (**d**) 30, and (**e**) 132 and (**f**) comprehensive comparison at 0 MPa.

3.2. Matrix Heterogeneity Parameter Analysis

Xi and Yao [66] (2002) presented a complete discussion of stratigraphic-scale random medium models, and Rao et al. [67] (2020) presented a pore-scale random medium model for heterogeneity simulation. In this study, the random medium model is used on the core scale to simulate matrix heterogeneity caused by structures such as the strips in Figure 6 and the ooze in Figure 7.

Equation (29) shows the expression for autocorrelation, where a and b (mm) are the autocorrelation lengths, n = 1 is the Gaussian autocorrelation function, and n = 0.5 is the exponential autocorrelation function, θ is particle alignment angle. The value of ε is 0.01. Figure 13 shows the different characteristics of the stochastic perturbation model produced by two different autocorrelation functions with different autocorrelation lengths a and b. The nonuniform anomalies presented in the correlation lengths and the type of autocorrelation function are shown in the Figure 13 show that the laminar nonuniformity is more pronounced when a = 20 mm and b = 5 mm, i.e., when the autocorrelation length is larger, the anomalous body is smaller, and the laminar nonuniformity is less pronounced when a = 10 mm and b = 2.5 mm. When n = 1, the autocorrelation function is Gaussian, the anomalies are larger, and the anomalous bodies are larger. The angle of the particles can be set to a specific angle, such as 135 degrees in Figure 9. If the nonuniform bodies are more concentrated, they can be added separately in layers.

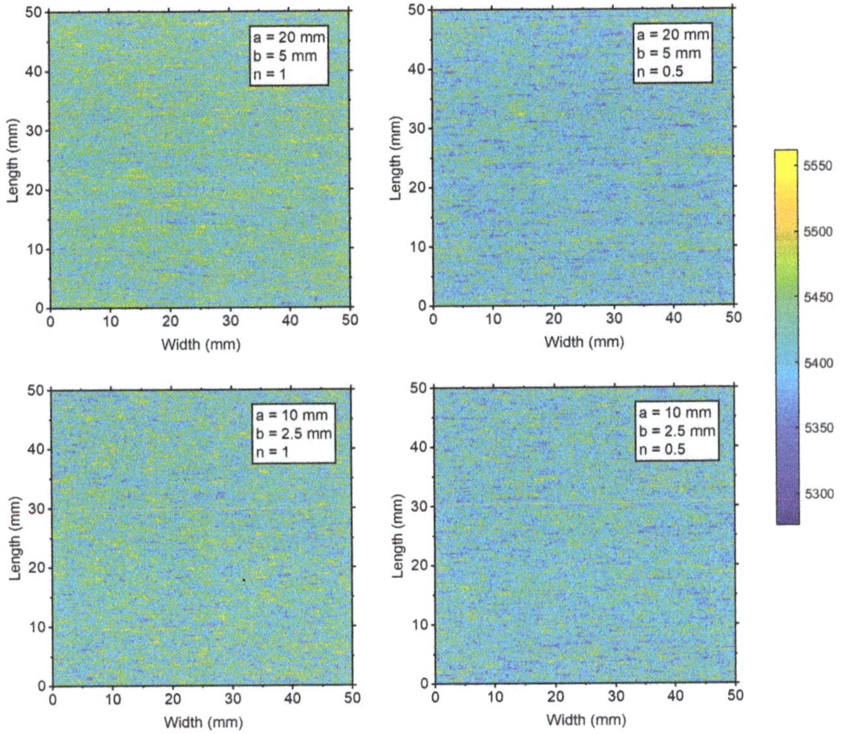

Figure 13. Core-scale random medium model: a and b are the autocorrelation lengths in the x and z directions, and n is the rounding coefficient.

3.3. Modified Squirt Model Application

In the previous section, the porosity distribution characteristics of five carbonate cores were obtained via inversion (Figure 12), and the P-wave and S-wave velocities were predicted via the obtained ultrasonic frequency squirt flow model considering multiple sets of microfractures with different aspect ratios. For comparison, Figure 14 shows measurements, predictions from the new model, calculations from the Biot model using the modulus of dry cores, and numerical simulation data from the new model. The simulation data suggest that the predictions of the new model, which combine the squirt flow model considering multiple microcracks and the mixed random medium model, agree well with the experimental data, and the accuracy is improved compared to the predicted values of the biot model.

In Figure 14, Modified-G is the "wet skeleton" modulus obtained from the dry core moduli using the modified squirt flow model and the prediction at saturation using the Gassmann model. The Biot model is the saturated prediction obtained by applying the dry core modulus to the Biot model. The simulations are the result of numerical simulation by adding the Modified-G predictions to the random medium model. The measurements are the saturated velocities measured in the ultrasonic experiments. As shown in Figure 14, the saturation velocities predicted by the modified squirt flow model at high frequencies agree well with the ultrasonic experimental measurements, whereas the predictions made by traditional Biot theory are grossly underestimated, especially for sample 26. The fit is slightly worse for dolomite sample 132, but the error is also much less than 5%, as shown in Figure 14. The difference lies first in the fact that all the limestones in this study contain solution pores and solution fractures, while limestone 132 is characterized by intergranular pores. The differences in pore size between large pores and small pores are also more significant, as are the differences in lithology. However, the distribution of the aspect ratio of microfractures obtained through inversion did not consider these differences. By controlling the size of the stiff pores and achieving a more uniform distribution of the matrix in the simulations, the simulated results became closer to the measured results. In Figure 14c,d, the curves plotted for both the calculated results of the new model and the measured data of the experiment are flatter, and the trends are more compatible. For the other three samples, the experimental data show a greater rate of change in the low-pressure section than that of the new model. Combined with the analyses of Figures 10–12, this suggests that the new model is more applicable to rocks with a uniform distribution of microporosity aspect ratios and a smaller and more uniformly distributed stiff pore radius.

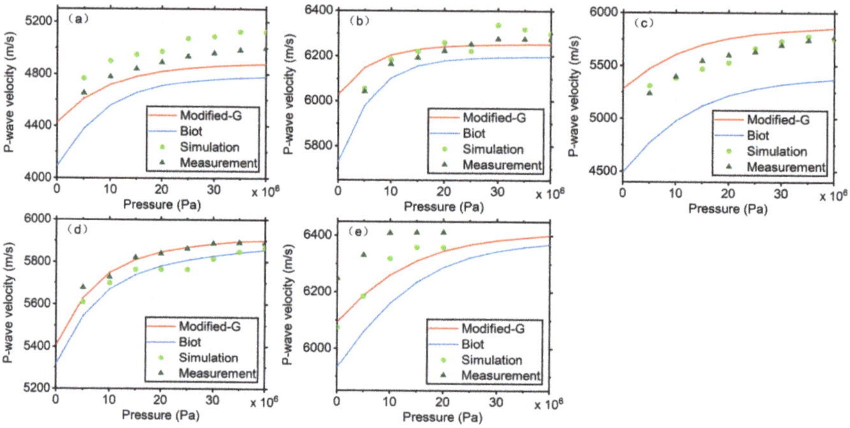

Figure 14. Ultrasonic velocity measurements and model predictions for water-saturated samples (**a**) 24, (**b**) 25, (**c**) 26, (**d**) 30 and (**e**) 132.

The errors of the predictions and saturated measurements for a variety of models are given in Figure 15. The dry and the saturated predictions of the modified squirt flow model are indicated by the blue and the red dots, and the predictions of the Gassmann and Biot models and the simulation results of the model considering the random medium are shown in green, purple, and orange, respectively, with different shapes representing different samples. In Figure 15a, the errors of the P-wave predictions obtained from all models are within 10%, and only the predictions of the modified squirt flow model in the dry case and the Biot model have errors of more than 5%. The predictions of the Modified-G model for the saturated cores (in red) are almost always close to both sides of y = x, with small errors. The new model, which combines the squirt flow model and random medium models, has smaller errors in the simulations (orange), which are closer to y = x. In Figure 15b, the S-wave predictions of the Modified-G model are the same as those of the Modified squirt flow model and similar to those of the Gassmann model; thus, the red colour is covered, and almost only the green and purple colours can be seen. The vast majority of the S-wave predictions are within 10%, which is a larger error than that of the P-wave, indicating that all of these model predictions are more accurate for the P-wave than they are for the S-wave.

The above study shows that the accuracy of the new model combines the modified squirt flow model and the mixed random medium model is greater for carbonates with a larger range of microfracture aspect ratios, but for carbonate sample 132 with a smaller range of microfracture aspect ratios, the effect of the size and morphology of the stiff pores on the experimental results gradually increases. Especially in dense carbonates, where the aspect ratio of microfractures is smaller and the range of variation is smaller, the morphology and size distribution of stiff pores in the samples have a more significant effect on the ultrasonic experimental results. Thus, future work will use 3D digital core techniques to construct the structure of stiff pores (at present, digital core technology is limited by resolution and is suitable for the extraction of large stiff pores), which, together with the squirt flow effect modelling of microfractures, will lead to more accurate and intuitive predictions of the ultrasonic elastic modulus.

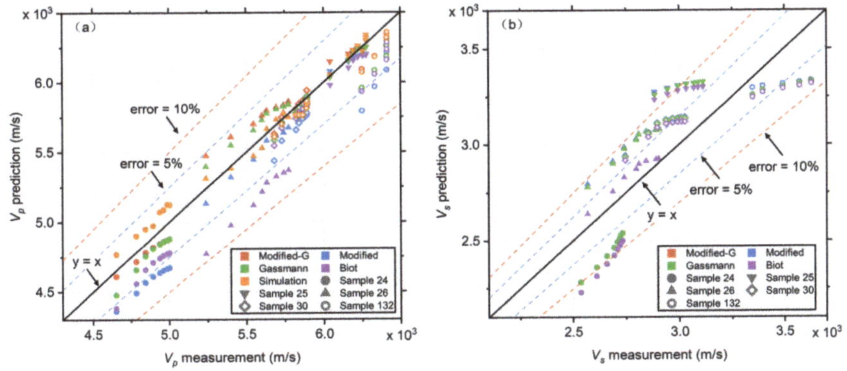

Figure 15. Error of the model calculation results and numerical simulation results for samples (**a**) V_P, (**b**) V_S.

4. Discussion

In core experiments, repeated pressurization and displacement can damage the core, so the experiments that can be conducted on each core are limited. Additionally, using multiple cores makes it impossible to discuss the impact of a single parameter on the results. The superiority of simulations lies in their repeatability and the ability to analyze any parameter individually. The key to the accuracy of the simulation is whether the model can capture the main characteristics of the core. The pore structure within rock cores is generally considered the primary factor affecting the attenuation and dispersion of waves. In addition, the description of the rock matrix is also one of the important factors influencing wave velocity. This paper proposes a comprehensive model that can flexibly adapt to each different rock core, making core analysis more accurate, comprehensive, and economical. This model can be applied to different types of rock and various types of fluid contained in the pores. Different types of rocks are simulated by considering the distribution of heterogeneous bodies and the distribution of pore structures obtained by inversion of pressure experiments. Depending on the simulation requirements, different fluids can consider different viscous forces, inertial forces, and even body forces.

In the theoretical part of the new model in the study, in addition to the basic parameters of the core, the parameters that control the characteristics of the core include the autocorrelation lengths a and b, the rounding coefficient n, the angle θ in the random medium model, and the compliant porosity ϕ_c in the multiple fracture squirt flow model. The above parameters set in the simulation are obtained through empirical estimation. The advantage of this method is that it can quickly obtain the parameters used for simulation based on the specific conditions of the core, and it has good accuracy. However, it has the drawbacks of relying on experience and lacking programmability. To address this drawback, the subsequent work involves using artificial intelligence to obtain the aforementioned parameters from the core samples and applying them in the simulation, which will further enhance the accuracy and standardization of the simulation.

5. Conclusions

In this study, the distribution of the pore aspect ratio of the microcracks was obtained based on the inversion of pressurized laboratory data according to the volume compression coefficient for each group of microcracks with different aspect ratios. The squirt flow effect between microfractures caused by their sequential closure with different aspect ratios is introduced into the Gurevich squirt flow model, and the modified Gurevich squirt flow model considers not only the local flow between compliant and stiff pores but also the local flow between microfractures with different aspect ratios. Meanwhile, the mixed random medium model was used to describe the heterogeneity of the matrix. Ultrasonic experimental data for four limestone core samples and one dolomite core sample were used for analysis, and the cumulative fracture density distribution and the microfracture porosity distribution extracted from the pressurized experimental data could well describe the microfracture ground pressure response law of the actual core. The simulation results of the new model are in good agreement with the ultrasonic experimental data, with a prediction error of less than 5% for the P-wave. The new model is more suitable for rocks with uniform micropore aspect ratio distribution, smaller pore radius, and more uniform distribution, such as samples 26 and 30, with an error of less than 2%.

Author Contributions: Conceptualization, Y.C.; Formal analysis, Y.C.; Investigation, Y.C.; Methodology, Y.C.; Project administration, P.D.; Software, Y.C.; Supervision, P.D.; Validation, Y.C.; Visualization, Y.C.; Writing—original draft, Y.C.; Writing—review & editing, Y.C. All authors have read and agreed to the published version of the manuscript.

Funding: This research was funded by the National Natural Science Foundation of China (Grant No. 42141009).

Data Availability Statement: The data and code that support the findings of this study are available in GitHub accessed on 4 December 2024 at https://github.com/YvetChen/Modelling-ultrasonic-propagation.git. Any additional information required to reanalyzed the data reported in this paper is available from the lead contact upon request.

Conflicts of Interest: The authors declare no conflicts of interest.

References

1. Vanorio, T.; Scotellaro, C.; Mavko, G. The effect of chemical and physical processes on the acoustic properties of carbonate rocks. *Lead. Edge* **2008**, *27*, 1040–1048. [CrossRef]
2. Nurmi, R.; Standen, E. Carbonates, the inside story. *Middle East Well Eval. Rev.* **1997**, *18*, 28–41.
3. Thompson, L.; Oftebro, C. Salt gets in your eyes: The geological challenges and solutions to sub-salt exploration. *First Break* **2011**, *29*. [CrossRef]
4. Wright, V.P. *Lacustrine Carbonates in Rift Settings: The Interaction of Volcanic and Microbial Processes on Carbonate Deposition*; Geological Society, London, Special Publications: London, UK, 2012; Volume 370, pp. 39–47. [CrossRef]
5. Teng, T.; Pan, J.; Zhang, H.; Liu, J. Prediction technique of carbonate reservoir in Tazhong area. *Lithol. Reserv.* **2010**, *22*, 14–19. [CrossRef]
6. Eshelby, J.D. The determination of the elastic field of an ellipsoidal inclusion, and related problems. *Proc. R. Soc. Lond. Ser. A Math. Phys. Sci.* **1957**, *241*, 376–396. [CrossRef]
7. Dunn, M.L.; Ledbetter, H. Poisson's ratio of porous and microcracked solids: Theory and application to oxide superconductors. *J. Mater. Res.* **1995**, *10*, 2715–2722. [CrossRef]
8. Wu, T.T. The effect of inclusion shape on the elastic moduli of a two-phase material. *Int. J. Solids Struct.* **1966**, *2*, 1–8. [CrossRef]
9. Kuster, G.T.; Toksöz, M.N. Velocity and attenuation of seismic waves in two-phase media: Part I. Theoretical formulations. *Geophysics* **1974**, *39*, 587–606. [CrossRef]
10. Kuster, G.T.; Toksoz, M.N. Velocity and attenuation of seismic waves in two-phase media; Part II, Experimental results. *Geophysics* **1974**, *39*, 607–618. [CrossRef]
11. Yan, J.; Li, X.; Enru, L. Effects of pore aspect ratios on velocity prediction from well-log data. *Geophys. Prospect.* **2002**, *50*, 289–300. [CrossRef]
12. Sun, S.; Stretch, S.; Brown, R. Comparison of borehole velocity-prediction models and estimation of fluid saturation effects: From rock physics to exploration problem. *J. Can. Pet. Technol.* **2004**, *43*, PETSOC-04-10-01. [CrossRef]
13. Cheng, C.H.; Toksöz, M.N. Inversion of seismic velocities for the pore aspect ratio spectrum of a rock. *J. Geophys. Res. Solid Earth* **1979**, *84*, 7533–7543. [CrossRef]
14. Berryman, J.G. Mixture theories for rock properties. In *Rock Physics and Phase Relations: A Handbook of Physical Constants*; American Geophysical Union: Washington, DC, USA, 1995; Volume 3, pp. 205–228. [CrossRef]
15. Mori, T.; Tanaka, K. Average stress in matrix and average elastic energy of materials with misfitting inclusions. *Acta Metall.* **1973**, *21*, 571–574. [CrossRef]
16. Norris, A.N. A differential scheme for the effective moduli of composites. *Mech. Mater.* **1985**, *4*, 1–16. [CrossRef]

17. Mavko, G.; Jizba, D. Estimating grain-scale fluid effects on velocity dispersion in rocks. *Geophysics* **1991**, *56*, 1940–1949. [CrossRef]
18. Dvorkin, J.; Nolen-Hoeksema, R.C.; Nur, A. The squirt-flow mechanism; macroscopic description. *Geophysics* **1994**, *59*, 428–438. [CrossRef]
19. Dvorkin, J.; Mavko, G.; Nur, A. Squirt flow in fully saturated rocks. *Geophysics* **1995**, *60*, 97–107. [CrossRef]
20. Mavko, G.; Nur, A. Melt squirt in the asthenosphere. *J. Geophys. Res.* **1975**, *80*, 1444–1448. [CrossRef]
21. Alkhimenkov, Y.; Quintal, B. An accurate analytical model for squirt flow in anisotropic porous rocks—Part 1: Classical geometry. *Geophysics* **2022**, *87*, MR85–MR103. [CrossRef]
22. Yang, Z.Q.; He, T.; Zou, C.C. A dynamic elastic model for squirt-flow effect and its application on fluid-viscosity-associated velocity dispersion in reservoir sandstones. *Geophysics* **2020**, *85*, MR201–MR212. [CrossRef]
23. Cheng, W.; Ba, J.; Fu, L.Y.; Lebedev, M. Wave-velocity dispersion and rock microstructure. *J. Pet. Sci. Eng.* **2019**, *183*, 106466. [CrossRef]
24. Han, X.; Wang, S.; Tang, G.; Dong, C.; He, Y.; Liu, T.; Zhao, L.; Sun, C. Coupled effects of pressure and frequency on velocities of tight sandstones saturated with fluids: Measurements and rock physics modelling. *Geophys. J. Int.* **2021**, *226*, 1308–1321. [CrossRef]
25. Chen, Y.; Dong, P. Modeling Squirt Dispersion and Attenuation of Ultrasonic in Fluid-Saturated Rocks at the Core Scale. In Proceedings of the International Geomechanics Symposium, ARMA-IGS-2023-0314, Al Khobar, Saudi Arabia, 30 October–2 November 2023; Volume All Days. [CrossRef]
26. de Paula, O.B.; Pervukhina, M.; Makarynska, D.; Gurevich, B. Modeling squirt dispersion and attenuation in fluid-saturated rocks using pressure dependency of dry ultrasonic velocities. *Geophysics* **2012**, *77*, WA157–WA168. [CrossRef]
27. Mavko, G.; Mukerji, T.; Dvorkin, J. *The Rock Physics Handbook*; Cambridge University Press: Cambridge, UK, 2020. [CrossRef]
28. Alkhimenkov, Y. Simulating squirt flow in realistic rock models using Graphical Processing Units (GPUs). *Geophys. J. Int.* **2024**, 238, 272–289. [CrossRef]
29. Alkhimenkov, Y.; Quintal, B. An accurate analytical model for squirt flow in anisotropic porous rocks Part 2: Complex geometry. *Geophysics* **2022**, *87*, MR291–MR302. [CrossRef]
30. Räss, L.; Utkin, I.; Duretz, T.; Omlin, S.; Podladchikov, Y.Y. Assessing the robustness and scalability of the accelerated pseudo-transient method. *Geosci. Model Dev.* **2022**, *15*, 5757–5786. [CrossRef]
31. Räss, L.; Licul, A.; Herman, F.; Podladchikov, Y.Y.; Suckale, J. Modelling thermomechanical ice deformation using an implicit pseudo-transient method (FastICE v1. 0) based on graphical processing units (GPUs). *Geosci. Model Dev.* **2020**, *13*, 955–976. [CrossRef]
32. Räss, L.; Duretz, T.; Podladchikov, Y. Resolving hydromechanical coupling in two and three dimensions: Spontaneous channelling of porous fluids owing to decompaction weakening. *Geophys. J. Int.* **2019**, *218*, 1591–1616. [CrossRef]
33. Deng, J.; Zhou, H.; Wang, H.; Zhao, J.; Wang, S. The influence of pore structure in reservoir sandstone on dispersion properties of elastic waves. *Chin. J. Geophys.* **2015**, *58*, 3389–3400. [CrossRef]
34. Duan, C.; Deng, J.; Li, Y.; Lu, Y.; Tang, Z.; Wang, X. Effect of pore structure on the dispersion and attenuation of fluid-saturated tight sandstone. *J. Geophys. Eng.* **2018**, *15*, 449–460. [CrossRef]
35. Todoeschuck, J.; Jensen, O.; Labonte, S. Gaussian scaling noise model of seismic reflection sequences: Evidence from well logs. *Geophysics* **1990**, *55*, 480–484. [CrossRef]
36. Müller, T.M.; Shapiro, S.A. Green's function construction for 2D and 3D elastic random media. In *SEG Technical Program Expanded Abstracts 1999*; Society of Exploration Geophysicists: Houston, TX, USA, 1999; pp. 1797–1800. [CrossRef]
37. Roth, M.; Korn, M. Single scattering theory versus numerical modelling in 2-D random media. *Geophys. J. Int.* **1993**, *112*, 124–140. [CrossRef]
38. Biswal, B.; Øren, P.E.; Held, R.; Bakke, S.; Hilfer, R. Stochastic multiscale model for carbonate rocks. *Phys. Rev. E* **2007**, *75*, 061303. [CrossRef] [PubMed]
39. Talukdar, M.; Torsaeter, O.; Ioannidis, M.; Howard, J. Stochastic reconstruction, 3D characterization and network modeling of chalk. *J. Pet. Sci. Eng.* **2002**, *35*, 1–21. [CrossRef]
40. Tahmasebi, P.; Hezarkhani, A.; Sahimi, M. Multiple-point geostatistical modeling based on the cross-correlation functions. *Comput. Geosci.* **2012**, *16*, 779–797. [CrossRef]
41. Khalili, A.D.; Arns, J.Y.; Hussain, F.; Cinar, Y.; Pinczewski, W.V.; Arns, C.H. Permeability Upscaling for Carbonates From the Pore Scale by Use of Multiscale X-Ray-CT Images. *SPE Reserv. Eval. Eng.* **2013**, *16*, 353–368. [CrossRef]
42. Gerke, K.M.; Karsanina, M.V.; Mallants, D. Universal stochastic multiscale image fusion: An example application for shale rock. *Sci. Rep.* **2015**, *5*, 15880. [CrossRef]
43. Ji, L.; Lin, M.; Cao, G.; Jiang, W. A core-scale reconstructing method for shale. *Sci. Rep.* **2019**, *9*, 4364. [CrossRef]
44. David, E.C.; Zimmerman, R.W. Pore structure model for elastic wave velocities in fluid-saturated sandstones. *J. Geophys. Res. Solid Earth* **2012**, *117*. [CrossRef]
45. Gurevich, B.; Makarynska, D.; de Paula, O.B.; Pervukhina, M. A simple model for squirt-flow dispersion and attenuation in fluid-saturated granular rocks. *Geophysics* **2010**, *75*, N109–N120. [CrossRef]
46. David, E.; Zimmerman, R.W. Elastic moduli of solids containing spheroidal pores. *Int. J. Eng. Sci.* **2011**, *49*, 544–560. [CrossRef]
47. Zimmerman, R.W. *Compressibility of Sandstones*; Elsevier: Amsterdam, The Netherlands, 1990.
48. Shapiro, S.A. Elastic piezosensitivity of porous and fractured rocks. *Geophysics* **2003**, *68*, 482–486. [CrossRef]

49. Ouyang, F.; Zhao, J.; Li, Z.; Xiao, Z.; He, Y.; Deng, J.; Zhao, H.; Ren, J. Inversion of pore aspect ration distribution based on effective medium theorie. *Chin. J. Geophys.* **2021**, *64*, 1016–1033. [CrossRef]
50. Gassmann, F. Uber die elastizitat poroser medien. *Vierteljahrsschr. Naturforschenden Ges. Zur.* **1951**, *96*, 1–23.
51. Birch, F. The velocity of compressional waves in rocks to 10 kilobars, part 2. *Elastic Prop. Equ. State* **1988**, *26*, 91–116. [CrossRef]
52. Ikelle, L.T.; Yung, S.K.; Daube, F. 2-D random media with ellipsoidal autocorrelation functions. *Geophysics* **1993**, *58*, 1359–1372. [CrossRef]
53. Wu, H.; Fu, L.; Lan, X. Analysis of reservoir heterogeneity based on random media models. *Prog. Geophys.* **2008**, *23*, 793–799.
54. Birch, F. The velocity of compressional waves in rocks to 10 kilobars: 1. *J. Geophys. Res.* **1960**, *65*, 1083–1102. [CrossRef]
55. Toksöz, M.; Johnston, D.H.; Timur, A. Attenuation of seismic waves in dry and saturated rocks: I. Laboratory measurements. *Geophysics* **1979**, *44*, 681–690. [CrossRef]
56. Best, A.; McCann, C.; Sothcott, J. The relationships between the velocities, attenuations and petrophysical properties of reservoir sedimentary rocks 1. *Geophys. Prospect.* **1994**, *42*, 151–178. [CrossRef]
57. Winkler, K.W.; Nur, A. Seismic attenuation: Effects of pore fluids and frictional-sliding. *Geophysics* **1982**, *47*, 1–15. [CrossRef]
58. King, M.S.; Marsden, J.R. Velocity dispersion between ultrasonic and seismic frequencies in brine-saturated reservoir sandstones. *Geophysics* **2002**, *67*, 254–258. [CrossRef]
59. Krishnamurthi, M.; Balakrishna, S. Attenuation of sound in rocks. *Geophysics* **1957**, *22*, 268–274. [CrossRef]
60. Reddy, J.N. *An Introduction to Continuum Mechanics*; Cambridge University Press: Cambridge, UK, 2013. [CrossRef]
61. Schenk, O.; Gärtner, K. Solving unsymmetric sparse systems of linear equations with PARDISO. *Future Gener. Comput. Syst.* **2004**, *20*, 475–487. [CrossRef]
62. Long, T. Rock Physics Modeling and Broad-Band Experimental Measurement of Carbonates. Ph.D. Thesis, China University of Petroleum-Beijing, Beijing, China, 2019. [CrossRef]
63. Choquette, P.W.; Pray, L.C. Geologic nomenclature and classification of porosity in sedimentary carbonates. *AAPG Bull.* **1970**, *54*, 207–250. [CrossRef]
64. Han, D. Velocity in carbonate rocks. In *Annual Report, Rock Physics and Fluid Consortium*; 2004.
65. Eberli, G.P.; Baechle, G.T.; Anselmetti, F.S.; Incze, M.L. Factors controlling elastic properties in carbonate sediments and rocks. *Lead. Edge* **2003**, *22*, 654–660. [CrossRef]
66. Xi, X.; Yao, Y. Simulations of Random Medium Model and Intermixed Random Medium. *Earth Sci.* **2002**, *27*, 67–71.
67. Rao, Y.; Fu, L.; Wu, Y.; Boateng, C.D. Heterogeneous characteristic analysis of shale based on multi-component and multi-scale rendom media method. *Chin. J. Geophys.* **2020**, *63*, 2800–2809. (In Chinese) [CrossRef]

Disclaimer/Publisher's Note: The statements, opinions and data contained in all publications are solely those of the individual author(s) and contributor(s) and not of MDPI and/or the editor(s). MDPI and/or the editor(s) disclaim responsibility for any injury to people or property resulting from any ideas, methods, instructions or products referred to in the content.

applied sciences

MDPI

Article

Spatially Constrained 1D Inversions of Common-Midpoint Marine Controlled-Source Electromagnetic Data to Create a 3D Electrical Model †

Jorlivan Lopes Correa [1,*] and Cícero Régis [2,3,*]

1 Petróleo Brasileiro S/A-Petrobras-EXP/TPGG/TG, Rio de Janeiro 20231-030, Brazil
2 Institute of Geosciences, Federal University of Pará–UFPA, Belém 66075-110, Brazil
3 National Institute of Science and Technology of Petroleum Geophysics–INCT-GP, Salvador 40110-909, Brazil
* Correspondence: jorlivan@petrobras.com.br (J.L.C.); cicero@ufpa.br (C.R.)
† This paper is an extended version of SEG 87th Annual Meeting, Houston, TX, USA, 24–29 September 2017.

Abstract: The proposed 3D stitching of spatially constrained Common Midpoint (CMP) 1D inversions is a method that integrates 1D laterally constrained inversion of marine Controlled-Source Electromagnetic (CSEM) data in the CMP domain to generate a cube of resistivities from 3D surveys. An interpretive model is built in the form of a set of columns of homogeneous cells that form a 3D resistivity grid. The proposed methodology is an iterative process that updates the entire model until it minimizes the data misfit between the real and synthetic data. We connect the model cells by smoothing regularization applied in all three directions, which generates stable solutions. Additionally, we evaluate the inversion result constrained by hard information, for example, well log resistivity data. Two applications to synthetic data and the inversion of a real data set illustrate the method. The synthetic data were generated from 3D models, one with two resistive targets at different depths and a second with a target inside a conductive layer over a resistive basement. The real data were gathered off the southeast coast of Brazil, in an area of gas hydrate accumulation. The results indicate that the method can generate useful approximations to the resistivity structures under the survey area in a much shorter time than that of a full 3D inversion.

Keywords: Marine CSEM; CMP; inversion

1. Introduction

The Marine Controlled Source Electromagnetic (MCSEM) method's rise in oil exploration since the 2000's brought forth an efficient tool for mapping resistive targets with good lateral definition [1–3].

Inversion of MCSEM data to resolve 3D structures is a highly demanding task in terms of computer resources. Our solution proposes a first approximation that may be generated by a fast program from layered earth models. Such a solution may be useful for allowing a quick first interpretation and also for the possibility of contributing a good bootstrap model to a true 3D inversion.

A common practice is to start the 3D inversion process with a homogeneous model and let the algorithm converge to a desired solution in a number of iterations. However, computer time can be saved by choosing a suitable initial model for the inversion. If a reasonable approximation to the geoelectrical structures under the survey lines can be built prior to the 3D inversion, then convergence may be achieved with fewer iterations.

Ref. [4] presented a 1D inversion scheme using the Common Midpoint (CMP) gathers approximation, which was able to recover 2D structures. In that method, the data for the 1D inversions are composed of measurements from all source-receiver offsets in the 2D survey that share the same common midpoint, and a separate 1D inversion is performed for each CMP. This method was improved by [5] with the introduction of analytical derivatives to calculate sensitivities, which made the process significantly faster, and lateral constraints [6–8] to allow the inversion of a CMP data set to influence the results of the surrounding CMPs, which improved the achieved approximations, as compared to the true 2D structures.

The same approach, with analytical derivatives and lateral constraints, is expanded here to invert data from 3D surveys. In a 3D survey, data on a single CMP are generated by source-receiver pairs from multiple survey lines and directions, incorporating a high amount of lateral information from broadside data into each 1D inversion.

From the data and analytical derivatives computed from each 1D CMP modeling, we formulate a least-squares problem that incorporates all columns of resistivities, which are used to build a 3D "cube". The columns are connected by the regularization, representing the volume underneath the survey area. This process allows the entire data set to be inverted in a single program execution.

The results show that our method, which integrates multiple electric field components and azimuthal information from broadside data, produces a final model of good accuracy. The solution is computed through an optimization process that updates an initial model to minimize the misfit between the integrated dataset and the synthetic data from the layered models. The final resistivity distribution achieves a sufficiently good approximation, enabling a preliminary 3D interpretation and the generation of an approximate 3D model for use as a first guess in a true 3D inversion.

2. Method

3D marine controlled-source electromagnetic (MCSEM) surveys collect data using a setup in which a dipole source is towed while a receiver mesh is spread on the seafloor. The data can be categorized as inline and broadside, each with three electric and three magnetic field components at various frequencies. The inline data are recorded when the source line is active and is towed along the receiver line, while the surrounding receivers record the lateral signal, known as broadside data. This coordinated effort helps map the lateral variation and anisotropy effects. The resulting setup (Figure 1) allows multiple source-receiver pairs to contribute to each CMP gather, which are then combined to formulate the mathematical problem and solve for a model that accounts for the lateral connection between the CMP columns.

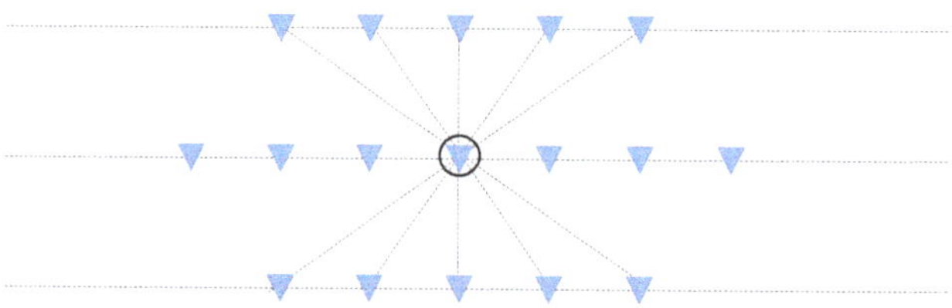

(**a**) CMP associated with a receiver position.

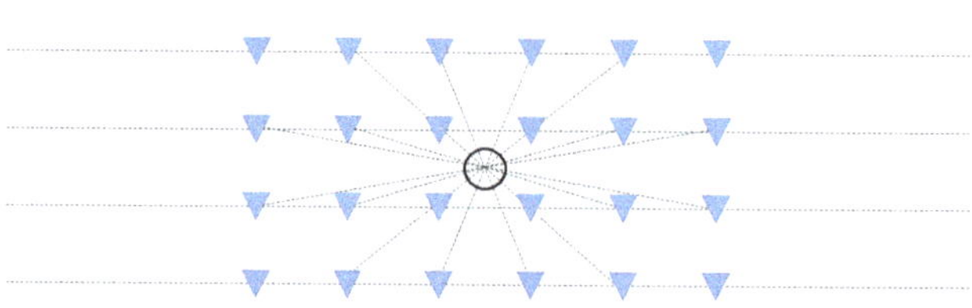

(**b**) CMP not associated with a receiver position.

Figure 1. Acquisition configurations. Triangles represent receivers. The transmitters are assumed to be over the appropriate receiver positions in each source-receiver pair.

The CMP inversion allows for building resistivity columns in points not associated with a receiver position, as illustrated in Figure 1b. Therefore, a more detailed discretization of the area can be achieved, when sufficient coverage is provided by the data, in comparison with more traditional common-source or common-receiver schemes. For a true layered earth, both approaches are equivalent, but when there are lateral resistivity variations, these are better mapped by organizing the data in CMP gathers, which receive information from multiple directions around the CMP.

In our approach, a maximum offset in a CMP gather is determined based on the cutoff level of the amplitude of the electric field. All source-receiver pairs that share the same CMP, and have offsets within this limit, define and contribute to that particular CMP. Therefore, the number of transmitter-receiver pairs defining a CMP varies in the area: CMPs close to the borders of the survey have the least number of data points, and those at the center of the survey area have contributions from the greatest number of data points.

This organization of the data for inversion is an efficient way to make use of most of the information that is available in a 3D survey, going beyond what is the common practice in published works performing 1D laterally constrained inversions, in which only the true in-line and broadside data are inverted [8–10]. For example, in the case of the Pelotas Basin survey presented in Section 4, the CMP with the biggest number of data points has contributions from 1573 transmitter-receiver pairs, and only 207 data points contribute to the one with the lowest number.

Although it is not common practice, it is possible to consider the case in which a common-receiver approach would use all data from transmitters around the receiver position, using the same maximum offset as in the CMP inversion. In this case, the data in the CMP inversion would be more closely localized around the CMP, because it is gathered in an area with a smaller radius. This closer concentration of the information around the central point benefits the inversion with lateral constraints because it tends to better separate the influences of localized resistive targets in the area, and also because over a given 3D target, the anomaly generated by the stacking of the CMP data is stronger than that generated by common-receiver data around the same point, as shown by the analysis of [5] for a model of a hydrocarbon reservoir.

To reduce the running time, only two electric field components are used in the data volume in this investigation, which provide sufficient information to generate quality inversion results, as the results show. The observed data are the logarithms of the horizontal components of the electric field (E_x, E_y), represented as the vector $\mathbf{d}$; the parameter vector with the resistivities of the interpretive model is $\mathbf{m}$, which is formed by the N_p values of the logarithms of the resistivities of each homogeneous layer in each CMP column (Figure 2), which are assumed to be isotropic; the 1D mathematical model describing the synthetic data as a function of the parameters is $\mathbf{F}(\mathbf{m})$. This function represents the layered Earth responses, which are calculated as shown by [5].

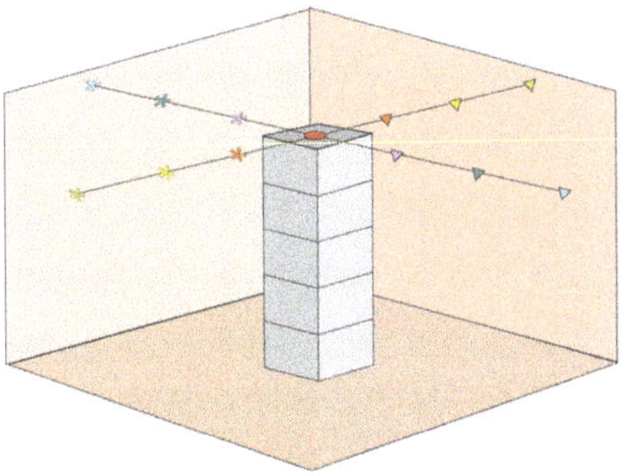

Figure 2. Resistivity column associated with one CMP.

In order to achieve stable solutions, we apply minimum gradient constraints [11] between adjacent parameters in both vertical and lateral directions. The vertical constraints are crucial for regularizing each CMP inversion, while the horizontal constraints enable the inversion of each CMP to impact the results of its neighboring columns. In this process, lateral smoothing is given more weight than vertical smoothing because, in reservoir environments, sharper variations are more common in the vertical direction than in the horizontal direction. Additionally, some parameters can be left unconstrained to allow for unexpected discontinuities, such as those at the borders of salt or igneous intrusions, when it is applied to complex geology regions.

In this application of Occam's inversion [11], the use of vertical constraints results in a smooth model that is able to generate synthetic data that fit observations gathered from layered environments, whereas the horizontal constraints contribute by enforcing lateral variations in the resistivity columns to approximate the observations under the influence of three-dimensional structures.

The smoothing constraints are introduced through the matrices $\mathbf{S}_x$, $\mathbf{S}_y$, and $\mathbf{S}_z$. Each row of a smoothing matrix defines which pairs of parameters are to be constrained to each other simply by placing a -1 and a 1 in the positions corresponding to those parameters and leaving zeroes in the remaining elements. When this matrix is applied to the parameter vector $\mathbf{m}$ in the functional to be minimized, it has the effect of imposing that the difference between those two parameters is as small as possible, by enforcing that the equation

$$\mathbf{Sm} = \mathbf{0} \tag{1}$$

be satisfied in a least squares sense. There will be as many rows in the matrix $\mathbf{S}$ as there are pairs of parameters to be constrained to each other.

Another kind of *a priori* information may be included through the use of absolute equality constraints [12], which are useful when reliable information exists about the resistivities in a given set of parameters, as when there are well logging data in the region under the survey. In this case, if there are N_c parameters to be constrained, the method defines a vector $\mathbf{m}_p$, with only the N_c values that are to be assigned based on the *a priori* information, and a matrix $\mathbf{P}_{N_c \times N_p}$, each row of which has a value of 1 in the position corresponding to the parameter to be constrained and zeroes elsewhere, such that

$$\mathbf{Pm} = \mathbf{m}_p. \tag{2}$$

To reiterate: in the inversion process, the equalities imposed by the constraint matrix Equations (1) and (2) are enforced in a least squares sense.

The functional to be minimized, then, includes a term for fitting the data, and one for each kind of constraint that enters the process. In its most general form, it is written as

$$\begin{aligned}\phi = [\mathbf{d} - \mathbf{F}(\mathbf{m})]\mathbf{W}_d[\mathbf{d} - \mathbf{F}(\mathbf{m})] + a_H\left(\mathbf{m}^T\mathbf{S}_x^T\mathbf{W}_x\mathbf{S}_x\mathbf{m} + \mathbf{m}^T\mathbf{S}_y^T\mathbf{W}_y\mathbf{S}_y\mathbf{m}\right) \\ +a_V\left(\mathbf{m}^T\mathbf{S}_z^T\mathbf{W}_z\mathbf{S}_z\mathbf{m}\right) + a_p(\mathbf{Pm} - \mathbf{m}_p)^T W_p(\mathbf{Pm} - \mathbf{m}_p).\end{aligned} \tag{3}$$

The regularization parameters a set the relative weighting for the different constraints in this objective function. $\mathbf{W}_d$ is a diagonal matrix whose elements depend on the standard deviations associated with the measurements. The diagonal weighting matrices $\mathbf{W}_x$, $\mathbf{W}_y$, $\mathbf{W}_z$, and $\mathbf{W}_p$ can be used to assign different degrees of constraining to different regions of the interpretive models by ascribing different weights to each constraining relation between pairs of parameters, i.e., each row in the matrix Equation (1).

The inversion is an implementation of the Gauss–Newton method, with Levenberg–Marquardt iterations [13,14]. In each iteration, the Marquardt's damping parameter μ is multiplied or divided by a constant factor, based on the success of the generated model in decreasing the value of the objective functional ϕ. In the examples shown here, the multiplicative factor is 2.

In the $(k+1)^{th}$ iteration, the regularized estimator is, then, written as

$$\mathbf{m}_{k+1} = \mathbf{m}_k + \left[\mathbf{J}_k^T\mathbf{W}_d\mathbf{J}_k + a_H(\mathbf{S}_x^T\mathbf{W}_x\mathbf{S}_x + \mathbf{S}_y^T\mathbf{W}_y\mathbf{S}_y) + a_V(\mathbf{S}_z^T\mathbf{W}_z\mathbf{S}_z) + a_p\mathbf{W}_p + \mu\mathbf{I}\right]^{-1}\left[(\mathbf{W}_d\mathbf{J}_k)^T(\mathbf{d} - \mathbf{F}(\mathbf{m}_k)) + a_H(\mathbf{S}_x^T\mathbf{W}_x\mathbf{S}_x\mathbf{m}_k + \mathbf{S}_y^T\mathbf{W}_y\mathbf{S}_y\mathbf{m}_k) + a_V(\mathbf{S}_z^T\mathbf{W}_z\mathbf{S}_z\mathbf{m}_k) + a_p\mathbf{W}_p(\mathbf{m}_k - \mathbf{m}_p)\right], \quad (4)$$

where $\mathbf{I}$ is the identity matrix and the elements of the Jacobian matrix $\mathbf{J}$ are the sensitivities

$$J_{ij} = \frac{\partial \log|E_i|}{\partial log(\rho_j)}. \quad (5)$$

Here, the logarithms are employed to reduce the range of values of the objective function since the field amplitudes, as well as the resistivities, can span several orders of magnitude. Reducing these ranges improves the process by allowing a more balanced fitting of the data set. The logarithms also have the advantage of preventing negative values to be attributed to the resistivities. In this application, all resistivities are isotropic; therefore, only one value (one parameter) is ascribed to each layer in a CMP column.

To take advantage of the analytical derivatives of the field ($\partial E_i/\partial\rho_j$), which are calculated as presented by [5], the logarithmic sensitivities, as defined in Equation (5), are calculated as

$$J_{ij} = \rho_j Re\left\{\frac{1}{E_i}\frac{\partial E_i}{\partial \rho_j}\right\}. \quad (6)$$

Starting with a suitable initial model, which can simply be a homogeneous earth (all layers with the same resistivity), the iteration represented by Equation (4) is repeated until the convergence criteria are satisfied. In this application, the stopping criteria are the minimum variation of the model and the minimum variation of data misfit from one iteration to the next.

3. Application to Synthetic Data

We apply the method to two synthetic data sets generated from distinct 3D models to demonstrate it. These models were designed to depict real-life features in reservoir environments, such as two simultaneous targets at different depths or a resistive target positioned above a resistive basement. These examples also showcase the use of two types of constraints to regulate the problem: relative equality constraints to ensure smooth solutions and absolute equality constraints applied to specific parameters to generate solutions where those parameters closely align with the imposed values [12].

3.1. Model 1

In the first example, we invert synthetic data generated by the 3D model shown in Figure 3. Two 100 m thick resistive targets (100 ohm-m) are embedded in a conductive (1 ohm-m) host, under a 1500 m thick water layer. The first target is a T shaped body placed at 900 m below the sea bed; the second is a slab at a depth of 1900 m below the sea bed.

The model covers a total area of 320 km^2. The survey configuration (Figure 4) has nine receiver lines, and seven source lines running at 50 m above the flat sea bed. The receiver lines are equally spaced, with a separation of 2000 m. Each line has 21 receivers spaced by 1 km. The source lines extend from 10 km to the left of the first receiver to 10 km to the right of the last receiver, between y coordinates -8000 m and 8000 m. The transmitter positions are at 100 m intervals.

Data were generated at three frequencies: 0.25 Hz, 0.75 Hz and 1.25 Hz.

The 3D synthetic observations were calculated with an Integral Equation code, based on the method presented by [15].

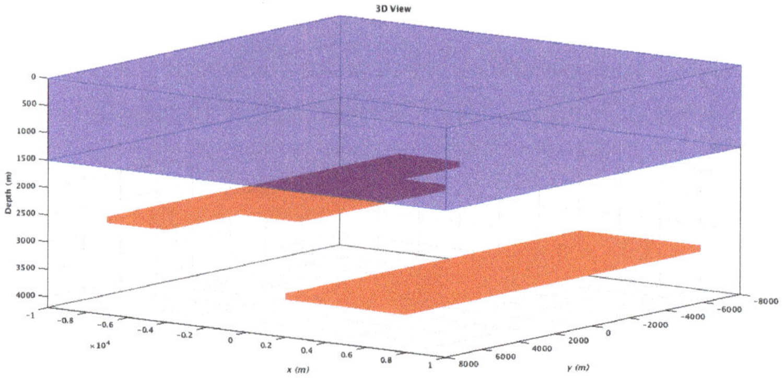

(**a**) 3D model view.

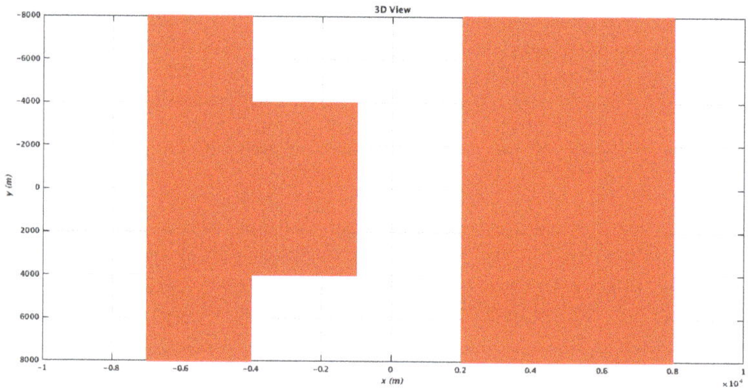

(**b**) Top view.

Figure 3. Test model: 1.5 km water layer with 0.3 ohm-m, sediment background with 1 ohm-m and two 100 ohm-m targets.

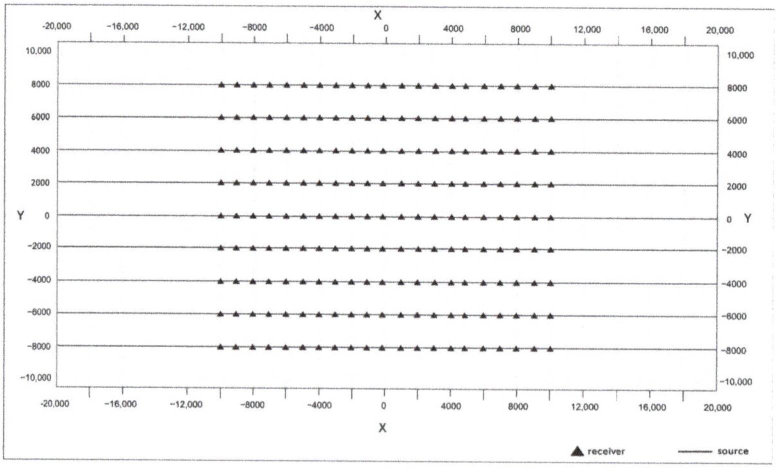

Figure 4. Acquisition layout for the simulated survey. The triangles represent the receivers and the black lines the towed source.

3.1.1. Interpretive Model

The interpretive model is formed by 410 CMP columns, distributed as a 41 × 10 grid in the xy plane. Each CMP column is composed of a stack of cells with horizontal sides of 500 m by 1500 m. The cells are narrow in the direction where the density of measurements is greatest, and wide in the direction perpendicular to the survey lines. Each 1D model has 89 layers. The first layer is 25 m thick. Thicknesses of the inversion layers increase with with depth by a factor of 1 % to compensate for the decrease in sensitivity with depth. The thickness of the last finite layer is 59.4 m. The starting model is homogeneous, all layers starting at a resistivity of 1 ohm-m.

3.1.2. MCSEM Data

The synthetic data were contaminated with gaussian noise with an amplitude of 3% of the amplitude of the field at the greatest offset in each frequency, which is representative of real world surveys, such as the one shown in Section 4.

Offsets ranging from 1 km to about 8 km were used. The cutoff of the data is at an amplitude of 10^{-15} V/Am^2. This cutoff corresponds to less than 4 km for the highest frequency and 8 km for the lowest frequency, for the E_y data, and slightly over those same distances for the E_x data, as shown in Figure 5.

3.1.3. Inversion Results

The first test runs the inversion without any absolute equality constraints. The process met the convergence criteria after 19 iterations for the data fit shown in Figure 5.

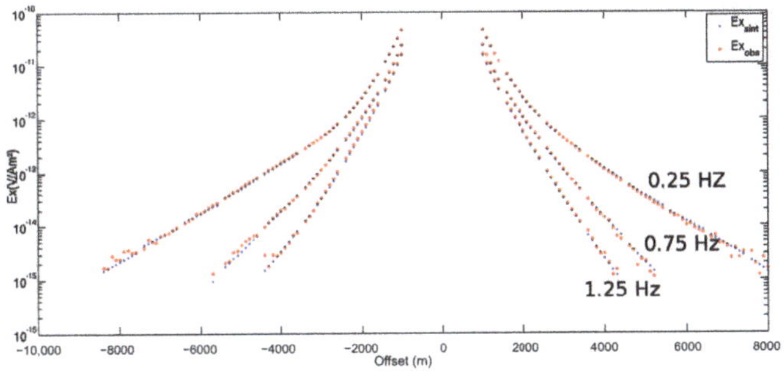

(a) E_x component.

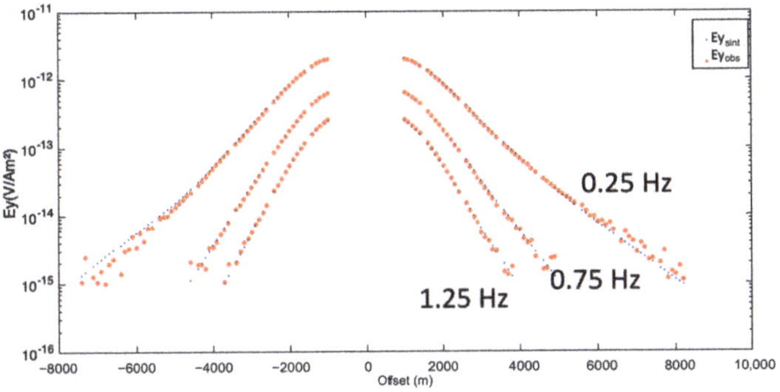

(b) E_y component.

Figure 5. Final data fit for the first model at the three frequencies, after 19 iterations.

Figures 6–10 show the results for this case. The smoothness regularization weight for the horizontal constraints is $a_H = 0.1$, and the one for the vertical constraints is $a_V = 0.05$. These parameters were chosen as the smallest values for which a stable solution was achieved.

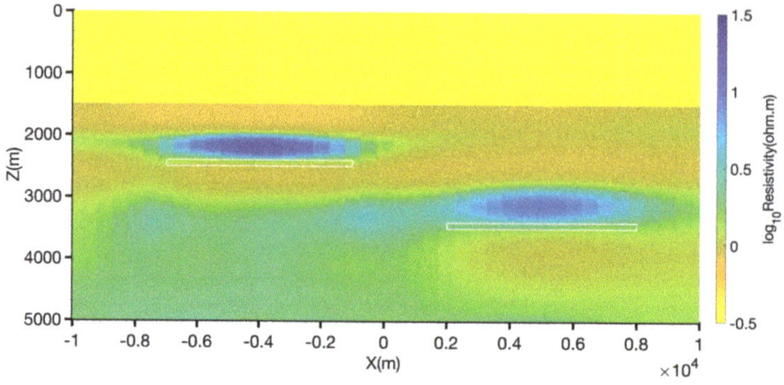

Figure 6. Slice XZ plane on $Y = 0$ m from the CMP inversion without absolute equality constraints. The white rectangles represent the real positions of the bodies.

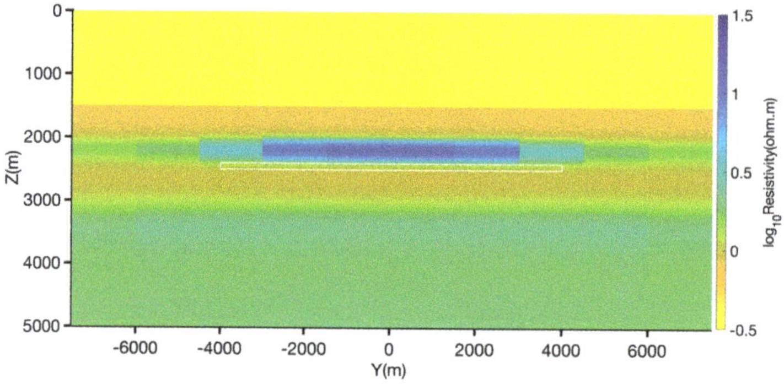

Figure 7. Slice YZ plane on $X = -2500$ m from the CMP inversion without absolute equality constraints. The white rectangle represents the real position of the body.

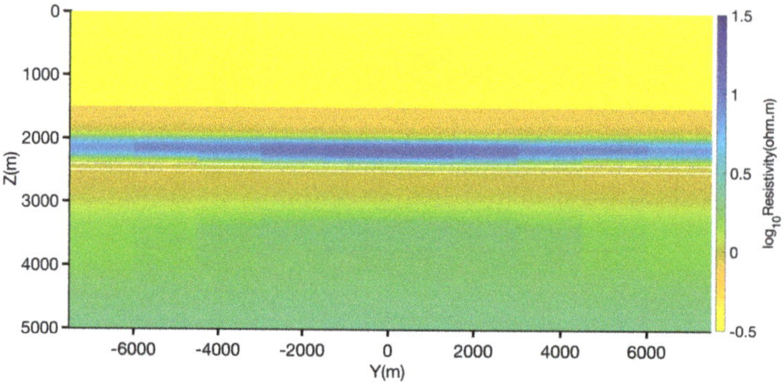

Figure 8. Slice YZ plane on $X = -5500$ m from the CMP inversion without absolute equality constraints. The white rectangle represents the real position of the body.

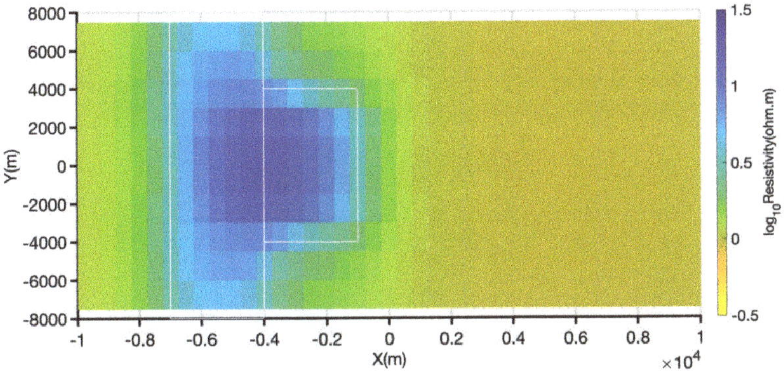

Figure 9. Depth slice on $Z = 2220$ m from the CMP inversion without absolute equality constraints. The white line is the projection of the true model shape at this depth.

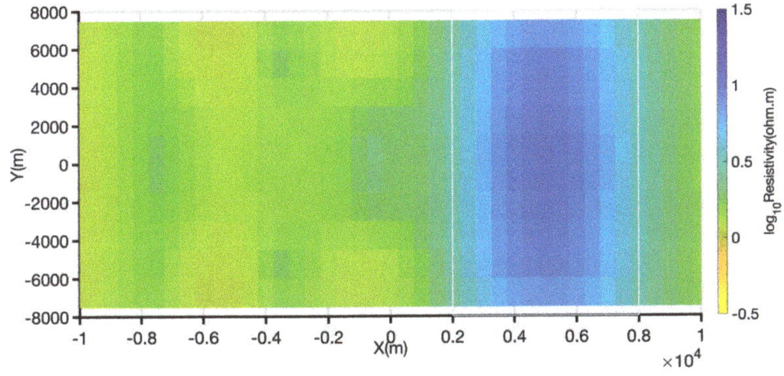

Figure 10. Depth slice on $Z = 3150$ m from the CMP inversion without absolute equality constraints. The white lines represent the real shape of the body.

Figure 6 shows a vertical section of the resistivity *cube* on the xz plane at $y = 0$, and Figures 7 and 8 show vertical slices in the yz plane on $x = -5500$ m and $x = -2500$ m, respectively, with the resistivity values recovered by the inversion process. The white lines show the true contours of the resistive targets. Both bodies appear in the final section positioned approximately 250 m above the true depths. The highest resistivity recovered is 15.5 ohm-m.

Figure 9 shows a horizontal slice at a depth of 2200 m, which is the depth at which the greatest resistivities were found in this case, underestimating the depth of the resistive target. The inversion has success in the delimitation of the left resistive body, albeit with overly smooth contours and at a depth above its true position. The delineation of the deeper resistive body is observed in the horizontal slice at 3150 m, as shown in Figure 10. Again, a good match with the true model is achieved, with smooth borders. The highest resistivities are found at the center of the bodies, which is also typically found in 3D inversions. In this case, this result is reinforced by the fact that in the points furthest from the borders, the behavior of the data is closest to the 1D approximation.

3.1.4. Inversion Including Well Constraints

In the exploration work, it is sometimes possible to use well logging information to drive the inversion process. In this example, this kind of information is input, in addition to the smoothing constraints, as the resistivity of a single cell at the true positions at the centers of both targets. The input resistivity is 100 ohm-m, and the weighing parameter $a_p = 1$.

Now, the true depths of both resistive bodies are enforced by the constraints; therefore, the fitting of the data is achieved by a model with better delineation of the targets, with the highest resistivity around 30 ohm-m, as shown in Figures 11–15.

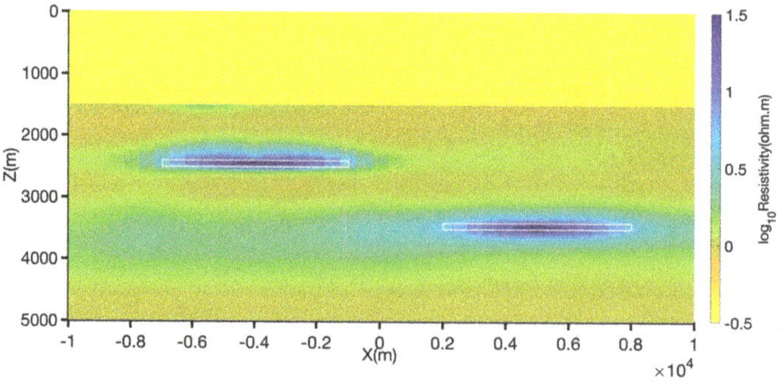

Figure 11. XZ slice on $Y = 0$ m from the constrained CMP inversion with absolute equality constraint from well log data. The results show the best fit with the real body in white.

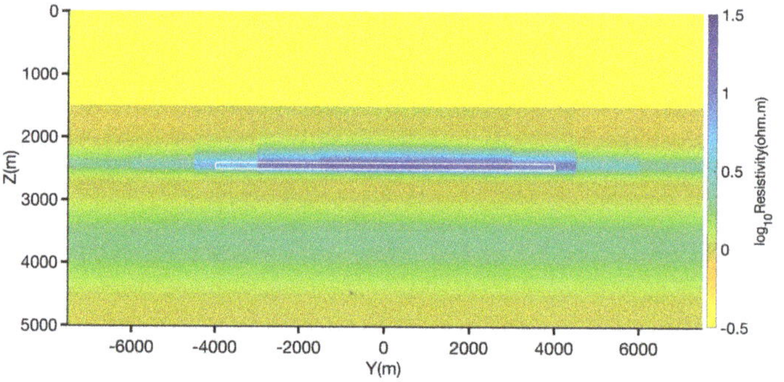

Figure 12. YZ slice on $X = -2500$ m from the constrained CMP inversion with absolute equality constraint from well log data. The results show the best fit with the real body in white.

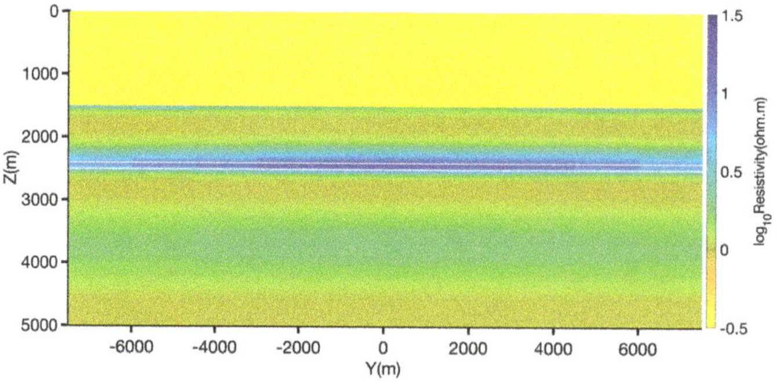

Figure 13. YZ slice on $X = -5500$ m from the constrained CMP inversion with absolute equality constraint from well log data. The results show the best fit with the real body in white.

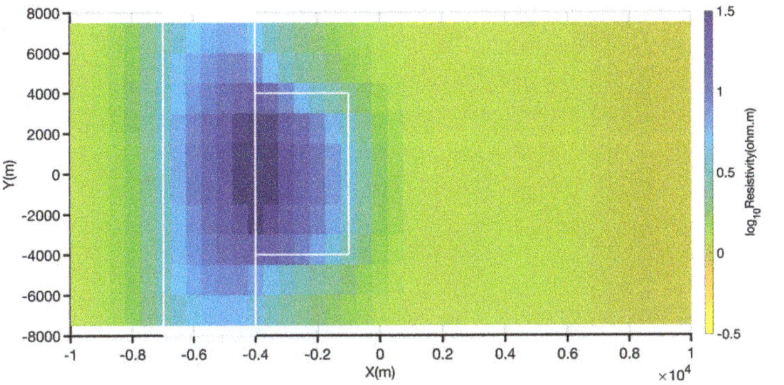

Figure 14. Depth slice at $Z = 2450$ m from the constrained CMP inversion with absolute equality constraint from well log data. The results show the best fit with shape of the real body in white.

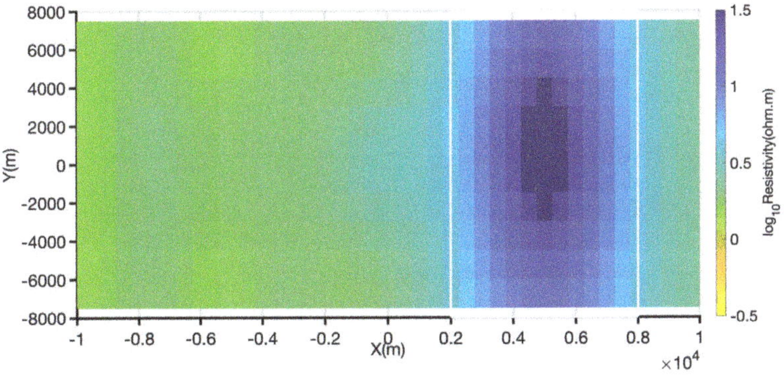

Figure 15. Depth slice at $Z = 3450$ m from the constrained CMP inversion with absolute equality constraint from well log data. The results show the best fit with the shape of the real body in white.

3.2. Model 2

The second example model illustrates the detection of a resistive target located above a resistive basement. The model is shown in Figure 16. One resistive body (100 ohm-m) with 100 m of thickness, at 900 m below the sea bed, is embedded in a layered model with 1, 2, and 3 ohm-m, over a 250 ohm-m basement. All layers are 1 km thick. The survey is run at the same frequencies and the same transmitter and receiver positions as in the previous example.

Inversion Results 2

The same inversion grid of the first example was used, but cropped in the y direction between -4500 m to 4500 m. Again, the resistivity start model was a homogenous layer of 1 ohm-m from the sea bed to 4500 m and the last layer with 250 ohm-m.

Figure 17 shows a vertical section on the xz plane at $y = 0$, and Figure 18 shows a vertical slice in the yz plane on $x = 0$ m, with the resistivity values recovered by the inversion process. The body appears in the final section well positioned in depth. The presence of the resistive basement is apparent, with its top interface located close to the true depth by the sharp vertical variation in the resistivity. The region below the body is detected with less sensitivity by the data and appears in the final model with lower resistivities than the true values.

A horizontal cut at a depth of 2500 m is shown in Figure 19.

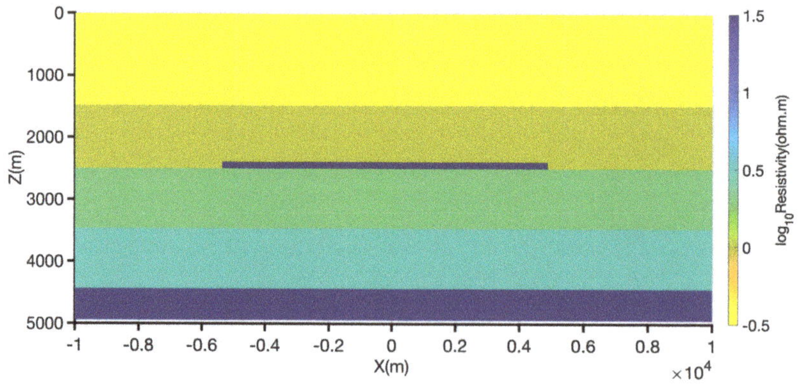

(**a**) Vertical section of model 2.

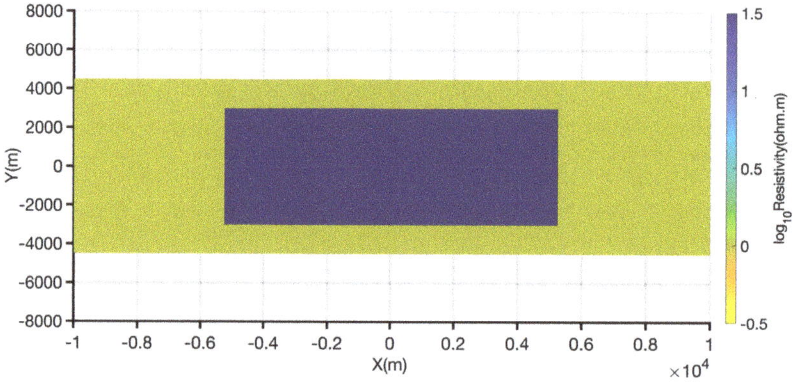

(**b**) Top view of model 2.

Figure 16. Model 2: 1.5 km water layer over layered sediments and a resistive target with 100 Ω·m.

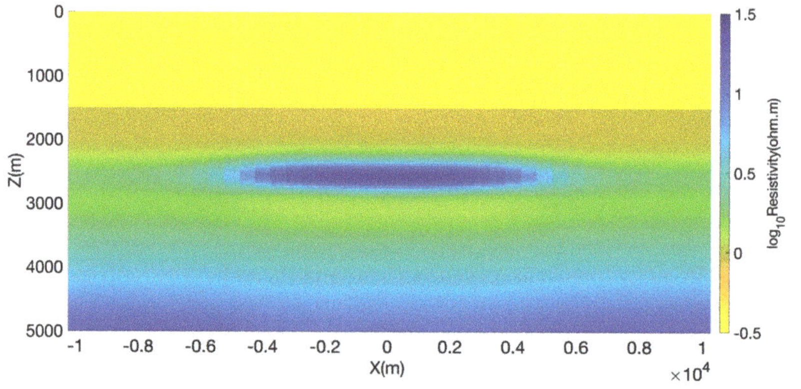

Figure 17. XZ slice on $Y = 0$ m from the CMP inversion. The real body has 100 m of thickness, starting on 2400 m of depth.

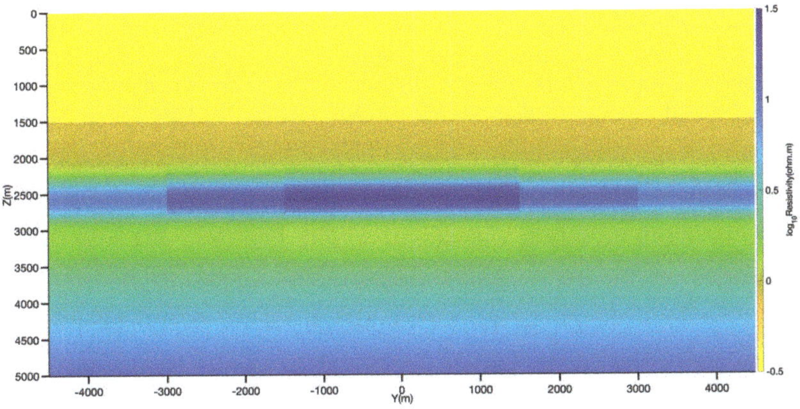

Figure 18. YZ slice plane on $X = 0$ m from the CMP inversion. The real body has 100 m of thickness, starting on 2400 m of depth.

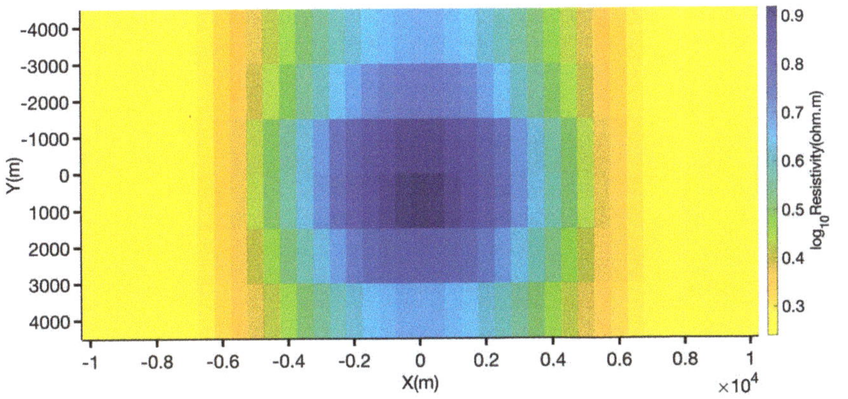

Figure 19. Depth slice on $z = 2500$ m from the CMP inversion.

4. Application to Real Data

The method was applied to data from a 3D MCSEM survey for gas hydrate deposits in the Pelotas Basin, in the southernmost rift margin in Brazil. The detailed discussion and interpretation of this dataset is presented in [16].

4.1. Geological Background

The Pelotas Basin is situated to the North, bounded by the Florianopolis High and the Santos Basin, and to the South, by the Punta Del Este Basin and the Polonio High in Uruguay. With a sedimentary section that is approximately 12,000 m thick and covering an area of more than 210,000 km^2, the structural framework of the Pelotas Basin is linked to extensional tectonism during the Late Jurassic to Early Cretaceous period, coinciding with the opening of the South Atlantic Ocean. The rift sequences in the Pelotas Basin started in the Mid-Cretaceous due to thermal subsidence and seafloor spreading, involving multiple stages of deposition and extensional faulting. As discussed by [17], significant volcanism is evident in certain sections of the early rift, with the basin displaying seaward dipping reflectors. Moving up in the section, a second late Miocene flooding surface signifies a stratigraphic transition from aggradational deposition to progradational depositional cycles that persist to the present day. Stratigraphic interpretations indicate that thickness variations are influenced by the Lagoa dos Patos, which contributes substantial sediment volumes to the Atlantic Ocean system.

The MCSEM survey was designed to cover part of the Rio Grande Cone. It is believed that organic material transported during this phase of deposition was well preserved in the intervals and likely matured to generate gas and gas-hydrates. Features related to the hydrates and free gas accumulations are identified in 2D seismic data as well as in previous studies in the Pelotas Basin [18].

On the Pelotas middle-lower continental slope, two high-amplitude reflectors extend almost parallel to the sea bottom surface and cut through the stratified Eocene–Miocene succession. These features triggered unstable conditions of the lower slope and deep-marine sediment immobilization, followed by debris deposition in the oceanic basin during the Neogene. According to [19], the layer with gas hydrate is under a layer of gaseous mud and has an average thickness of 400 to 500 m. Ideal conditions of temperature and pressure allow for the freezing of the water, trapping the gas.

4.2. CMP Inversion

The geometry of the receiver positioning is illustrated in Figure 20a. In this survey each source line passes between two receiver lines, therefore there are no purelly inline data. This configuration obligates us to apply algorithms that handle broadside data as a full 3D inversion, as well as our proposed method. The survey was conducted in an area for which several seismic sections were already available. Figure 20b shows the locations of the seismic lines in relation to the MCSEM receivers. The frequencies used in the inversion were 1.3 and 5.0 Hz.

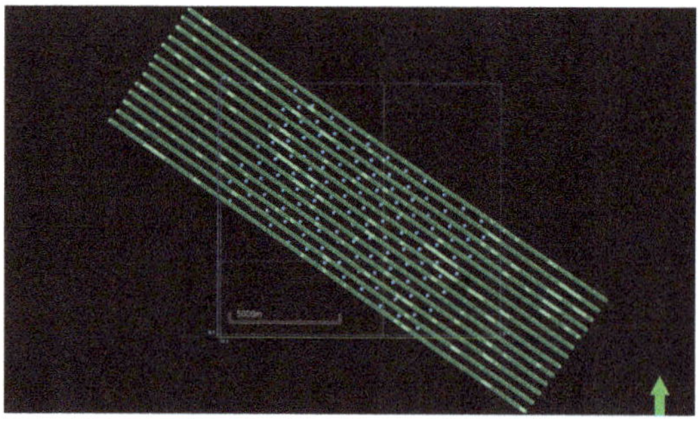

(**a**) The positioning of the MCSEM receivers in blue with the source navigation lines in green.

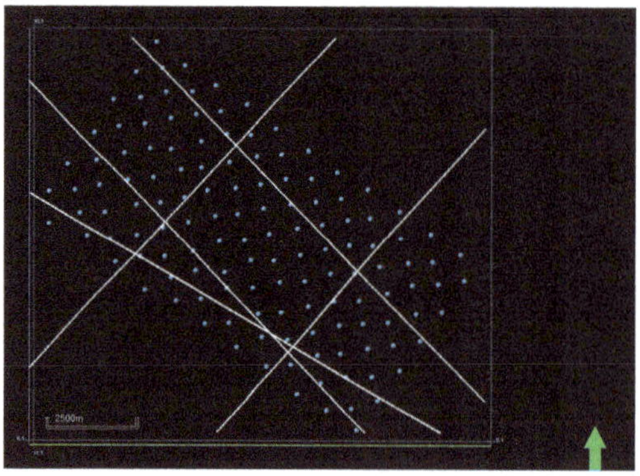

(**b**) Seismic lines in the MCSEM survey area. The blue points represent the MCSEM receiver and the white lines represent the 2D seismic lines.

Figure 20. Geometry of the MCSEM survey.

Due to the sparsity of data points in the area, the columns corresponding to the CMPs have 300 m sides. This implies low spatial resolution and limits the capacity of the inversion to detect sub-vertical bodies, such as resistivity variations associated with fractures. However, it allows for the identification of regional tendencies in the resistivity structures. The mesh in depth has increasing thickness to account for variations in sensitivity, and the bathymetry is addressed through regularization that connects the lateral cells.

A data sample illustrating the small noise level in this dataset and the fit obtained by the generated synthetic data is shown in Figure 21, which shows curves for the E_x and E_y components in one receiver for one transmitter pass. Observe that the signal to noise ratio is large for many of the curves, but that the noise overwhelms the signal in the longest offsets. The synthetic data closely matches the field data in areas where the earth is layered, but less so in areas strongly influenced by 3D structures.

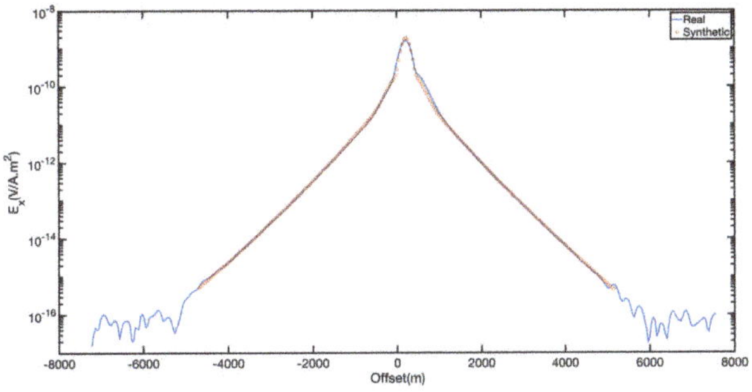

(**a**)

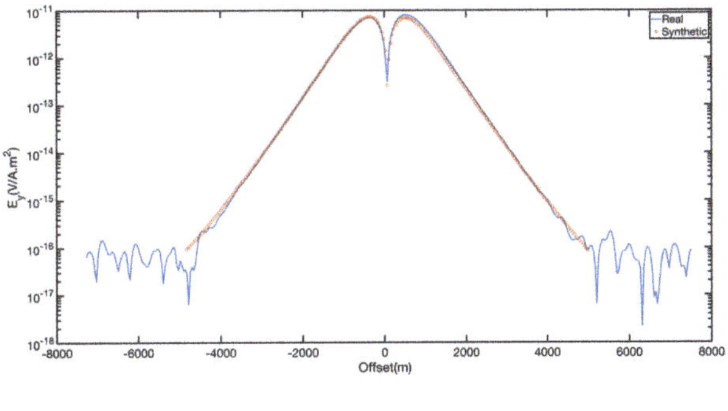

(**b**)

Figure 21. Example of the data in a receiver superimposed with the synthetic data obtained from the inverted model. (**a**) E_x and (**b**) E_y.

The inversion results are integrated to 2D seismic sections for comparison with the stratigraphy, with the structural behavior of the area, and with the positioning of the *Bottom Simulating Reflector* (BSR), which is usually associated with the base of the hydrate accumulation zone.

The CMP inversion was able to locate a resistive layer below the BSR and a shallow zone of resistive bodies. These two anomalous resistive zones are clearly delineated in the inverted resistivity sections shown in Figure 22, superimposed on the corresponding seismic sections indicated by white lines. In the first interpretation, the shallow and less resistive layer is associated with gas hydrate accumulation and the intermediate zone is associated with free gas accumulation underneath the BSR. The results match the stratigraphy overall, and we can highlight some higher anomalies close to the faults.

This inversion runs in parallel in 62 core processors with 128 GB/node. The execution time was approximately 10 hours with 45 iterations, which is about the same time as two iterations of a true 3D Gauss–Newton inversion for this data set in the same machine.

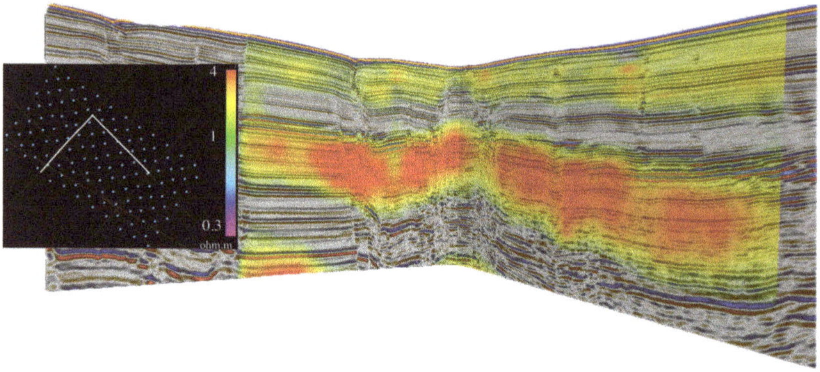

Figure 22. CMP Inverted resistivity section superimposed on the seismic section under the same survey line.

5. Conclusions

The 1D inversion of 3D MCSEM data in the CMP domain with spatial constraints has been able to produce a quick first approximation to the geoelectrical structures under a survey area. The use of regularization constraints not only contributes to the achievement of stable solutions, but also helps to create a better delineation of the targets by allowing the data from one CMP to exert influence on its surrounding CMP columns.

The 3D models that result from this inversion are intended to be useful as a first interpretation tool, and also as initial models in a true 3D inversion, with the possibility of saving computing time by reducing the number of iterations.

The application of this methodology needs to take into account the characteristics of the geologic setting; the more complex the setting, the less reliable the inverted model can be.

In the presented example of the application to real MCSEM data for hydrates, the method was able to produce a good resistivity model because of the mostly layered structure of the geology of the area. In this case, the presence of faults imposes localized lateral variations in the resistivity sections.

Author Contributions: Conceptualization, J.L.C. and C.R.; methodology, J.L.C. and C.R.; software, J.L.C.; validation, J.L.C.; writing—original draft preparation, J.L.C.; writing—review and editing, J.L.C. and C.R.; supervision, C.R. All authors have read and agreed to the published version of the manuscript.

Funding: This research was funded by Petrobras.

Data Availability Statement: The data and all details from the synthetic examples are available through contact with the authors. The real data are confidential and cannot be shared.

Acknowledgments: The authors thank the National Institute of Science and Technology of Petroleum Geophysics – INCT-GP for its support. Régis thanks the National Council for Scientific and Technological Development–CNPq for the financial support. We also thank Emanuel Capechi for his support in implementing the original MATLAB code in C (Java). We are also grateful to PETROBRAS for permission to publish this work, particularly Adriano Viana. Finally, we thank José Cupertino for his support of this research.

Conflicts of Interest: Author Jorlivan Lopes Correa was employed by the company Petróleo Brasileiro S/A-Petrobras-EXP/TPGG/TG. The remaining authors declare that the research was conducted in the absence of any commercial or financial relationships that could be construed as a potential conflict of interest.

References

1. Constable, S. Ten years of marine CSEM for hydrocarbon exploration. *Geophysics* **2010**, *75*, 75A67–75A81. [CrossRef]
2. Buonora, M.P.P.; Correa, J.L.; Martins, L.S.; Menezes, P.T.; Pinho, E.J.; Silva Crepaldi, J.L.; Ribas, M.P.; Ferreira, S.M.; Freitas, R.C. mCSEM data interpretation for hydrocarbon exploration: A fast interpretation workflow for drilling decisioninterpretation for hydrocarbon exploration: A fast interpretation workflow for drilling decision. *Interpretation* **2014**, *2*, SH1–SH11. [CrossRef]
3. Barker, N.; Baltar, D. CSEM anomaly identification. *First Break* **2016**, *34*, 47–50. [CrossRef]
4. Mittet, R.; Brauti, K.; Maulana, H.; Wicklund, T.A. CMP inversion and post-inversion modelling for marine CSEM data. *First Break* **2008**, *26*.. [CrossRef]
5. Crepaldi, J.L.S.; Buonora, M.P.P.; Figueiredo, I. Fast marine CSEM inversion in the CMP domain using analytical derivatives. *Geophysics* **2011**, *76*, F303–F313. [CrossRef]
6. Dell'Aversana, P.; Vivier, M. Expanding the frequency spectrum in marine CSEM applications. *Geophys. Prospect.* **2009**, *57*, 573–590. [CrossRef]
7. Tseng, H.W.; MacGregor, L.; Ackermann, R.V. 2D inversions of 3D marine CSEM data. In *SEG Technical Program Expanded Abstracts 2012*; Society of Exploration Geophysicists: Houston, TX, USA, 2012; pp. 1–5.
8. MacGregor, L.; Tomlinson, J. Marine controlled-source electromagnetic methods in the hydrocarbon industry: A tutorial on method and practice. *Interpretation* **2014**, *2*, SH13–SH32. [CrossRef]
9. Key, K. 1D inversion of multicomponent, multifrequency marine CSEM data: Methodology and synthetic studies for resolving thin resistive layers. *Geophysics* **2009**, *74*, F9–F20. [CrossRef]
10. Luz, E.d.C.; Régis, C.R.T.; dos Santos, W.G. Delineation of anisotropic layers through 1D inversion of marine CSEM data. *Braz. J. Geophys.* **2016**, *34*, 543–558. [CrossRef]
11. Constable, S.C.; Parker, R.L.; Constable, C.G. Occam's inversion: A practical algorithm for generating smooth models from electromagnetic sounding data. *Geophysics* **1987**, *52*, 289–300. [CrossRef]
12. Medeiros, W.E.; Silva, J.B.C. Geophysical inversion using aproximate equality constraints. *Geophysics* **1996**, *61*, 1678–1688. [CrossRef]
13. Marquardt, D.W. An algorithm for least-squares estimation of nonlinear parameters. *J. Soc. Ind. Appl. Math.* **1963**, *11*, 431–441. [CrossRef]
14. Auken, E.; Christiansen, A.V. Layered and laterally constrained 2D inversion of resistivity data. *Geophysics* **2004**, *69*, 752–761. [CrossRef]
15. Zhdanov, M.S. *Geophysical Electromagnetic Theory and Methods*; Elsevier: Amsterdam, The Netherlands, 2009.
16. Tharimela, R.; Augustin, A.; Ketzer, M.; Cupertino, J.; Miller, D.; Viana, A.; Senger, K. 3D controlled-source electromagnetic imaging of gas hydrates: Insights from the Pelotas Basin offshore Brazil. *Interpretation* **2019**, *7*, SH111–SH131. [CrossRef]
17. Stica, J.M.; Zalán, P.V.; Ferrari, A.L. The evolution of rifting on the volcanic margin of the Pelotas Basin and the contextualization of the Paraná–Etendeka LIP in the separation of Gondwana in the South Atlantic. *Mar. Pet. Geol.* **2014**, *50*, 1–21. [CrossRef]
18. Miller, D.J.; ao Marcelo Ketzer, J.; Viana, A.R.; Kowsmann, R.O.; Freire, A.F.M.; Oreiro, S.G.; Augustin, A.H.; Lourega, R.V.; Rodrigues, L.F.; Heemann, R.; et al. Natural gas hydrates in the Rio Grande Cone (Brazil): A new province in the western South Atlantic. *Mar. Pet. Geol.* **2015**, *67*, 187–196. [CrossRef]
19. Fontana, R.L.; Mussumeci, A. Hydrates Offshore Brazil. *Ann. N. Y. Acad. Sci.* **1994**, *715*, 106–113. [CrossRef]

Disclaimer/Publisher's Note: The statements, opinions and data contained in all publications are solely those of the individual author(s) and contributor(s) and not of MDPI and/or the editor(s). MDPI and/or the editor(s) disclaim responsibility for any injury to people or property resulting from any ideas, methods, instructions or products referred to in the content.

applied sciences

MDPI

Article

CSEM Optimization Using the Correspondence Principle

Adriany Valente [1,†], Deivid Nascimento [2,*,†] and Jessé Costa [1,3,†]

1 Faculty of Geophysics, Pará Federal University, Av. Augusto Correa, 01, Belém 66070-110, PA, Brazil; adrianytmreis@gmail.com (A.V.); jesse.ufpa@gmail.com (J.C.)
2 PETROBRAS, Petrobras Research Center—CENPES, Rio de Janeiro 21941-915, RJ, Brazil
3 INCT-GP/CNPq—National Institute of Science and Technology for Petroleum Geophysics—CNPq, Brasília 70070-010, DF, Brazil
* Correspondence: sndeivid@gmail.com or sndeivid@petrobras.com.br
† These authors contributed equally to this work.

Abstract: Traditionally, 3D modeling of marine controlled-source electromagnetic (CSEM) data (in the frequency domain) involves high-memory demand, requiring solving a large linear system for each frequency. To address this problem, we propose to solve Maxwell's equations in a fictitious dielectric medium with time-domain finite-difference methods, with the support of the correspondence principle. As an advantage of this approach, we highlight the possibility of its implementation for execution with GPU accelerators, in addition to multi-frequency data modeling with a single simulation. Furthermore, we explore using the correspondence principle to the inversion of CSEM data by calculating the gradient of the least-squares objective function employing the adjoint-state method to establish the relationship between adjoint fields in a conductive medium and their counterparts in the fictitious dielectric medium, similar to the approach used in forward modeling. We validate this method through 2D inversions of three synthetic CSEM datasets, computed for a simple model consisting of two resistors in a conductive medium, a model adapted from a CSEM modeling and inversion package, and the last one based on a reference model of turbidite reservoirs on the Brazilian continental margin. We also evaluate the differences between the results of inversions using the steepest descent method and our proposed momentum method, comparing them with the limited-memory BFGS (Broyden–Fletcher–Goldfarb–Shanno) algorithm (L-BFGS-B). In all experiments, we use smoothing by model reparameterization as a strategy for regularizing and stabilizing the iterations throughout the inversions. The results indicate that, although it requires more iterations, our modified momentum method produces the best models, which are consistent with results from the L-BFGS-B algorithm and require less storage per iteration.

Keywords: CSEM inversion; correspondence principle; optimization methods

1. Introduction

The controlled source electromagnetic method (CSEM) is a consolidated method in the off-shore industry [1,2]. It is used not only to reduce the exploratory risk [3], helping to decide well placement, but also in applications aimed at reservoir monitoring [4,5], given the significant role of electrical resistivity in characterizing rock porosity and fluid saturation.

Based on measurements of the electric and magnetic fields made by receivers on the seabed (while a ship towed the submerged electromagnetic source), CSEM allows for obtaining a model of electrical conductivity in the subsurface through an inversion process, which iteratively minimizes the misfit between observed and modeled data.

Several possible approaches exist to perform forward modeling in the context of the inversion of CSEM data [6]. Some of these approaches are based on solving Maxwell's equations for conductive media directly in the frequency domain and require solving a linear system for each frequency. Furthermore, the Gauss–Newton approximation for the model update is common during inverse problem iterations [7,8], which usually require solving large linear systems, especially in 3D problems. Another critical aspect of CSEM inversion is the regularization strategy applied to stabilize the iterative process [9–12]. We propose modeling CSEM data through the numerical time-domain finite-difference solution of Maxwell's equations for a fictitious dielectric medium, following the prescription of [13]. The correspondence principle allows the evaluation of the fields for the corresponding conductive medium in the frequency domain [14]. In the context of inversion, we present the formulation for gradient calculation, obtained using the adjoint-state method [15,16], based on adjoint modeling following the same formulation as the forward modeling. This is the approach presented by [17] but with the correction to the equations that define the mapping between the fictitious and diffusive domains. To stabilize the inverse problem, we use model reparameterization smoothing [18,19], which does not require trial-and-error selection of a regularization parameter.

We also evaluate two optimization algorithms: the steepest descent (using a simple line search strategy) and a momentum method [20], parameterized according to the Adam algorithm [21], known as a low memory demand method. We also compared the results with those obtained using L-BFGS-B [22,23], which is traditionally used in minimization problems. Our numerical experiments show that both algorithms successfully fit the observations and correctly localize the inversion targets, but our momentum approach has low storage demand.

2. Methodology

2.1. Csem Forward Modeling

Maxwell's equations (in the frequency domain) for a conductive medium at low frequencies are as follows:

$$-\nabla \times \mathbf{H}(\mathbf{x},\omega) + \sigma(\mathbf{x})\mathbf{E}(\mathbf{x},\omega) = -\mathbf{J}(\mathbf{x},\omega), \tag{1}$$
$$\nabla \times \mathbf{E}(\mathbf{x},\omega) - i\omega\mu\mathbf{H}(\mathbf{x},\omega) = \mathbf{0}, \tag{2}$$

where $\mathbf{E}(\mathbf{x},\omega)$ and $\mathbf{H}(\mathbf{x},\omega)$ are the electric and magnetic field vectors, respectively, $\sigma(\mathbf{x})$ represents the conductivity tensor, μ is the magnetic permeability, and $\mathbf{J}(\mathbf{x},t)$ is the electric current density vector, which represents the injection source distribution.

In the same context, Maxwell's equations in the time domain are given as follows:

$$-\nabla \times \mathbf{H}(\mathbf{x},t) + \sigma(\mathbf{x})\mathbf{E}(\mathbf{x},t) = -\mathbf{J}(\mathbf{x},t), \tag{3}$$
$$\nabla \times \mathbf{E}(\mathbf{x},t) + \mu\partial_t\mathbf{H}(\mathbf{x},t) = \mathbf{0}, \tag{4}$$

since we can obtain Equations (1) and (2) from Equations (3) and (4) using the Fourier transform:

$$F(\omega) = \int_{-\infty}^{+\infty} dt f(t)e^{i\omega t}, \tag{5}$$
$$f(t) = \frac{1}{2\pi}\int_{-\infty}^{+\infty} d\omega F(\omega)e^{-i\omega t}. \tag{6}$$

In a non-conductive dielectric medium, Maxwell's equations in the time domain are:

$$-\nabla \times \mathbf{H}(\mathbf{x},t) + \epsilon(\mathbf{x})\frac{\partial\mathbf{E}(\mathbf{x},t)}{\partial t} = -\mathbf{J}(\mathbf{x},t), \tag{7}$$
$$\nabla \times \mathbf{E}(\mathbf{x},t) + \mu\frac{\partial\mathbf{H}(\mathbf{x},t)}{\partial t} = \mathbf{0}, \tag{8}$$

and in the frequency domain they are the following:

$$-\nabla \times \mathbf{H}(\mathbf{x},\omega) - i\omega\epsilon(\mathbf{x})\mathbf{E}(\mathbf{x},\omega) = -\mathbf{J}(\mathbf{x},\omega), \tag{9}$$
$$\nabla \times \mathbf{E}(\mathbf{x},\omega) - i\omega\mu\mathbf{H}(\mathbf{x},\omega) = \mathbf{0}, \tag{10}$$

where $\epsilon(\mathbf{x})$ is the dielectric permittivity tensor.

Defining the fictitious dielectric permittivity,

$$\sigma(\mathbf{x}) \equiv 2\omega_0 \epsilon'(\mathbf{x}), \tag{11}$$

we can rewrite Equations (1) and (2):

$$-\nabla \times \mathbf{H}'(\mathbf{x}, \omega') - i\omega' \epsilon'(\mathbf{x}) \mathbf{E}'(\mathbf{x}, \omega') = -\mathbf{J}'(\mathbf{x}, \omega'), \tag{12}$$

$$\nabla \times \mathbf{E}'(\mathbf{x}, \omega') - i\omega' \mu \mathbf{H}'(\mathbf{x}, \omega') = \mathbf{0}, \tag{13}$$

where

$$\omega' \equiv (1+i)\sqrt{\omega\omega_0}, \tag{14}$$

$$\mathbf{E}'(\mathbf{x}, \omega') \equiv \mathbf{E}(\mathbf{x}, \omega), \tag{15}$$

$$\mathbf{H}'(\mathbf{x}, \omega') \equiv \sqrt{\frac{-i\omega}{2\omega_0}} \mathbf{H}(\mathbf{x}, \omega), \tag{16}$$

$$\mathbf{J}'(\mathbf{x}, \omega') \equiv \sqrt{\frac{-i\omega}{2\omega_0}} \mathbf{J}(\mathbf{x}, \omega). \tag{17}$$

Comparing with Equations (9) and (10), we obtain the equations in the fictitious time domain:

$$-\nabla \times \mathbf{H}'(\mathbf{x}, t') + \epsilon'(\mathbf{x}) \frac{\partial \mathbf{E}'(\mathbf{x}, t')}{\partial t'} = -\mathbf{J}'(\mathbf{x}, t'), \tag{18}$$

$$\nabla \times \mathbf{E}'(\mathbf{x}, t') + \mu \frac{\partial \mathbf{H}'(\mathbf{x}, t')}{\partial t'} = \mathbf{0}. \tag{19}$$

The mapping defined in Equations (14)–(17) warrants this construction, which establishes a direct relationship between the electromagnetic fields in a conductive medium and their corresponding in a fictitious dielectric model, as established by the correspondence principle [14]. We solve Equations (18) and (19) using finite difference, following the prescriptions of [13], and obtain the components E_i of the electric field in the conductive medium using Equation (15):

$$E_i(\mathbf{x}, \omega) = \int_0^T dt' E_i'(\mathbf{x}, t') e^{i\omega' t'} = \int_0^T dt' E_i'(\mathbf{x}, t') e^{-\sqrt{\omega\omega_0} t'} e^{i\sqrt{\omega\omega_0} t'}, \tag{20}$$

whose numerical evaluation is stable in consequence of the exponential decay imposed by the term $e^{-\sqrt{\omega\omega_0} t'}$, regardless of the duration of the signal recorded during the simulation in the fictitious time domain.

The injection of the source component $J'(\mathbf{x}, t')$ uses the band-limited spatial Delta function approximation:

$$J'(\mathbf{x}_s, t') = \delta(\mathbf{x} - \mathbf{x}_s) J'(t'), \tag{21}$$

with a signature spectrum given as follows:

$$J(\omega) = \sqrt{\frac{-2\omega_0}{i\omega}} \int_0^T dt' J'(t') e^{-\sqrt{\omega\omega_0} t'} e^{i\sqrt{\omega\omega_0} t'}. \tag{22}$$

In practice, the choice of the source signature for modeling in the fictitious domain is flexible since the data obtained after performing the CSEM survey corresponds to the field normalized by the source at each frequency.

2.2. Gradient Calculation for CSEM Inversion

The estimate of subsurface resistivity is an ill-posed non-linear problem and inversion of CSEM data usually applies interactive algorithms, where update of the resistivity model is a step along the gradient of the objective function relative to the model parameters. The adjoint state method allows for the efficient computation of the gradient [16].

For a conductive model, σ, the objective function for inversion of CSEM data is the least squares mismatch between the observed and modeled data:

$$\chi(\mathbf{E},\sigma) = \sum_s \sum_\omega \sum_r W^E(\mathbf{x}^r,\mathbf{x}^s) \left\| \mathbf{E}^{obs}(\mathbf{x}^r,\omega;\mathbf{x}^s) - \mathbf{E}(\mathbf{x}^r,\omega;\mathbf{x}^s|\sigma) \right\|^2, \tag{23}$$

where index s indicates the summation over all sources, index r indicates the summation over the receiver position, and index ω indicates the summation over all recorded frequencies. The weighting factors W^E penalize the residuals between the modeled and observed field components [24]. The observed electric field at the receiver position, denoted by $\mathbf{x}^r$, due to a source located at $\mathbf{x}^s$, for a given frequency, ω, is represented by the vector $\mathbf{E}^{obs}(\mathbf{x}^r,\mathbf{x}^s,\omega)$. The field vector $\mathbf{E}(\mathbf{x}^r,\mathbf{x}^s,\omega)$ represents the corresponding modeled field for a given conductivity model, σ.

We derived the gradient of $\chi(\mathbf{E},\sigma)$ relative to the conductivity model through the adjoint state method, which is an extension of the Lagrange multipliers method. The starting point is the extended Lagrangian, containing the sum of the least-squares data misfit plus the state equations that enforce the modeled fields and medium properties should honor the Maxwell equations, i.e.,

$$\begin{aligned}\mathcal{L}(\mathbf{E},\mathbf{H},\sigma,\boldsymbol{\lambda}_E,\boldsymbol{\lambda}_H) =& \sum_s \sum_\omega \sum_r W^E(\mathbf{x}^r,\omega;\mathbf{x}^s) \left\| \mathbf{E}^{obs}(\mathbf{x}^r,\omega;\mathbf{x}^s) - \mathbf{E}(\mathbf{x}^r,\omega;\mathbf{x}^s|\sigma) \right\|^2 \\ &+ \sum_s \sum_\omega \Re\left(\int_\Omega d\Omega \boldsymbol{\lambda}_E^*(\mathbf{u})[-\nabla\times\mathbf{H}(\mathbf{u}) + \sigma(\mathbf{x})\mathbf{E} + \mathbf{J}(\mathbf{u})] \right) \\ &+ \sum_s \sum_\omega \Re\left(\int_\Omega d\Omega \boldsymbol{\lambda}_H^*(\mathbf{u})[\nabla\times\mathbf{E}(\mathbf{u}) - i\omega\mu\mathbf{H}(\mathbf{u})] \right)\end{aligned} \tag{24}$$

where $\Re(z)$ indicates the real part of its complex argument and $\mathbf{u} \equiv (\mathbf{x},\omega;\mathbf{x}^s)$. $\boldsymbol{\lambda}_E(\mathbf{x},\omega;\mathbf{x}^s)$, and $\boldsymbol{\lambda}_H(\mathbf{x},\omega;\mathbf{x}^s)$ are the adjoint-state fields.

The stationarity of the first variation of the Lagrangian functional relative to the electromagnetic fields produces the system of equations for the adjoint-state fields [15,16], $\boldsymbol{\lambda}_E$ and $\boldsymbol{\lambda}_H$,

$$\nabla\times\boldsymbol{\lambda}_H^*(\mathbf{u}) + \sigma(\mathbf{x})\boldsymbol{\lambda}_E^*(\mathbf{u}) = \sum_r \delta(\mathbf{x}-\mathbf{x}^r) W^E(\mathbf{u}) \left[\mathbf{E}^{obs}(\mathbf{u}) - \mathbf{E}(\mathbf{u}) \right], \tag{25}$$

$$-\nabla\times\boldsymbol{\lambda}_E^*(\mathbf{u}) - i\omega\mu\boldsymbol{\lambda}_H^*(\mathbf{u}) = \mathbf{0}, \tag{26}$$

and the gradient of the objective function relative to the conductivity tensor,

$$\left.\frac{\partial\chi}{\partial\sigma}\right|_{\mathbf{x}} = \sum_s \sum_\omega \Re\{\boldsymbol{\lambda}_E^*(\mathbf{x},\omega;\mathbf{x}^s)\mathbf{E}(\mathbf{x},\omega;\mathbf{x}^s)\}. \tag{27}$$

Therefore, to evaluate the gradient of the objective function, we need to solve the direct electromagnetic fields equation's and the adjoint-state fields equation's.

To solve the system Equations (25) and (26), we apply the correspondence principle proposed by [13].

Our derivation demonstrates that the corresponding adjoint-field equations in the fictitious dielectric medium in the frequency domain are as follows:

$$-\nabla\times\boldsymbol{\lambda}'_H(\mathbf{x},\omega') + (-i\omega')\boldsymbol{\epsilon}'(\mathbf{x})\boldsymbol{\lambda}'_E(\mathbf{x},\omega') = -\sum_r \delta(\mathbf{x}-\mathbf{x}^r)\mathbf{J}'(\mathbf{x},\omega'), \tag{28}$$

$$\nabla\times\boldsymbol{\lambda}'_E(\mathbf{x},\omega') + (-i\omega')\mu\boldsymbol{\lambda}'_H(\mathbf{x},\omega') = \mathbf{0}. \tag{29}$$

The relationship between the adjoint-state equations in a conductive medium and the corresponding adjoint-state equations in a fictitious dielectric medium are:

$$\omega' \equiv (1+i)\sqrt{\omega\omega_0}, \tag{30}$$
$$\boldsymbol{\epsilon}'(\mathbf{x}) \equiv \sigma(\mathbf{x})/(2\omega_0), \tag{31}$$
$$\boldsymbol{\lambda}'_E(\mathbf{x},\omega') \equiv \boldsymbol{\lambda}^*_E(\mathbf{x},\omega), \tag{32}$$
$$\boldsymbol{\lambda}'_H(\mathbf{x},\omega') \equiv -\sqrt{\frac{-i\omega}{2\omega_0}}\boldsymbol{\lambda}^*_H(\mathbf{x},\omega), \tag{33}$$
$$\mathbf{J}'(\mathbf{x},\omega') \equiv -\sqrt{\frac{-i\omega}{2\omega_0}}\Delta\mathbf{E}^*(\mathbf{x},\omega). \tag{34}$$

The Equations from (30) to (34) map the time-domain adjoint-state fields to the corresponding frequency-domain adjoint-state fields in the conductive medium. The system of Equations (28) and (29) in the time domain is:

$$-\nabla\times\boldsymbol{\lambda}'_H(\mathbf{x},t') + \boldsymbol{\epsilon}'(\mathbf{x})\partial'_t\boldsymbol{\lambda}'_E(\mathbf{x},t') = -\sum_r \delta(\mathbf{x}-\mathbf{x}^r)\mathbf{J}'(\mathbf{x},t'), \tag{35}$$
$$\nabla\times\boldsymbol{\lambda}'_E(\mathbf{x},t') + \mu\partial'_t\boldsymbol{\lambda}'_H(\mathbf{x},t') = \mathbf{0}. \tag{36}$$

To solve these equations through finite-difference, we use the algorithm proposed by [13]. The current density vector at each receiver position is as follows (34):

$$\mathbf{J}'(\mathbf{x}^r,\omega) = \int_0^T dt'\mathbf{J}(\mathbf{x}^r,t')e^{-\sqrt{\omega\omega_0}t'}e^{i\sqrt{\omega\omega_0}t'}. \tag{37}$$

Finally, the adjoint-field $\boldsymbol{\lambda}^*_E(\mathbf{x},\omega)$ is evaluated on the fly at each finite-difference time update, applying the correspondence relation in Equation (32):

$$\boldsymbol{\lambda}^*_E(\mathbf{x},\omega) = \int_0^T \boldsymbol{\lambda}'_E(\mathbf{x},t')e^{-\sqrt{\omega\omega_0}t'}e^{i\sqrt{\omega\omega_0}t'}dt'. \tag{38}$$

2.3. Adjoint Source

The source injected at a receiver during adjoint modeling in the fictitious time domain must satisfy the following equation:

$$\Delta E^*(\mathbf{x}_r,\omega) = -\sqrt{\frac{-2\omega_0}{i\omega}}\int_0^T dt' J'(\mathbf{x}_r,t')e^{-\sqrt{\omega\omega_0}t'}e^{i\sqrt{\omega\omega_0}t'}. \tag{39}$$

Obtaining the adjoint source (in fictitious time domain) from $\Delta E^*(\mathbf{x}_r,\omega)$ is an underdetermined problem, which allows a high degree of freedom in the choice of $J'(\mathbf{x}_r,t')$ [13]. For example, one can assume an estimated pulse, $\hat{J}'(\mathbf{x}_r,t')$, formed from the sum of pulses corresponding to the second derivative of a Gaussian:

$$\hat{J}'(\mathbf{x}_r,t') = \sum_{m=1}^{N_\omega} A_m\partial^2_{t'}\Gamma(t'-\tau_m), \tag{40}$$
$$\Gamma(t'-\tau_m) = \sqrt{\frac{\beta_m}{\pi}}e^{-\beta_m(t'-\tau_m)^2}, \tag{41}$$

where N_ω is the number of frequencies, and $\beta_m = \pi^3/\tau_m^2$.

With this choice, the adjoint source depends on the $2N_\omega$ parameters, A_m and τ_m, determined by minimizing the functional defined by the difference between the spectra of the residuals and the sources (calculated from Equation (40)) in the least-squares sense.

Following another strategy, we chose to estimate the adjoint source by solving the linear system corresponding to the discrete form of Equation (39), with equations given as follows:

$$\Delta E^*(\mathbf{x}_r, \omega_q) = \sum_{n=1}^{N_t} A_{qn} J'(\mathbf{x}_r, t'_n), \tag{42}$$

$$A_{qn} = -\Delta t' \sqrt{\frac{-2\omega_0}{i\omega_q}} e^{-\sqrt{\omega_q \omega_0} t'_n} e^{i\sqrt{\omega_q \omega_0} t'_n}, \tag{43}$$

where q is the index associated with the frequencies, i.e., $q \in \{1, \cdots, N_\omega\}$, and N_t is the number of time samples. To stabilize the system solution, we added the minimum norm regularization with coefficient λ [25], and we also used preconditioning via reparameterization:

$$\mathbf{J}' = \mathbf{Pm}, \tag{44}$$

where the preconditioning operator $\mathbf{P}$ has the form of a Ricker pulse with peak frequency f_p, expressed as follows:

$$P_{nk} = \left(1 - 2\pi^2 f_p^2 (t_n - t_k)^2\right) e^{-\pi^2 f_p^2 (t_n - t_k)^2}. \tag{45}$$

Finally, we obtain the adjoint source (in the fictitious time domain) by solving the linear system:

$$\begin{pmatrix} \mathbf{AP} \\ \lambda \mathbf{I} \end{pmatrix} \mathbf{m} = \begin{pmatrix} \Delta \mathbf{E}^* \\ \mathbf{0} \end{pmatrix}. \tag{46}$$

We obtain the solution through the conjugate gradient algorithm [19].

2.4. Gradient Preconditioning

The gradient calculated according to Equation (27) is subjected to a preconditioning process [26], used to compensate for its attenuation with depth. We implement two different approaches. The first approach, preconditioning 1, involves calculating the ratio of the gradient to the squared modulus of the electric field in each model sample, requiring calibration to avoid amplifying the gradient at the edges. In practice, at each frequency, the division is performed by the sum of the normalized field with a constant:

$$g_{precond} = \frac{g}{\left(\frac{E}{\max(E)} + \delta\right)}, \tag{47}$$

where $E = E_x^2 + E_y^2 + E_z^2$, and δ can be chosen by evaluating the square modulus of the field in the region of interest.

The second strategy (referred to as preconditioning 2) involves essentially applying a depth gain factor by multiplying the gradient by an exponential [27]:

$$g_{precond} = g \times e^{kz/\delta}, \tag{48}$$

where k is a dimensionless parameter (close to 1.0), z is the depth related to the seabed, and δ is the skin-depth [28] at the frequency of the gradient calculation.

2.5. Model Regularization

We used the model reparameterization method proposed by [18] to promote the model's smoothness. In this case, we assume the following:

$$\sigma = \mathbf{S}_2 \mathbf{S}_1 \mathbf{p}, \tag{49}$$

where S_1 and S_2 are exponential filters [29] that depend on the size of the previously defined smoothing window and operate in the directions where the model parameters vary.

We can obtain the gradient relative to the parameter vector, $\mathbf{p}(\mathbf{x})$, using the chain rule:

$$\frac{\partial \chi}{\partial \mathbf{p}} = \frac{\partial \chi}{\partial \sigma}\frac{\partial \sigma}{\partial \mathbf{p}} = S_2 S_1 \frac{\partial \chi}{\partial \sigma}\,. \tag{50}$$

2.6. Steepest Descent and Momentum Methods

CSEM inversion aims to obtain a subsurface conductivity model representative of the investigated region and honors the measurements of electromagnetic fields collected by receivers on the seabed. It is an iterative process that seeks to minimize the objective function defined in Equation (23) through successive steps in the parameter space, using the direction opposite to the gradient of the objective function calculated from the model obtained in each iteration. This update form is the main feature of steepest descent methods [20], which leads to the following expression:

$$\sigma_{k+1} = \sigma_k + \alpha \mathbf{d}_k, \tag{51}$$

where α is the step size, determined by a simple backtracking line search, and the update direction, $\mathbf{d}_k$, is given by the preconditioned gradient, $\mathbf{g}_k$, of $\chi(\sigma)$ relative to the model conductivity:

$$\mathbf{d}_k = -\frac{\mathbf{g}_k}{||\mathbf{g}_k||}. \tag{52}$$

We use the most straightforward line search strategy, starting with a step size equal to 1.0 in Equation (51). We calculate the value of the objective function using the updated model, and if there is a decrease, we proceed to the next iteration. Otherwise, we discard the update and compute a new model with a step size that is half of the previous step. This process iterates until the objective function decreases or we reach the maximum number of iterations allowed, in which case the inversion stops. In our experiments, this maximum number is 20.

Another inversion strategy used involves updating the model with a momentum vector, $\mathbf{s}$, defined from the gradient of the objective function and the momentum calculated in the previous iteration as follows:

$$\sigma_{k+1} = \sigma_k - \alpha \hat{\mathbf{s}}_k, \tag{53}$$

$$\hat{\mathbf{s}} = \frac{\sqrt{1-\beta_2^k}}{1-\beta_1^k}\mathbf{s}_k; \quad \mathbf{s}_k = \beta_1 \mathbf{s}_{k-1} + (1-\beta_1)\mathbf{d}_k, \tag{54}$$

where we adopt the parameters $\beta_1 = 0.9$ e $\beta_2 = 0.999$, according to the usual prescription of the Adam algorithm [21,30], formulated initially with a second momentum (which we do not use here). Indeed, our momentum update is very similar to the formulation of the Adam algorithm, where we essentially remove the term involving the second momentum from the equation. Furthermore, although it is not common practice, we combine momentum with the line search strategy already adopted in the context of steepest descent in an attempt to obtain more accurate models.

3. Numerical Experiments

3.1. Gradient Calculation

We validate our implementation to calculate the gradient of the objective function using the data from a 3D isotropic model measuring 18 km in the x direction, 10 km in the y direction, and 3 km deep (in the z direction), inspired by [31]. This model simulates a 2D distribution of conductivities consisting of a resistive body with a resistivity of 25 ohm.m, in a homogeneous background medium with 1 ohm.m, and a water layer of 800 m, with its central section shown in Figure 1. We also use a uniform grid with cell dimensions of

100 m (vertical) × 200 m × 200 m (in this work, we use a 3D modeling code, and to ensure compatibility with the 2D inversion codes, we adopt the strategy of replicating the models obtained during the optimization process along the third dimension).

With the model illustrated in Figure 1, we simulate, using reciprocity, the survey along its central section to use the horizontal component of the electric field (at frequencies of 0.25 Hz, 0.50 Hz, and 0.75 Hz) measured by the receivers on the seafloor with a regular spacing of 1 km, from coordinates 6 km to 12 km. The sources, located 100 m above the receivers, are distributed at intervals of 200 m, between 1.1 km and 16.9 km. Figure 2 illustrates the data associated with the central receiver at the three frequencies (since we use reciprocity, which allows the interchange between sources and receivers, the data illustrated in Figure 2 correspond to the field measured at the source positions for a source attributed to the position of the central receiver).

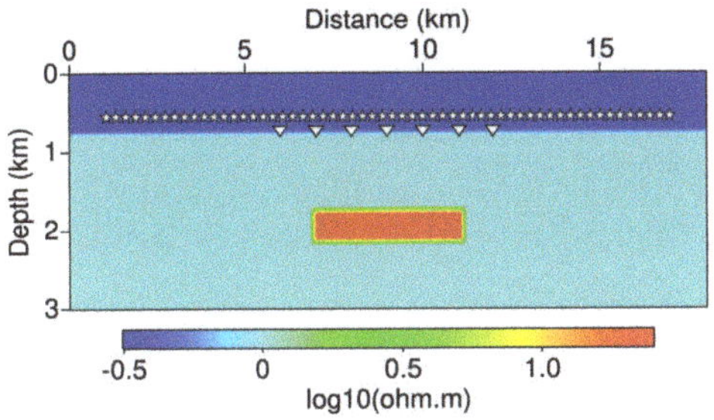

Figure 1. Resistivity model used to obtain the synthetic data to validate the gradient calculation. Triangles represent the receivers, and stars represent the sources used in the simulation. Not all sources are represented, and their height relative to the seafloor is illustrative.

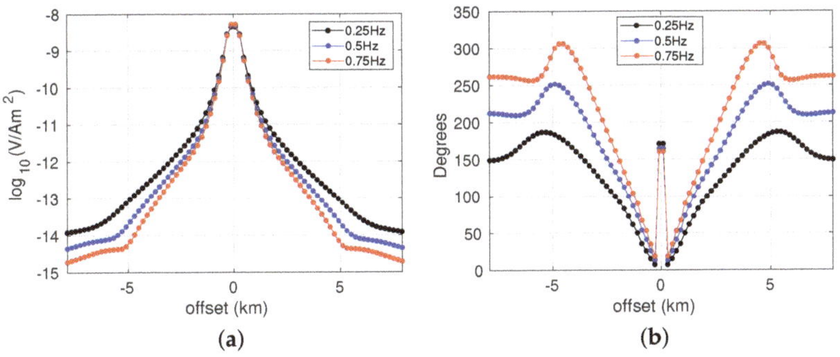

Figure 2. (**a**) Amplitude and (**b**) phase curves of the in-line electric field at frequencies 0.25 Hz, 0.50 Hz, and 0.75 Hz associated with the central receiver in the data from the model shown in Figure 1.

The estimation of the adjoint source involves solving the system of Equation (46) for the residual (right-hand side of the equation) at each offset, such that at the end of the process, the conjugate gradient routine provides a pulse in the (fictitious) time domain (vector **m**) as the solution. An important aspect of the approach we adopted is related to the need to fit the residuals at each frequency separately, in contrast to the more intuitive strategy of simultaneously fitting the residuals at the frequencies where the gradient is to be calculated, as we can see in [17]. The adoption of this approach is essential to obtain the results that will be presented below.

Next, we analyze the accuracy of our adjoint source estimation, considering the residuals between the observed data and the data calculated with a model consisting only of the 1 ohm.m background present in the model shown in Figure 1. Figure 3 illustrates the estimated pulses (at each frequency) for source 60 and the data from receiver number 4 and their amplitude spectra, along which we indicate the residuals (in the frequency domain) adjusted during obtaining adjoint sources.

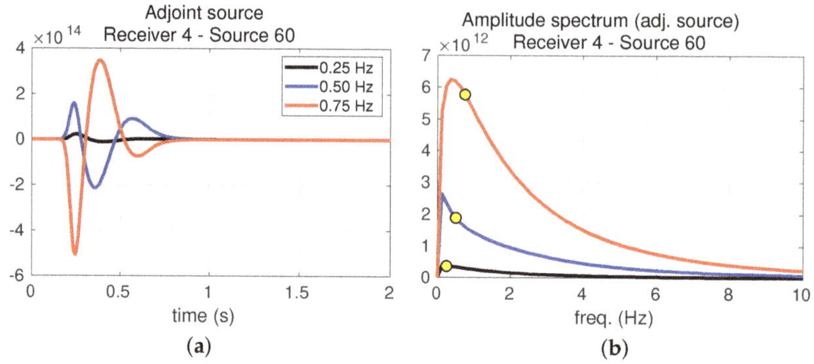

Figure 3. (**a**) Estimated adjoint source pulses derived from residuals fitted separately for each frequency; (**b**) the yellow circles indicate the amplitude spectra of the pulses, along with the residuals at each frequency.

We also illustrate the gradient calculation relative to the vertical component of the conductivity tensor based on data for receiver 1 at a frequency of 0.50 Hz. Figure 4 shows the absolute value (in log scale) of the E_z component of the electric field normalized by the source and the field λ_{E_z}, computed using forward and adjoint modeling, respectively, as well as their product, calculated according to Equation (27). To improve the visualization, we apply muting in a depth range corresponding to five samples below the seafloor.

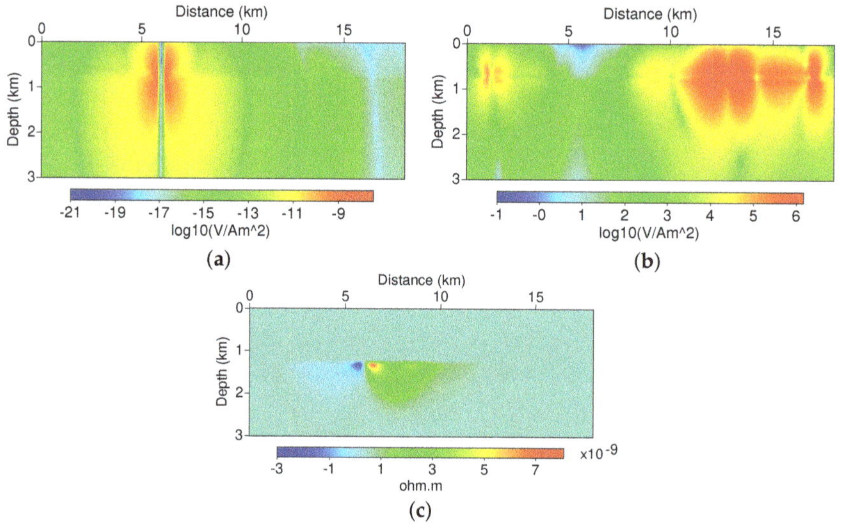

Figure 4. (**a**) Absolute values (log scale) of the E_z component of the electric field, normalized by the source, computed by forward modeling with the source at the position of receiver 1 (at frequency of 0.50 Hz); (**b**) absolute values (log scale) of the λ_{E_z} field obtained by adjoint modeling with the sources estimated from data from receiver 1 (at frequency of 0.50 Hz); (**c**) gradient calculated from the product of the fields from forward modeling and the adjoint modeling.

The "complete" gradient is obtained by summing the gradients calculated from the data of each receiver. Figure 5 shows the results for frequencies of 0.25, 0.50, and 0.75 Hz, which have similar characteristics in shape but different depth ranges. Figure 6 illustrates the effect of the preconditioning strategies discussed in Section 2.4, where we show the result of their application to the gradient obtained with the 0.25 Hz data. As we can see, precondition 1 (performed with $\delta = 10^{-4}$) positions the anomaly deeper while precondition 2 (performed with $k = 1.0$) increases the depth range of the anomaly.

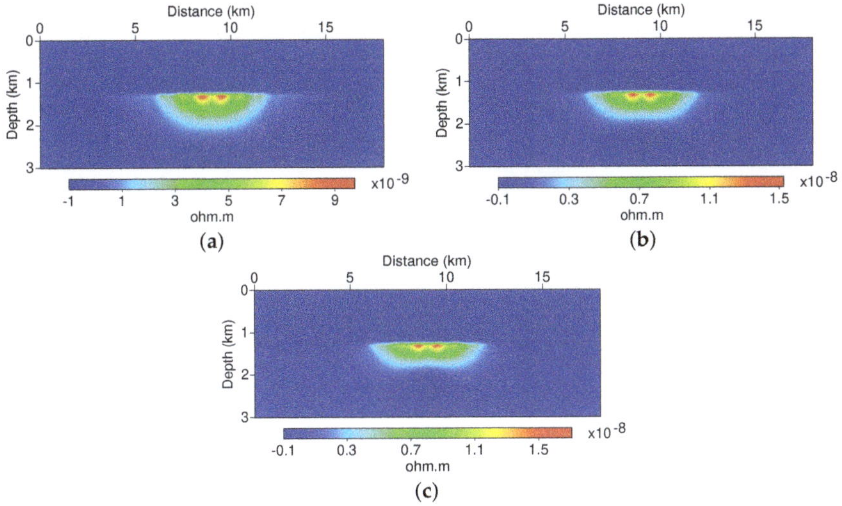

Figure 5. Sum of the gradients associated with each receptor, obtained from the product between forward modeling and adjoint modeling fields. (**a**) Gradient at frequency of 0.25 Hz; (**b**) gradient at frequency of 0.50 Hz; (**c**) gradient at the frequency of 0.75 Hz.

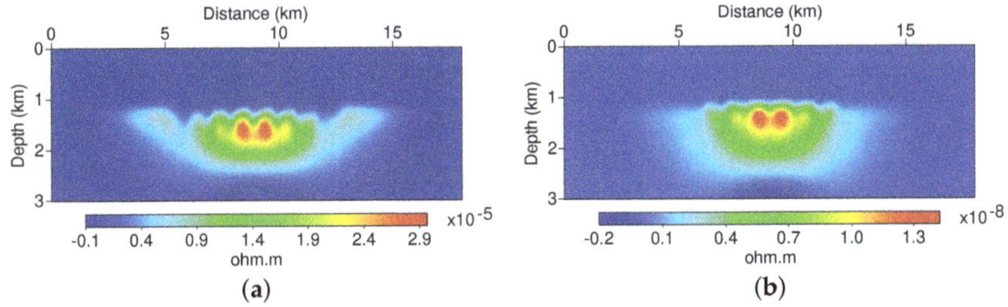

Figure 6. (**a**) Gradient relative to the vertical conductivity (at frequency of 0.25 Hz), preconditioned with precondition 1 (obtained with $\delta = 10^{-4}$); (**b**) gradient relative to vertical conductivity (at frequency of 0.25 Hz), preconditioned with precondition 2 (obtained with $k = 1.0$).

3.2. Inversion Results

In this section, we will discuss the inversion results of the data obtained under three different scenarios. The first is similar to the model used in Section 3.1, consisting of two resistors located at varying depths in a homogeneous background medium. The second is an adaptation of one of the demonstration models from the MARE2DEM modeling and inversion package (available at http://mare2dem.ucsd.edu acessed on 11 December 2023) [32], and the third corresponds to an adaptation of the synthetic model of the Marlim field (on the Brazilian continental margin) [33], constructed from seismic sections and information from the main horizons in the area. The optimizers used will be the steepest descent and the momentum method described in Section 2.6, as well as L-BFGS-B [22] (computed using a limited-memory matrix using the previous five gradients).

3.2.1. Model with Two Resistors

The model with two resistors has the same background as in Section 3.1, with the same resistivity and water layer thickness. The difference is that there are two resistors (not just one) located at depths of 1.8 km and 2.1 km (Figure 7a). The data correspond to the simulated horizontal electric field measured between 3 and 15 km, with sources every 200 m, in the interval of 1.1 km to 16.9 km. The data from each receiver were restricted to offsets between 1 and 6 km. Figure 7b illustrates the gradient at the frequency of 0.25 Hz, relative to the vertical conductivity, computed in a homogeneous medium with a resistivity of 1 ohm.m. As we can see, there is consistency in the regions where the gradient is prominent, considering the position of the resistors in the model used to generate the observed data. The higher value in the left portion is a consequence of the shallower depth of the resistor located in that region.

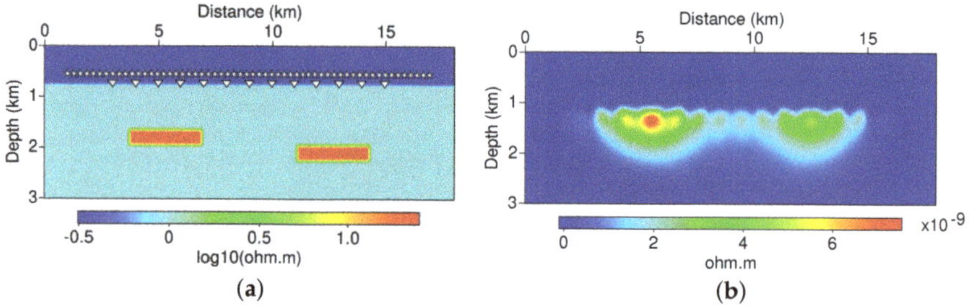

Figure 7. (**a**) Resistivity model formed by two 25 ohm.m bodies at depths of 1.8 and 2.1 km in a homogeneous medium with 1 ohm.m resistivity. Triangles represent the receivers, and stars represent the sources used in the simulation. Not all sources are represented, and their height relative to the seafloor is illustrative. (**b**) Gradient relative to the vertical conductivity obtained from 0.25 Hz data in a homogeneous medium with a resistivity of 1 ohm.m.

The inversions used data at a frequency of 0.25 Hz on a mesh with a grid interval of 100 m (along the vertical) and 200 m (along the horizontal) and a homogeneous initial model with a resistivity of 1 ohm.m. Below, we present the best results obtained with each tested method. In inversions with the steepest descent and momentum, we used preconditioning 1 ($\delta = 10^{-4}$), and in inversion with L-BFGS-B, we used preconditioning 2 ($k = 1.0$). Figure 8 shows the curves of the objective function's evolution during the inversions, highlighting the greater efficiency of the steepest descent method, which takes just over 20 iterations to reach a value comparable to that obtained by the momentum inversion after 60 iterations. The L-BFGS-B method is similar to the steepest descent, although it takes more iterations to reach the minimum of the objective function value.

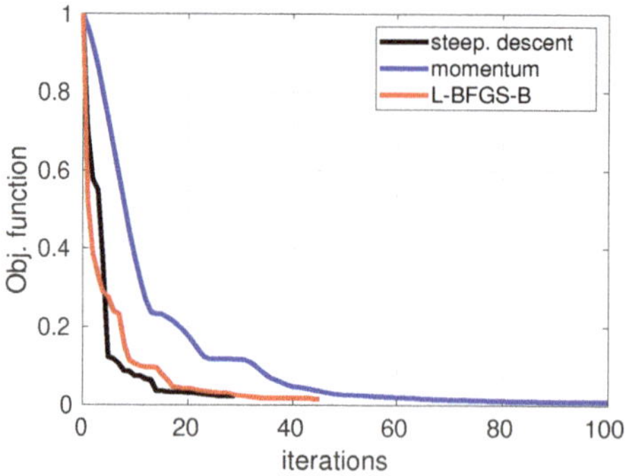

Figure 8. Curves showing the evolution of the objective function during the inversions for the model with two resistors, using the steepest descent method (black curve), momentum (blue curve), and the L-BFGS-B (red curve).

The models obtained from the inversions are similar (Figure 9). They are indicated by the two resistive bodies at different depths, consistent with the resistors' positions in the model used to produce the observed data. In the cases of the steepest descent and the momentum method, we observe that the deeper resistor is slightly above the desired position, unlike the model obtained with the L-BFGS-B method, in which this resistor is better positioned.

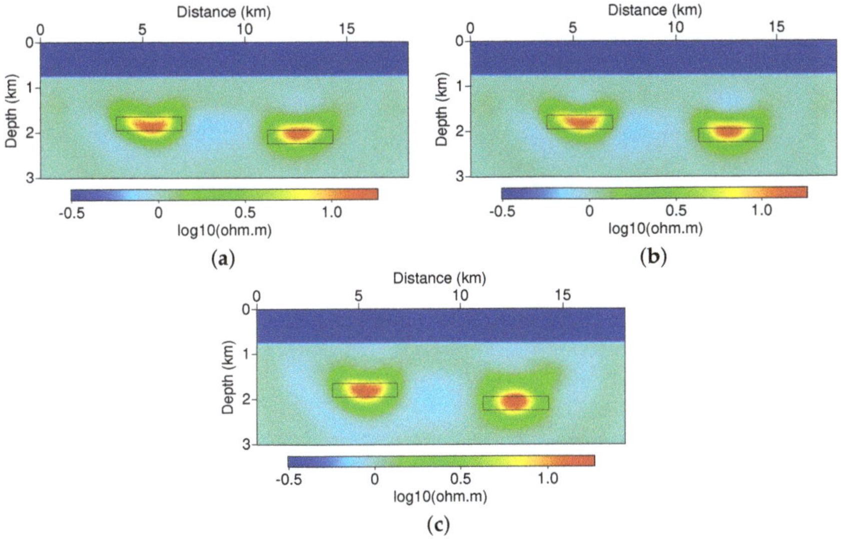

Figure 9. Results of the inversions with (**a**) the steepest descent, (**b**) the momentum method, and (**c**) the L-BFGS-B method uses the observed data simulated from the model with two resistors. Black boxes represent the edge of the resistors.

3.2.2. MARE Model

The model we will call MARE consists of an adaptation of the isotropic conductivity model available in one of the versions of the MARE2DEM package for modeling and inversion (the MARE2DEM package [32] is a CSEM 2.5D data modeling and inversion software suite based on finite elements. The code was initially developed as part of the SEMC consortium (Seafloor Electromagnetic Methods Consortium) at Scripps Institution of Oceanography, University of California, San Diego, USA). It simulates the presence of two bodies: a smaller one with a resistivity of 50 ohm.m and a larger one, to the left of the first, of 4 ohm.m. These resistors are embedded in a medium formed by conductive layers and an extended structure with 100 ohm.m, located in the lower portion of the model. The adaptation involved replacing the bathymetry present in the original model with a horizontal seabed at a depth of 1.2 km, as we can see in Figure 10.

The acquisition used in the simulation of observed data consisted of 25 receivers positioned on the seabed, regularly spaced between 1 and 25 km, and sources located 50 m above the receivers, from coordinate 0 to coordinate 26 km, with sampling every 200 m. We computed the synthetic data using the MARE2DEM modeling code [34]. The field used corresponds to the inline electric field. The receiver data were limited to offsets greater than 1 km and those with field amplitude greater than 10^{-15} V/m. The inversions used 0.25 Hz data.

To perform the inversions, we limited the vertical extension of the model to a depth of 6 km on a grid with a spacing of 100 m (vertical) by 200 m (horizontal), as seen in the initial model (shown in Figure 11a), constructed by smoothing the background formed by the conductive layers present in the reference model. Figure 11b illustrates the gradient relative to vertical conductivity at a frequency of 0.25 Hz, calculated with the model presented in

Figure 11a. There is a strong correlation between the regions where the gradient stands out and the locations where the resistors are present in the reference model that produced the observed data.

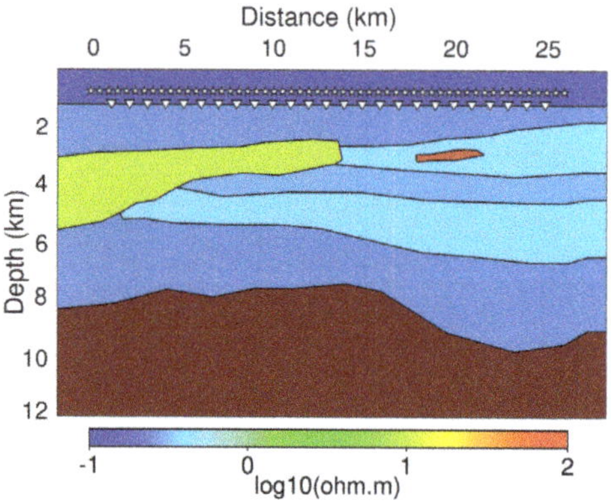

Figure 10. The MARE model used to simulate observed data in the inversions. Triangles represent the receivers, and stars represent the sources used in the simulation. Not all sources are represented, and their height relative to the seafloor is illustrative.

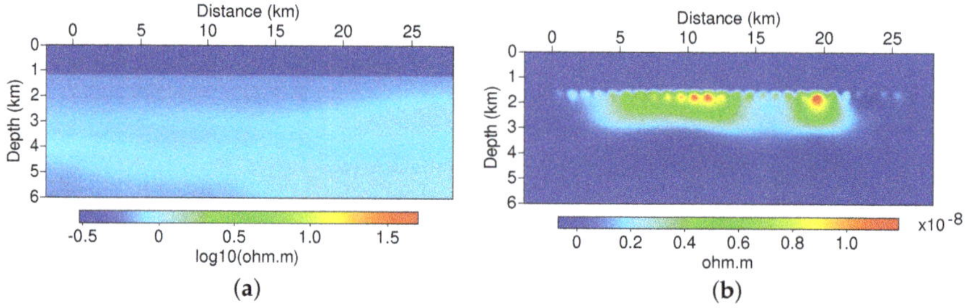

Figure 11. (**a**) Initial model used in the inversions of data modeled from the MARE model. (**b**) Gradient relative to vertical conductivity at a frequency of 0.25 Hz, obtained from the initial model used in the inversions.

Next, we present the best results obtained with each methodology presented. In the case of the steepest descent with line search (Figure 12a) and the inversion with momentum (Figure 12b), we used the gradient with precondition 2. In contrast, in the inversion with L-BFGS-B (Figure 12c), the best result was obtained using precondition 1. Figure 13 shows the convergence curves for the objective function. As we can see, the steepest descent and L-BFGS-B present the best convergence behavior compared to the inversion with the momentum algorithm. The estimated models are very similar, especially concerning the smaller and more resistive body on the right, which is well-defined by the three inversions (with resistivity values slightly closer to the reference model in the results obtained with momentum and L-BFGS-B). Regarding the larger resistor on the left, it is possible to notice that there is an overestimation of resistivity in the three results, with a slightly pronounced effect in the result of the inversion with the steepest descent and L-BFGS-B.

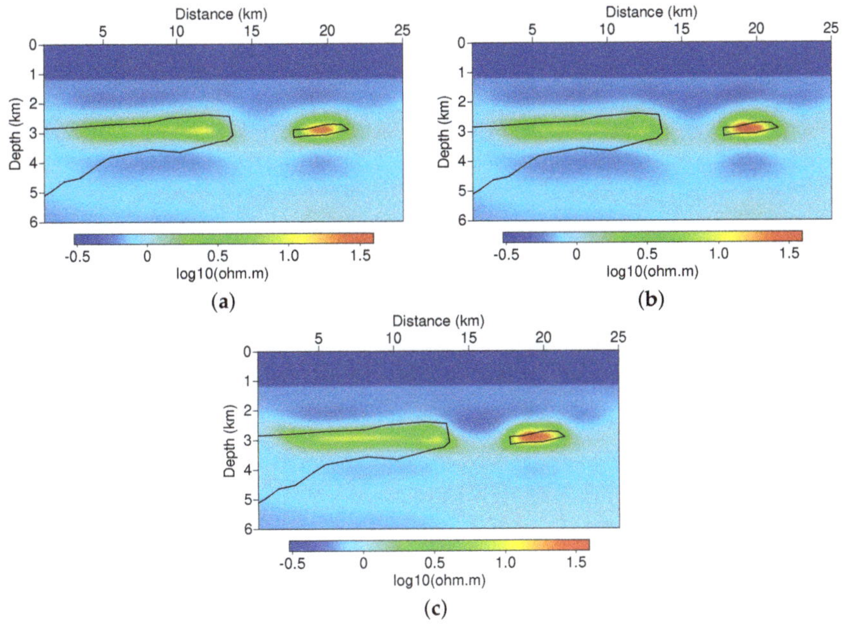

Figure 12. Results of the inversions with (**a**) the steepest descent, (**b**) the momentum method, and (**c**) the L-BFGS-B method, performed using observed data simulated from the MARE model. Black outlines indicate the boundaries of the resistors.

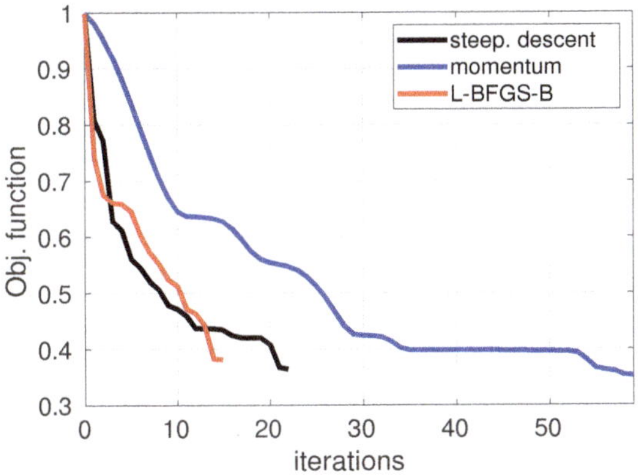

Figure 13. Curves showing the evolution of the objective function during the inversions for the MARE model, using the steepest descent method (black curve), momentum (blue curve), and the L-BFGS-B (red curve).

3.2.3. Marlim Model

The model we will refer to as Marlim is an adaptation of the MR3D resistivity model constructed by [33] from 3D seismic and well-log data from the Marlim field (Brazilian continental margin). The features of interest in using electromagnetic (EM) methods include a heterogeneous reservoir embedded in a conductive background and a salt body with a resistivity of 1000 ohm.m. It is a good representation of the type of scenario where the use of the CSEM method might be required, which was also described by [35] when discussing the forward modeling performed from the MR3D model.

The modifications made relative to the original MR3D model consist of removing the effects of bathymetry and using only the vertical resistivity of the original model (in this case, we considered the model to be isotropic). Figure 14 illustrates the central section of the model (Figure 14a) from which we simulated the observed data through modeling that used a grid of 20 m (vertical) by 100 m (horizontal) and recorded the inline electric field measured by 20 receivers regularly arranged along the line that coincides with the section in Figure 14a (in the interval from 11 km to 30 km). The sources are distributed between 0 to 41 km every 200 m. We discarded data with offsets smaller than 1 km and amplitudes smaller than 10^{-15} V/m, and the inversions used single-frequency data at 0.25 Hz. The initial model is homogeneous, with a resistivity of 2.8 ohm.m (which corresponds to the average background resistivity in the region where the reservoir is in the Marlim model), the same used in calculating the gradient relative to vertical conductivity (Figure 14b) on a mesh with a grid spacing of 100 m (along the vertical) and 200 m (along the horizontal). Note the asymmetry in the gradient shape associated with the location of the resistor in the reference model.

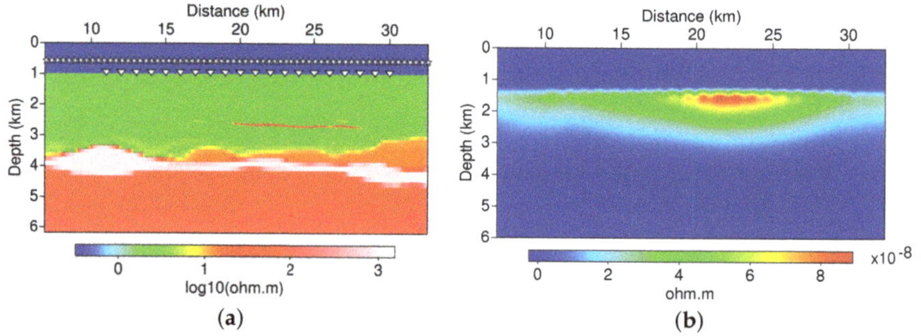

Figure 14. (**a**) Central section of the Marlim model used to simulate the observed data utilized in the inversions. Triangles represent the receivers, and stars represent the sources used in the simulation. Not all sources are represented, and their height relative to the seafloor is illustrative. (**b**) Gradient (at a frequency of 0.25 Hz) relative to vertical conductivity, calculated from the homogeneous medium with a resistivity of 2.8 ohm.m.

Using the data and the initial model described, the inversions performed produced the results that will be illustrated below. Using the steepest descent method with gradient precondition 1, we obtained the model shown in Figure 15a. In the inversion with momentum, where we used precondition 2, we obtained the model in Figure 15b, and the inversion with L-BFGS-B (performed using the gradient with precondition 1), resulting in the model shown in Figure 15c. All results indicate the presence of a resistor in the region where the reservoir is located in the Marlim model, each of them presenting its own characteristics. The models obtained with the steepest descent and L-BFGS-B are similar, as they are characterized by an anomaly with higher resistivity in the case of the L-BFGS-B inversion. The result obtained with momentum is better distributed, covering a larger reservoir extent. The curves showing the evolution of the objective function during the inversions are shown in Figure 16, indicating a slightly more effective reduction rate in the case of L-BFGS-B, especially in the first iterations. In contrast, the others show similar behavior, with the momentum method reaching a slightly lower residual value.

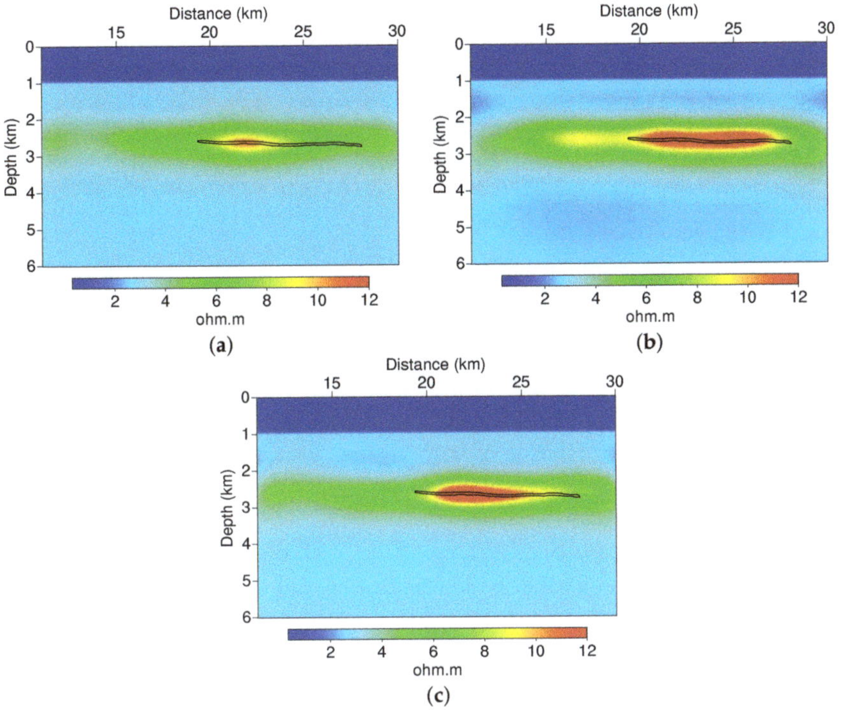

Figure 15. Resistivity models obtained from the inversions using (**a**) steepest descent, (**b**) momentum, and (**c**) L-BFGS-B. The black outline represents the boundary of the resistor present in the Marlim model.

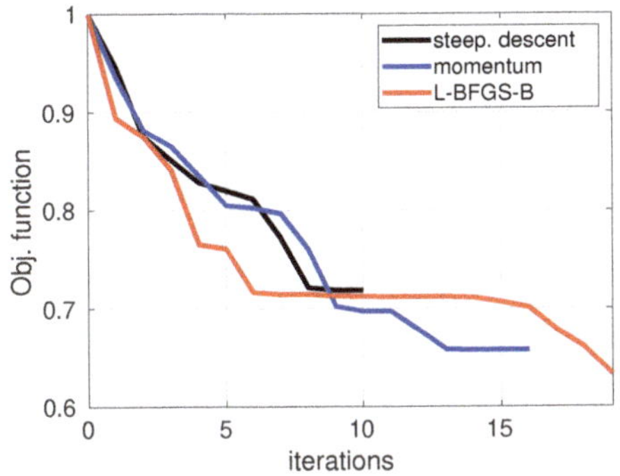

Figure 16. Curves showing the evolution of the objective function during the inversions for the Marlim model, using the steepest descent method (black curve), momentum (blue curve), and L-BFGS-B (red curve).

4. Discussion

The results illustrated in the previous section consolidate the use of the adjoint-state method and the correspondence principle to obtain the gradient (relative to the model conductivities) of the objective function defined in Equation (23). The fictitious domain modeling strategy, already established in the context of solving the forward problem, also proved efficient in obtaining the adjoint fields. For the calculation of the gradient consistent with the perturbation observed in the data due to the presence of a resistor in the model, we opted for the approach based on obtaining the adjoint sources for each frequency separately, unlike other authors who address the topic, as seen in the work of [36].

We validated the quality of the gradient calculated with the proposed methodology, preconditioning, and regularization using model parameterization through the inversions of simulated data from three different models: the model consisting of two resistors in a conductive background, the MARE model, an adaptation of the example model from the MARE2DEM package, and the Marlim model, adapted from the MR3D model, constructed by [33]. We used three optimization methods: steepest descent, our proposed modified momentum algorithm, and L-BFGS-B.

Regardless of the dataset we used and the inversion methodology employed, we observed interesting aspects in the inversion results, which produced models with well-positioned resistors at depth and without artifacts near the receiver positions, commonly seen in CSEM inversion results. We attribute these achievements to the use of gradient preconditioning and smoothing by reparameterization.

For the model with two resistors, all three methods successfully obtained results that included the bodies at different depths, with the positioning of the deeper resistor being slightly better in the case of L-BFGS-B (Figure 9). From the perspective of the number of iterations needed for the objective function to reach its minimum value, L-BFGS-B was slightly slower than the steepest descent, with the momentum inversion requiring the most iterations (Figure 8).

The inversions for the MARE model were obtained using synthetic data produced with the MARE2DEM package, and the analysis of the objective function evolution curves during the inversions indicated slower convergence for the momentum optimization, which reached its lowest level after iteration 30, in contrast to the performance of steepest descent and L-BFGS-B, which did so at around iterations 20 and 15, respectively (Figure 13). The results obtained were similar but presented a subtle difference in the resistivity of the larger resistor, which was slightly overestimated in some cases (Figure 12). This feature was somewhat more pronounced in the model obtained with the steepest descent and less noticeable in the model obtained with momentum.

In the inversions for the Marlim model, we observed similar behavior in the convergence curve of the objective function during the first iterations with the steepest descent and momentum; the L-BFGS-B showed a sharper reduction followed by stagnation that lasted for about ten iterations, after which the objective function began to decrease again (Figure 16). The models obtained with the three optimizers differed (Figure 15). While the steepest descent result indicated the presence of a modest and localized resistor, the result obtained with momentum was more consistent with the overall geometry of the reservoir. The result produced with L-BFGS-B was somewhat similar to that obtained with the steepest descent but slightly better as it identified a larger and more resistive resistor.

The analysis of the results produced with the different optimization methodologies indicates that the momentum method tends to fit the observed data slightly better compared to the steepest descent method (as observed in the experiments conducted with the data from the two-resistor model and the MARE model). In the inversion for the Marlim model, this approach not only fit the data better but also recovered a model that was more consistent with the reference model.

5. Conclusions

In this work, we implemented CSEM inversion without solving large linear systems, which can be extremely useful in the context of 3D problems, where this task might become prohibitive. Our implementation used the correspondence principle to derive the forward modeling and gradient calculation. To improve the conditioning of the inversion, we used gradient preconditioning and model reparameterization smoothing. As a result, we obtained very satisfactory models from the inversion of data produced from three different scenarios, using the steepest descent and the momentum algorithm and comparing them to the results from L-BFGS-B. In the two-resistor and the MARE models, the superior performance of the steepest descent and L-BFGS-B was evident, as they required fewer iterations to achieve results similar to those obtained by the momentum method (which needed more iterations). In the case of the Marlim model, the performances were comparable, with better models achieved by the momentum method and L-BFGS-B. Our proposal for using momentum is based on its update from two parameters, similar to what is done in the formulation of the Adam algorithm. However, unlike this one, we used only one momentum term. As we observed, the results from the algorithm we proposed are comparable to the models produced with L-BFGS-B, with the advantage of requiring less memory consumption and fitting the observed data slightly better.

Author Contributions: Methodology, A.V., D.N. and J.C.; validation, A.V., D.N. and J.C.; numerical experiments, A.V., D.N. and J.C.; writing—review and editing, A.V., D.N. and J.C. All authors have read and agreed to the published version of the manuscript.

Funding: This research was funded by the Coordenação de Aperfeiçoamento de Pessoal de Nível Superior-Brasil (CAPES)-grant number 88882.445206/2019-01 (A.V.), by the Conselho Nacional do Desenvolvimento Científico e Tecnológico-CNPq-grant number 313011/2022-2 and the Instituto Nacional de Ciência e Tecnologia-Geofísica do Petróleo-INCT-GP (J.C.) and by PETROBRAS (D.N.).

Data Availability Statement: The data used in this research can be requested directly from the authors.

Conflicts of Interest: The authors declare no conflicts of interest.

References

1. Constable, S. Ten years of marine CSEM for hydrocarbon exploration. *Geophysics* **2010**, *75*, 75A67–75A81. [CrossRef]
2. Menezes, P.T.L.; Ferreira, S.M.; Correa, J.L.; Menor, E.N. Twenty Years of CSEM Exploration in the Brazilian Continental Margin. *Minerals* **2023**, *13*, 870. [CrossRef]
3. Buonora, M.; Correa, J.; Martins, L.; Menezes, P.; Pinho, E.; Crepaldi, J.; Ribas, M.; Ferreira, S.; Freitas, R. mCSEM data interpretation for hydrocarbon exploration: A fast interpretation workflow for drilling decision. *Interpretation* **2014**, *2*, SH1–SH11. [CrossRef]
4. Menezes, P.T.L.; Correa, J.L.; Alvim, L.M.; Viana, A.R.; Sansonowski, R.C. Time-Lapse CSEM Monitoring: Correlating the Anomalous Transverse Resistance with SoPhiH Maps. *Energies* **2021**, *14*, 7159. [CrossRef]
5. Ettayebi, M.; Wang, S.; Landrø, M. Time-Lapse 3D CSEM for Reservoir Monitoring Based on Rock Physics Simulation of the Wisting Oil Field Offshore Norway. *Sensors* **2023**, *23*, 7197. [CrossRef]
6. Alumbaugh, D.L.; Newman, G.A.; Prevost, L.; Shadid, J.N. Three-dimensional wideband electromagnetic modeling on massively parallel computers. *Radio Sci.* **1996**, *31*, 1–23. [CrossRef]
7. da Piedade, A.A.; Régis, C.; Nunes, C.M.B.; da Silva, H.F. Computational cost comparison between nodal and vector finite elements in the modeling of controlled source electromagnetic data using a direct solver. *Comput. Geosci.* **2021**, *156*, 104901. [CrossRef]
8. Crepaldi, J.L.S.; Buonora, M.P.P.; Figueiredo, I. Fast marine CSEM inversion in the CMP domain using analytical derivatives. *Geophysics* **2011**, *76*, F303–F313. [CrossRef]
9. Hansen, K.; Panzner, M.; Shantsev, D.; Mittet, R. TTI inversion of marine CSEM data. In *SEG Technical Program Expanded Abstracts 2016*; Society of Exploration Geophysicists: Houston, TX, USA, 2016; pp. 1014–1018.
10. Meju, M.A.; Mackie, R.L.; Miorelli, F.; Saleh, A.S.; Miller, R.V. Structurally tailored 3D anisotropic controlled-source electromagnetic resistivity inversion with cross-gradient criterion and simultaneous model calibration. *Geophysics* **2019**, *84*, E387–E402. [CrossRef]
11. Hoversten, G.M.; Mackie, R.L.; Hua, Y. Reexamination of controlled-source electromagnetic inversion at the Lona prospect, Orphan Basin, Canada. *Geophysics* **2021**, *86*, E157–E170. [CrossRef]
12. Cai, H.; Long, Z.; Lin, W.; Li, J.; Lin, P.; Hu, X. 3D multinary inversion of controlled-source electromagnetic data based on the finite-element method with unstructured mesh. *Geophysics* **2021**, *86*, E77–E92. [CrossRef]
13. Mittet, R. High-order finite-difference simulations of marine CSEM surveys using a correspondence principle for wave and diffusion fields FDTD simulation of marine CSEM surveys. *Geophysics* **2010**, *75*, F33–F50. [CrossRef]
14. de Hoop, A.T. A general correspondence principle for time-domain electromagnetic wave and diffusion fields. *Geophys. J. Int.* **1996**, *127*, 757–761. [CrossRef]
15. Plessix, R.E. A review of the adjoint-state method for computing the gradient of a functional with geophysical applications. *Geophys. J. Int.* **2006**, *167*, 495–503. [CrossRef]
16. Chavent, G. *Nonlinear Least Squares for Inverse Problems—Theoretical Foundations and Step-by-Step Guide for Applications;* Number 1 in Scientific Computation; Springer: Dordrecht, The Netherlands, 2010; p. 360. [CrossRef]

17. Valente, A.; Nascimento, D.; Costa, J.C. CSEM inversion using the correspondence principle: The adjoint-state approach. In Proceedings of the 17th International Congress of the Geophysical Society, Online, 8–11 November 2021; Sociedade Brasileira de Geofísica: Rio de Janeiro, Brazil, 2021.
18. Harlan, W.S. Regularization by model reparameterization. *Citeseer* **1995** .
19. Claerbout, J.F.; Fomel, S. *Image Estimation by Example: Geophysical Soundings Image Construction—Multidimensional Autoregression;* Stanford University: Stanford, CA, USA, 2003.
20. Kochenderfer, M.J.; Wheeler, T.A. *Algorithms for Optimization;* Massachusetts Institute of Technology: Cambridge, MA, USA, 2019.
21. Kingma, D.P.; Ba, J. Adam: A method for stochastic optimization. *arXiv* **2014**, arXiv:1412.6980.
22. Zhu, C.; Byrd, R.H.; Lu, P.; Nocedal, J. Algorithm 778: L-BFGS-B: Fortran subroutines for large-scale bound-constrained optimization. *ACM Trans. Math. Softw. (Toms)* **1997**, *23*, 550–560. [CrossRef]
23. Nguyen, A.K.; Nordskag, J.I.; Wiik, T.; Bjørke, A.K.; Boman, L.; Pedersen, O.M.; Ribaudo, J.; Mittet, R. Comparing large-scale 3D Gauss–Newton and BFGS CSEM inversions. In *SEG Technical Program Expanded Abstracts 2016;* Society of Exploration Geophysicists: Houston, TX, USA, 2016; pp. 872–877.
24. Plessix, R.E.; Mulder, W. Resistivity imaging with controlled-source electromagnetic data: Depth and data weighting. *Inverse Probl.* **2008**, *24*, 034012. [CrossRef]
25. Menke, W. *Geophysical Data Analysis: Discrete Inverse Theory;* Academic Press: Cambridge, MA, USA, 2018.
26. Shin, C.; Jang, S.; Min, D.J. Improved amplitude preservation for prestack depth migration by inverse scattering theory. *Geophys. Prospect.* **2001**, *49*, 592–606. [CrossRef]
27. Plessix, R.; van der Sman, P. 3D CSEM modeling and inversion in complex geologic settings. In *SEG Technical Program Expanded Abstracts 2007;* Society of Exploration Geophysicists: Houston, TX, USA, 2007; pp. 589–593. [CrossRef]
28. Chave, A.D.; Constable, S.C.; Edwards, R.N. Electrical Exploration Methods for the Seafloor. In *Electomagnetic Methods in Applied Geophysics, Vol. 2, Application;* Nabighian, M.N., Ed.; SEG: Houston, TX, USA, 1987; Volume 2, Investigations in Geophysics, Chapter 12; pp. 931–966.
29. Hale, D. My Favorite Ten-Line Computer Program. 2012. Available online: https://inside.mines.edu/~dhale/notebook.html (accessed on 30 May 2021).
30. Liu, L.; Yang, B.; Zhang, Y.; Xu, Y.; Peng, Z.; Wang, F. Solving Electromagnetic Inverse Problem Using Adaptive Gradient Descent Algorithm. *IEEE Trans. Geosci. Remote Sens.* **2023**, *61*, 5902415. [CrossRef]
31. Støren, T.; Zach, J.; Maaø, F. Gradient calculations for 3D inversion of CSEM data using a fast finite-difference time-domain modelling code. In Proceedings of the 70th EAGE Conference and Exhibition incorporating SPE EUROPEC 2008, Rome, Italy, 9–12 June 2008.
32. Key, K. MARE2DEM: A 2-D inversion code for controlled-source electromagnetic and magnetotelluric data. *Geophys. J. Int.* **2016**, *207*, 571–588. [CrossRef]
33. Carvalho, B.R.; Menezes, P.T.L. Marlim R3D: A realistic model for CSEM simulations-phase I: Model building. *Braz. J. Geol.* **2017**, *47*, 633–644. [CrossRef]
34. Key, K.; Ovall, J. A parallel goal-oriented adaptive finite element method for 2.5-D electromagnetic modelling. *Geophys. J. Int.* **2011**, *186*, 137–154. [CrossRef]
35. Correa, J.L.; Menezes, P.T.L. MR3D: A realistic model for CSEM simulations—Phase II: The CSEM dataset. In *SEG Technical Program Expanded Abstracts 2018;* SEG: Houston, TX, USA, 2018; pp. 959–963. [CrossRef]
36. Yang, P. 3D fictitious wave domain CSEM inversion by adjoint source estimation. *Comput. Geosci.* **2023**, *180*, 105441. [CrossRef]

Disclaimer/Publisher's Note: The statements, opinions and data contained in all publications are solely those of the individual author(s) and contributor(s) and not of MDPI and/or the editor(s). MDPI and/or the editor(s) disclaim responsibility for any injury to people or property resulting from any ideas, methods, instructions or products referred to in the content.

applied sciences

Article

Testing the Efficacy of Indirect Methods on Characterization of Sedimentary Basins by Correlation of Direct Data and Geophysical Techniques

Javier Rey [1], Rosendo Mendoza [2], M. Carmen Hidalgo [1,*] and Bruna Marinho [1]

1 Department of Geology and CEACTEMA, Higher Polytechnic School of Linares, Technological Scientific Campus, University of Jaen, 23700 Linares, Spain; jrey@ujaen.es (J.R.); bmb00013@red.ujaen.es (B.M.)
2 Department of Mechanical and Mining Engineering and CEACTEMA, Higher Polytechnic School of Linares, Technological Scientific Campus, University of Jaen, 23700 Linares, Spain; rmendoza@ujaen.es
* Correspondence: chidalgo@ujaen.es

Abstract: The information obtained from direct data (geological mapping and boreholes) and indirect techniques (reflection seismology, time-domain electromagnetics and magnetometry) is combined to analyse the northern limit of the Bailén basin (southeastern Spain). This Triassic–Neogene basin is confined by a graben-type structure, limited by two normal faults in the SW–NE direction (the Baños de la Encina-La Carolina fault and Guarromán fault). The movement of these faults was complex, with different pulses occurring over time. Therefore, the subsidence of the basin and the sedimentary filling of the graben were different, giving rise to lateral changes in the facies and thicknesses. This study focuses on the Baños de la Encina fault, chosen as the experimental site to analyse the effectiveness and accuracy of these geophysical techniques to reveal the basement structure and geometry. Seismic reflection allows to detect two faults that caused the subsidence of the eastern sector of the graben. The TDEM method made it possible to calculate the depth of the Palaeozoic basement, as well as reveal the presence of the two aforementioned faults. Magnetic total field data highlight variations in the basement depth that can be used to infer previously unknown fractures, in this case, in the NW–SE direction.

Keywords: fault detection; seismic reflection; time-domain electromagnetics; magnetometry; Spain

1. Introduction

There are many examples of the application of different geophysical techniques in geological-mining research. The most commonly used techniques include gravimetric, magnetic, electrical, electromagnetic and seismic methods. However, these indirect techniques include a data inversion process, which generates uncertainty in the interpretation of these data (inverse problem). The solution to the inverse problem is not unique, since the same set of field data supports a domain of valid mathematical solutions, and modifications can be made in the measured property (e.g., resistivity) and in the thicknesses of the geological unit with the same measurement results [1]. For this reason, it is advisable to combine indirect techniques that measure different properties and whose results have a good geological significance depending on the direct information available [2–5].

In this work, the capacities of different geophysical techniques to study geological structures are analysed. For this purpose, the northern boundary of the detrital aquifer of Bailén (southeastern Spain, Figure 1a), characterized by the presence of normal faults in a NE–SW direction with several hundred meters of vertical displacement, was selected as the experimental site. The following two geophysical techniques that measure different properties (elasticity and conductivity) and offer good resolution at several hundred metres depth are used together: seismic and electromagnetic techniques (time-domain electromagnetic (TDEM) method). As a complementary geophysical technique, the magnetic method was chosen. Although its use in the study of fractures is not as frequent in the literature as the previous ones, it was expected that it will provide additional information, such as the strike of the fractures existing in the subsoil and covered by recent materials. In addition, direct field information and groundwater catchment data were available, providing verification and correlation tools for the indirect techniques used in this work.

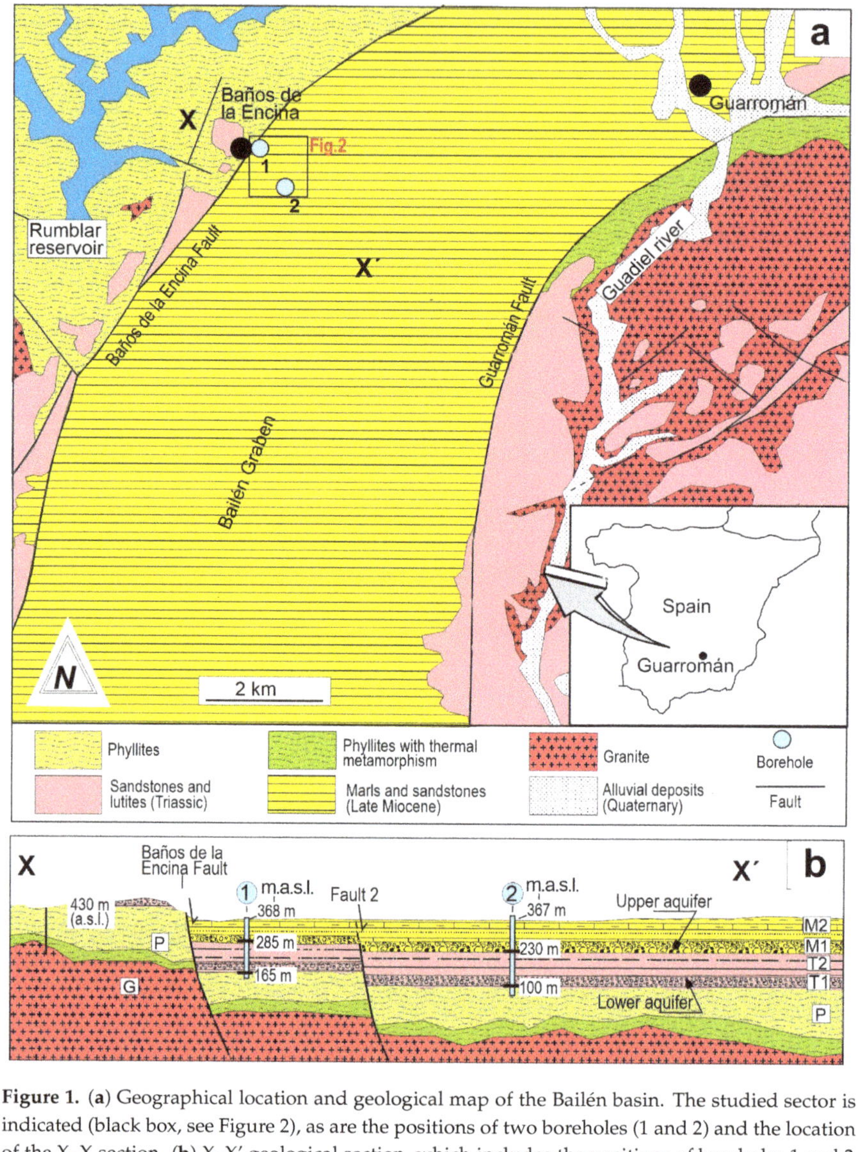

Figure 1. (**a**) Geographical location and geological map of the Bailén basin. The studied sector is indicated (black box, see Figure 2), as are the positions of two boreholes (1 and 2) and the location of the X–X section. (**b**) X–X′ geological section, which includes the positions of boreholes 1 and 2. G: granite, P: phyllites; T1: conglomerates and sandstones of the basal Triassic (lower aquifer); T2: lutites of the upper part of the Triassic; M1: rudites and calcarenites from the basal Miocene unit (upper aquifer); M2: marls from the upper part of the Miocene.

First of all, the information provided by seismic reflection is analysed and compared with stratigraphic data obtained in the field. Seismic methods are the most important geophysical technique in terms of accuracy, resolution and penetration depth [1]. The main objective of seismic reflection is to provide information on the geological structures of the subsoil and to identify different boundaries through changes in the acoustic impedance of the materials passed through [6]. Reflection seismology is the most commonly used tool for establishing the fundamental structural and depositional framework, therefore it has been widely used in the analysis of sedimentary basins and in hydrocarbon exploration [7–10].

Second, the time-domain electromagnetic method (TDEM) is applied in this study. The TDEM operating principle consists of circulating an electric current through a transmitting coil for short periods of time. When the flow of current is abruptly interrupted, a variation in the primary magnetic field occurs, which induces a variable electric current. This, in turn, generates a transient secondary magnetic field. These currents flow in closed paths and migrate in depth, decreasing in intensity over time. Changes in the secondary magnetic field over time induce a transient voltage in the receiving coil. The decay of this voltage provides information on the distribution of the subsoil conductivity, which can be used for its characterization [11,12]. This geophysical prospecting technique is used for many tasks. Although this method was originally used in the prospecting of deep mineral deposits [12], its ability to detect changes in conductivity associated with variations in the salinity of aquifers was later confirmed. This led to its widespread use in the investigation of marine intrusions in coastal areas [13–16] or in the detection of pollution plumes in environmental studies [17]. It has also been used in the analysis of sedimentary basins, in particular, to detect the depth of the base under sedimentary coverings [3,4,18,19].

In addition, the information that magnetic techniques can provide in this geological context is analysed and compared with the above-mentioned methods. Magnetic prospecting is based on the detection of anomalies in the Earth's magnetic field. These anomalies are caused by materials that present high magnetic susceptibility; that is, they can be magnetically polarized in the presence of an inductive field [1]. Before these measurements can be interpreted, the influence of external fields (basically the diurnal variations in the magnetic field) must be considered. This methodology has been extensively described in the literature [1,6,20] and it is widely used in different fields of Earth science. In mining, this method has proven useful in prospecting for ferromagnetic materials [21–25] and in hydrocarbon prospecting, providing information on deep structures that can function as traps [26]. It has also been used as a complementary tool for geological mapping [27,28], locating faults in covered areas [29,30].

In this study, the information provided by the surface geological survey and the data from boreholes is compared with the results obtained through geophysical prospecting. This will allow, on the one hand, to complement the geological information of this sedimentary basin, as well as the characterization of the related aquifers. On the other hand, the comparison of the results obtained with the different indirect techniques used (seismic reflection, TDEM and magnetic) allow us to analyse the effectiveness of each of them in this geological context.

2. Geological Setting

2.1. Regional Geological Context

In the studied region, the following two geological units can be distinguished: the Palaeozoic basement and the sedimentary cover (Figure 1). The first unit is basically composed of intensely folded phyllites intruded by granite in the last stages of the Hercynian Orogeny [31,32]. Post-Hercynian sedimentary cover is deposited unconformably on this basement, arranged sub-horizontally. This cover is composed of Triassic, Miocene and Quaternary materials. The basal Triassic deposits are characterized by the presence of levels of fluvial conglomerates and sandstones that are wedged laterally, with abundant quartzite and sandstone clasts, that do not exceed 20 m in thickness (T1 in Figure 1b), according to the information obtained in the geological research carried out in the nearby outcrops, as well as the data provided in previous works [31]. Above these facies, there is a set of red lutites with some sandstone intercalations whose thickness ranges between

60 and 120 m, depending on the sector (T2 in Figure 1b) [31]. Above this set and separated by a stratigraphic discontinuity is the Late Miocene (Messinian) unit, with thicknesses of approximately 100–150 m in the sector but sometimes reaching 200–300 m towards the centre and south of the basin. At the base, calcarenitic breccias or polygenic conglomerates (M1 in Figure 1b) appear, which change laterally and vertically to marl and limestone marl with fine calcarenitic intercalations (M2 in Figure 1b) [31]. The Quaternary deposits are associated only with the filling of river channels and are characterized by the presence of silt, sand and gravel [31–33].

There are two subparallel normal faults (the Baños fault and Guarromán fault, Figure 1a), oriented NE–SW, that delimit a horst/graben structure. The differential erosion of these structures results in the sedimentary cover being conserved almost exclusively in the depressed sector (graben structure), which is known as the "Bailén Basin".

The Bailén Basin is characterized by the presence of two superimposed aquifer units, which coincide with the basal Triassic materials (T1 in Figure 1b) and basal Late Miocene (M1 in Figure 1b). The lower aquifer offers low pumping flow rates (between 1–2 L/s), and the upper aquifer offers somewhat higher pumping rates (between 1 and 5 L/s). Laterally, these aquifers are limited by the two faults mentioned above. The movement of these faults was complex, with different pulses occurring over time [4]. Therefore, the subsidence and sedimentary filling of the basin were different, with lateral changes in facies and thicknesses, which in turn influenced both the hydrogeological characteristics of the units described and the discharge rates of the exploitation boreholes [33,34].

2.2. Local Geological Data

This study focused on the northern boundary of the Bailén Basin, which is in the vicinity of Baños de la Encina (Figures 1 and 2). At some points in this locality, the contact between the base of the Triassic and the Palaeozoic basement outcrops, in the horst block of the fault (Figures 1b and 2), at an elevation of 430 m (above sea level (asl)). In this area, geological information is provided by two groundwater exploitation boreholes. In addition to the information on the lithologies traversed, the thickness of the geological units and the heights of the lithological changes, they are used to calculate the velocities in the seismic profile analysed.

Towards the east, in the graben block and at an elevation of 368 m asl, a borehole was drilled (borehole 1 in Figures 1 and 2), in which the contact between the basal Miocene and the Triassic was reached at 83 m of drilling (285 m asl), and the base of the Triassic was reached at 203 m of drilling (165 m asl). From the difference in the elevation of the basal sedimentary cover on both sides of the Baños de la Encina fault in the vicinity of this locality, we estimate a throw of 265 m for this structure.

Approximately 700 m to the SE, another borehole was drilled for groundwater withdrawal (borehole 2 in Figures 1 and 2). In borehole 2, the contact between the sedimentary cover and the top of the Palaeozoic basement was reached at a depth of 267 m. Taking into account that the starting point was 367 m asl, the base of the Triassic unit was located at 100 m asl. By comparing the position of this level in the two soundings, the contact between the sedimentary cover and the Palaeozoic basement is deduced to be located 55 m below the previous one, which could be associated with a new fault (fault 2 in Figures 1b and 2) that would sink the SE sector.

The presence of these faults would be consistent with the aforementioned graben model and would justify the increase in thickness of the sedimentary cover towards the centre and south of the Bailén basin mentioned in some studies [32,33]. However, in the loamy facies (M2) that emerges on the surface, the presence of this structure has not been verified; therefore, this structure does not appear on geological maps.

In this work, different geophysical techniques (seismic reflection, time-domain electromagnetics and magnetometry) are used to corroborate the existence of this structure, as deduced from the direct data of the aforementioned surveys.

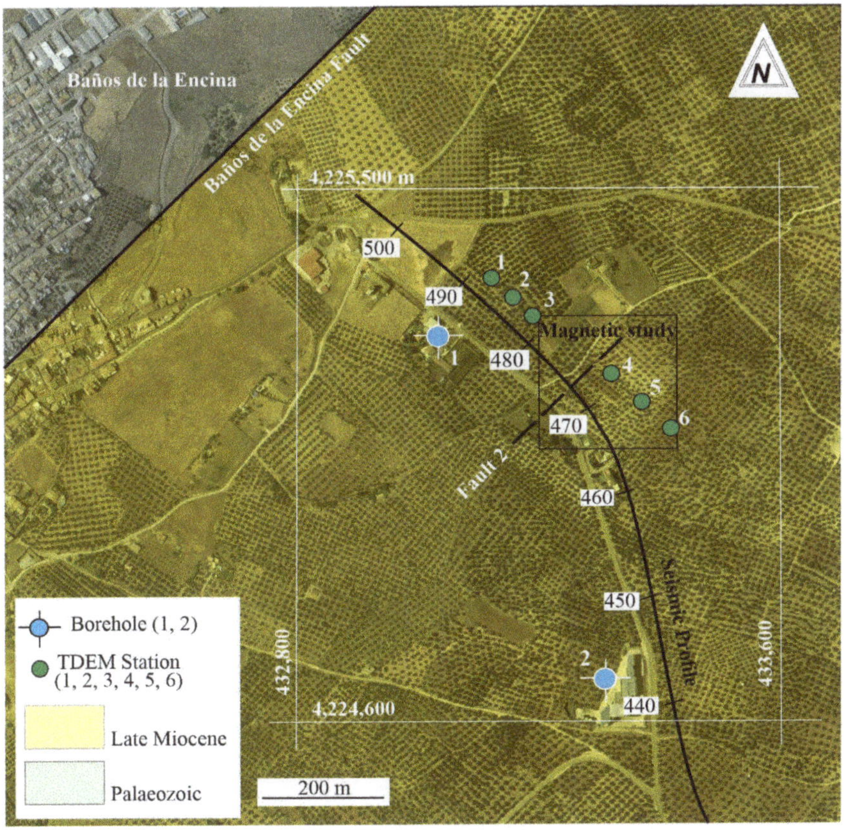

Figure 2. Details of the studied sector (black box in Figure 1a). The positions of fault 2, the boreholes (1 and 2), the different time-domain electromagnetic stations (1, 2, 3, 4, 5, 6) and the seismic profile with the positions of its traces are shown. Coordinates are in m.

3. Geophysical Survey

3.1. Seismic Reflection Transect

Seismic reflection profiles collected in the 1980s by the Geological and Mining Institute of Spain [35] were analysed. A Mini-Sosie vibrator was used as the seismic source, and model SM-7GR geophones with 30 Hz sensors were used to record the signals. An array of 18 geophones per trace was used, grouped in two lines parallel to the orientation of the profile, with 9 geophones each and with a separation between them of 1.85 m (the measurement was made by means of a double geophone in each position). The spacing between the traces was 15 m.

At each firing point, 2100 hits were generated. To fix the shooting distance, tests were carried out at 30, 45, 60 and 75 m. The best reflections were obtained at distances of 45 m.

The information was processed by the Compagnie Génèrale de Géophysique (CGG). The processing flow followed a standard sequence of processing, including collection of points in the "slalom line", deconvolution prior to stacking, static corrections from ground level at the referred point to mean ground level (time origin), continuous velocity analysis, dynamic velocity correction, automatic adjustment of residual static corrections (SATAN), gains, bandpass filtering (3–100 Hz), coherence improvement, migration and analogical output of the information [35].

In this work, one of these seismic lines is used, namely, line 1 (see location in Figure 2). For the double-time/depth transformation, interval velocities of 1900 m/s were considered to the base of the Miocene unit, and an interval velocity of 2200 m/s were considered to the base of the Triassic unit. This velocity was determined in the processing of the line and was verified based on the known thicknesses of the different units according to mechanical probing performed in the area.

3.2. Time-Domain Electromagnetic (TDEM) Survey

The TDEM equipment used in this work was the AIE-2 model of the ELGEO Research & Production Company. This device has a maximum output power of 200 W and a current intensity of up to 10 A. The TDEM receiver is based on a 16-bit analogue–digital converter with a signal processor that provides an analogue–digital conversion immune to input voltage noise and real-time signal pre-processing [36].

In the field campaign, a total of six TDEM measurement stations were established in the NW–SE direction, coinciding with the seismic line analysed. The points located in Figure 2 indicate the centres of the reading stations for square loops of 100 and 200 m on each side. At each of these points, measurements were taken with central-loop and single-loop devices, and the measurement parameters (intensity, time and voltage) were varied, allowing the efficiency and stability of the different configurations to be compared. Thus, to reach research depths on the order of 200 m (at the first three stations), transmitter loops of 100 × 100 m were used. For the last three stations, which were designed to study the graben block of the fault, larger transmitter loops (200 × 200 m) were used since a greater Palaeozoic basement depth was expected. In all cases, a 6 mm^2 section cable was used, which reduces its resistance and, therefore, allows a higher effective current intensity to be achieved. At those stations in which a central loop device was used, the receiving coil had dimensions of 20 × 20 m and a loop of five turns. For the visualization and processing of the different curves, the software TEMBIN version 1.1 was used. The modelling and inversion process was carried out with the software ZondTEM1D version 1.1.

3.3. Magnetic Survey

In this study, field work was carried out with a GEM GSMP-35G v8.0 portable potassium magnetometer (GEM SYSTEMS, Markham, ON, Canada). This potassium magnetometer obtains high data quality due to its high sensitivity (0.0003 nT), minimum reading error, resolution (0.0001 nT) and high precision (+/−0.05 nT). It has the capacity to take up to 20 readings per second, which provides a remarkable sampling density while walking [37]. Nineteen continuous profiles in the NW–SE direction were made in walking mode, with a length between 250–350 m and a separation between profiles of 10 m. Given the stability of the equipment used, measurements were taken every 50 ms. The georeferencing was carried out with a NovAtel GPS device mounted on the top of the mobile device.

To calculate the main magnetic field, the International Geomagnetic Reference Field (IGRF, 13th) calculator of the British Geological Survey (https://geomag.bgs.ac.uk/data_service/home.html, accessed on 24 March 2023) was used. Regarding the effect of external fields, diurnal correction was carried out. For this, data from the San Pablo Observatory of the National Geographic Institute (https://www.ign.es/web/ign/portal/gmt-area-geomagnetismo, accessed on 24 March 2023) were used.

To obtain the total field magnetic anomaly, the value of the main magnetic field (obtained with the IGRF model) was subtracted from the contributions of the external field (diurnal correction) and the values measured with the magnetometer at each point. Daytime correction and despiking were carried out using the software GEMLink version 5.4. The software Surfer was used for data interpolation and 2D representation. The data were analysed with different interpolation methods (kriging, cokriging, natural neighbour, nearest neighbour, etc.). Ultimately, kriging was selected as the interpolation method. The corresponding variogram showed the best fit with the maximum distance.

4. Results and Discussion

4.1. Seismic Reflection Survey

The trace of the seismic profile is perpendicular to the mapped Baños fault (Figures 1 and 2) that bound the Bailén Basin. Therefore, it is expected that it will allow the detection of other possible subparallel structures associated with the same fracturing system. In addition, the different mechanical behaviour of the lithologies studied should allow differentiating the described units, as well as deducing lateral thickness changes in the profile.

In the upper part of the seismic profile in Figure 3, materials attributable to the Miocene are represented in two shades of yellow (M1 + M2). At the base of this unit, there are levels that contain moderate-amplitude reflectors (M1), which are interpreted as sandstones and are related to the upper aquifer (Figure 3). The thicknesses obtained using an interval velocity of 1900 m/s [35] fit the depths to the base of the Miocene found in boreholes 1 and 2.

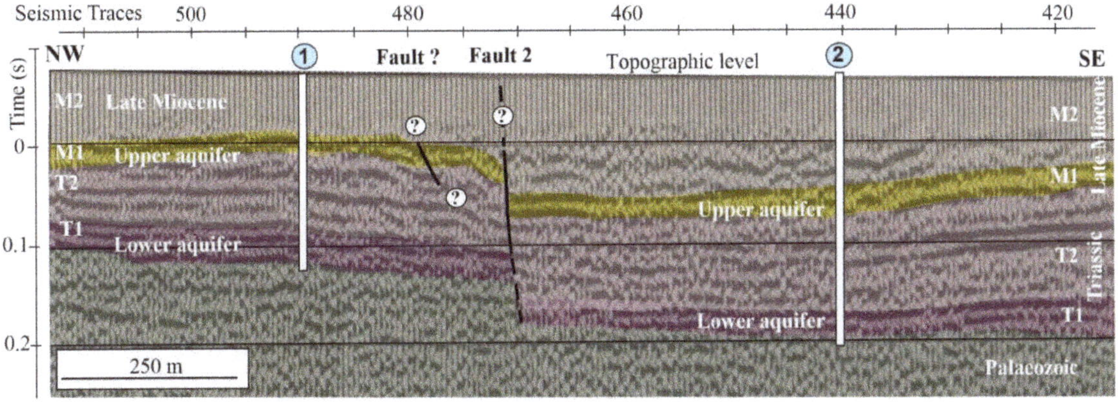

Figure 3. Seismic reflection profile and interpretation (see location in Figure 2). The positions of boreholes 1 and 2 are indicated. T1: conglomerate and sandstones of the lower Triassic (lower aquifer); T2: lutites of the upper Triassic; M1: rudites and calcarenites from the basal Miocene unit (upper aquifer); M2: marls from the upper Miocene.

Underlying the Miocene unit, the Triassic unit appears, represented in shades of pink in the seismic profile (Figure 3). One hundred seismic traces are represented, corresponding to 1500 m on the horizontal axis of the profile. This short distance is sufficient to detect changes in the thickness of the Triassic unit. At the NW end (trace 510), it covers a two-way time of 0.08 s, which corresponds to a thickness of 88 m (with an estimated interval velocity of 2200 m/s). In contrast, at the SE end (trace 420), the Triassic unit covers a greater two-way time, approximately 0.15 s, which is correlated with a thickness of 165 m. The estimated interval velocity to the base of the Triassic (2200 m/s) is verified in boreholes 1 and 2, where the thicknesses are 120 and 130 m, respectively.

Within the Triassic unit, maximum increase in wave amplitudes are also detected in the seismic trace, which are correlated with levels of sandstones and conglomerates. They are especially abundant at the base of the Triassic unit but also appear in the middle part (SE sector). The increase in wave amplitude associated to the Triassic base corresponds to the lower aquifer (T1 in Figure 3). Laterally, this base reflector may be lost due to changes in thickness or even due to the wedging of the basal conglomerates (between traces 460–470), as seen in outcrops in the northwestern sector of the Baños fault. In the lower part of the profile, the acoustic basement consists of the folded phyllites of the Palaeozoic basement, which are characterized by a discontinuous reflectivity, which generates chaotic features and disorganized seismic facies (Figure 3).

The seismic reflection profile in Figure 3 also shows a series of faults that affect the basement and the sedimentary cover. The most obvious difference is between seismic traces 470 and 475 (Fault 2 in Figure 3). The reflectors sink towards the SE by approximately 0.05 s, which corresponds to the 50 m throw detected in the two boreholes. This fracture affects the base and the sedimentary cover. Although the faults affect the Late Miocene and are, therefore, of a later age, the thickness changes in the Triassic suggest previous pulses that affected the subsidence and the sedimentation rate already in the Triassic. Furthermore, it should be noted that the lack of information in the first 0.06 s of the seismic profile makes it impossible to know if the highest part of the Miocene is also affected or, on the contrary, fossilizes the structure. This creates some uncertainty about the age of the last pulse of the fault.

In addition to this fault, the morphology of the reflectors suggests the presence of another fault with a smaller throw around seismic trace 490 (Figure 3). This structure affects the materials attributed to the upper part of the Triassic and to the base of the Miocene (M1), which sink slightly towards the SE. The lack of resolution in the upper part of the profile makes it difficult, also in this case, to stablish whether the fracture affects the most recent materials and, therefore, whether it was before or after them. It is noteworthy that the materials of the lower part of the Triassic and the basement do not suffer vertical displacements associated with this fracture, which would not be coherent if it were a normal fault.

4.2. *Time-Domain Electromagnetic (TDEM) Method*

The TDEM campaign was designed after detecting fault 2 in the seismic profile and calculating the depths of the different units in both fault blocks. Three stations were positioned on each block of the fault (Figure 2). In the footwall block, loops of 100 × 100 m were used (stations 1, 2 and 3 in Figure 2); in the hanging wall block it was necessary to increase the size to 200 × 200 to be able to detect the Palaeozoic basement, since it is located at a greater depth (stations 4, 5 and 6 in Figure 2).

Figure 4 shows the curves generated at the six TDEM stations carried out in this study. The induced voltage curve measured in the field and that generated by the ZONDTEM1D version 1.1 software model are represented in blue. The red line represents the apparent resistivity curve obtained as a function of time and, likewise, that generated according to the software model. In addition, the resistivity model generated by the inversion software is represented by a thick red line. According to a comparison of the induced voltage curves at the different stations, a better fit is obtained for Stations 4, 5 and 6, where both curves coincide (the one generated with the field data and the one from the model proposed by the software). This is also the case for the apparent resistivity curves obtained with measured data and modelled data. The fit is worse for Stations 1, 2 and 3, where the two voltage curves coincide poorly, especially in the final part. This is reflected in the root-mean-square (RMS) error, with values of 0.5%, 0.7% and 0.3% for the measurements made at Stations 4, 5 and 6, respectively. In contrast, for Stations 1, 2 and 3, the RMS error values are high, on the order of 8%, 17% and 9%, respectively.

For Stations 1, 2 and 3, several sections are differentiated in the resistivity curves obtained after the inversion. These first measurements are associated with the M2 section, where the marly facies generate decreasing resistivity values. Occasionally, within this section, abrupt increases in resistivity are associated with calcarenitic intercalations in this unit. This is detected, for example, at Station 3 at a depth of about 50 m (Figure 4). The contact between the marly facies of M2 and calcarenites, breccias or polygenic conglomerates (M1) is very well evidenced in all curves by a sharp increase in resistivity at depths of 78 m, 79 m and 116 m at Stations 1, 2 and 3, respectively (Figure 4). The basal detrital section of the Miocene and the Triassic levels cannot be differentiated, so the set is represented as M1 + T in Figure 4. The last section of the curves corresponds to the Palaeozoic basement. In this case, the contact between the sedimentary cover and the basement is detected at all stations as a sudden increase in resistivity, at depths of 228 m, 232 m and 228 m at Stations 1, 2 and 3, respectively (Figure 4). If the elevation of the ground surface at each station and the depth measurements are taken into account, the elevations above sea level of this contact are 145 m, 140 m and 141 m at Stations 1, 2 and 3, respectively (Figure 5). These results are very similar to the elevation of this same contact in borehole 1 (165 m asl). If these values are compared with the elevation of the Triassic base in the uplifted block (430 m asl), a vertical displacement of about 285 m is deduced for the Baños de la Encina fault. This value is similar to that calculated with the data from borehole 1 (265 m).

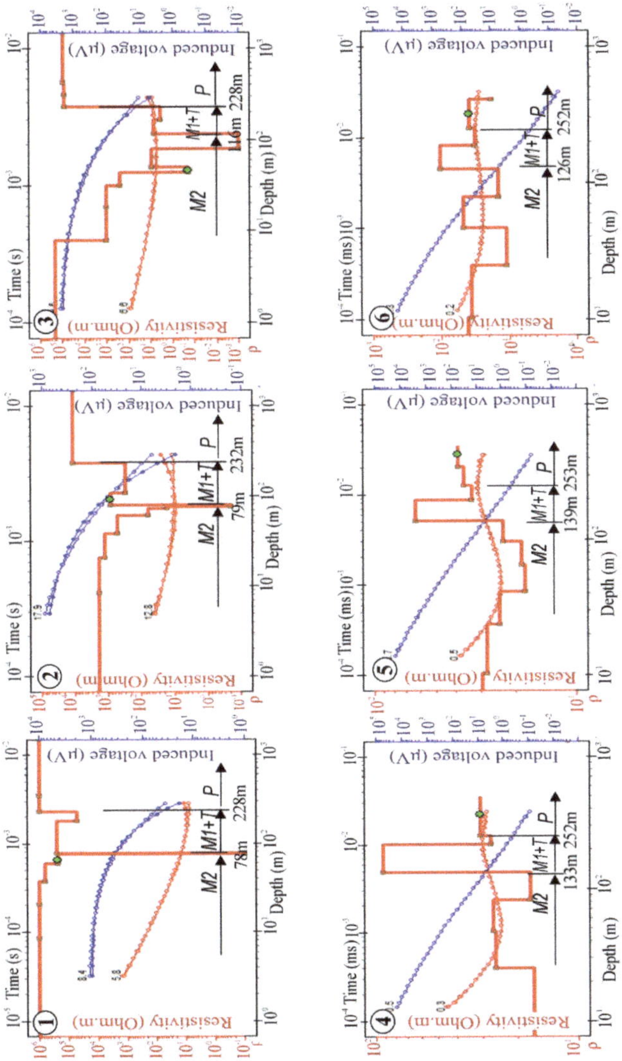

Figure 4. Induced voltage curves (blue lines) and apparent resistivity (red lines) as a function of time at the six TDEM stations (1, 2, 3, 4, 5 and 6). Two voltage lines are represented: one defined by circles filled in blue (field measured values) and the other by unfilled circles (model proposed by the software). In turn, two red lines are represented for apparent resistivity: one with circles filled (field measured values) and the other with unfilled circles (model proposed by the software). The interpretation after the inversion of the data is also included: the red straight lines represent the resistivity as a function of depth. P: Palaeozoic phyllites; T: Triassic; M1: rudites and calcarenites from the basal Miocene unit (upper aquifer); M2: marls from the upper Miocene unit.

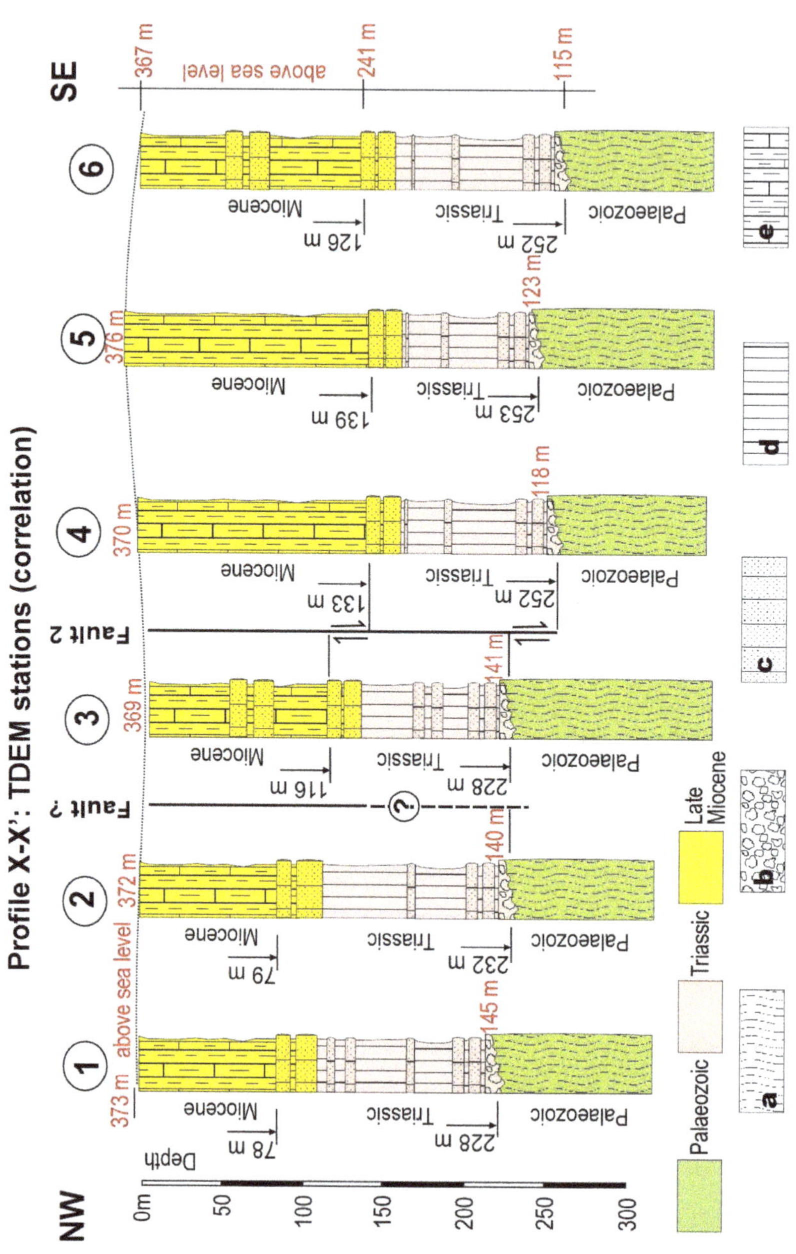

Figure 5. Stratigraphic columns deduced from the TDEM method information at each of the measurement stations (1, 2, 3, 4, 5 and 6). The numbering in black corresponds to the depths at which the reference levels were found (top of the basal Late Miocene calcarenite/conglomerate and base of the Triassic). The numbering in red corresponds to the elevation of the land surface at the point where the TDEM measurement was taken and to that of the two reference levels. a: folded phyllites; b: breccias or conglomerates; c: sandstones; d: lutites; e: marls and marly limestones.

At Stations 4, 5 and 6, the contact between assemblies M2 and M1 is characterized by a significant increase in resistivity associated with the top of the Miocene calcarenites, breccias or polygenic conglomerates. This contact was found at depths of 133 m, 139 m and 126 m at Stations 4, 5 and 6, respectively (Figure 4). If the values between Stations 1, 2 and 3 and Stations 4, 5 and 6 are compared, a vertical movement of approximately 50 m is obtained, which is consistent with the information provided by hydrogeological boreholes and the seismic profile. The contact of the sedimentary cover with the Palaeozoic basement is also associated, in this case, with a rise in resistivities at depths of 252 m, 253 m and 252 m (at the different stations shown in Figure 4). If the elevation of the terrain at each station and the depth measurements are considered, the elevations above sea level of this contact would be 118 m, 123 m and 115 m at Stations 4, 5 and 6, respectively (Figure 5). The elevation obtained for this contact using this geophysical technique is very similar to that derived from direct data in borehole 2 (100 m). Additionally, if the elevation data of the basement at the six stations are compared, the presence of fault 2 is also deduced by means of the TDEM method.

In Figure 5, the lithological columns deduced for each of these TDEM stations are correlated. They indicate the depth measurements up to the reference levels (in black), as well as the elevation of these surfaces above sea level (in red). If the TDEM X–X′ profile (Figure 5) and the seismic profile of Figure 3 are compared, it can be noticed that fault 2 is detected by both techniques. Also, in both cases, another structure (Fault ? in Figures 3 and 5) is inferred, which is characterized by a deepening of the Miocene base (M1) that does not affect the height of the Palaeozoic basement.

4.3. Magnetic Prospecting

The magnetic profiles were made in the SE-NW direction, parallel to the seismic profile and perpendicular to the SW-NE faults, with the aim of detecting them (Figure 6). After the field campaign was performed and the diurnal correction of the data was applied, despiking was carried out to eliminate environmental noise in both sectors. To obtain the total field magnetic anomaly, the value of the main magnetic field (obtained through the IGRF model) was subtracted. The result is represented in Figure 6.

In the magnetic anomaly map of the total field, a magnetic anomaly direction of N30E is detected, on the order of 20 nT, which coincides with the trace of fault 2 deduced in the seismic profile. These are medium-wavelength magnetic anomalies (Figure 7). In other regions, equivalent anomalies have been associated to a mean-depth origin in the sedimentary basin [20]. Thus, as described in other studies [27–30], fractures affecting sedimentary basins bring in contact basement and sedimentary cover materials with different magnetic susceptibility values, which would justify these linear anomalies.

In addition to fault 2, another alignment (N60E) that corresponds to a minor order anomaly is detected (approximately 10 nT), which could coincide with the secondary fault detected in the seismic profile, as well as with the direction of the faults mapped in the Palaeozoic basement in sectors adjacent to that studied [31].

Furthermore, in the surface representation of Figure 6, isolated anomalies of small dimensions and short wavelengths can be seen, suggesting very shallow bodies [20]. They are especially abundant towards the NE of the sector studied, in the vicinity of the road (Figure 6). All these anomalies are associated with the existence on the surface of anthropic remains with ferromagnetic elements (observed in the outcrop during measurements) that generate these small magnetic dipoles.

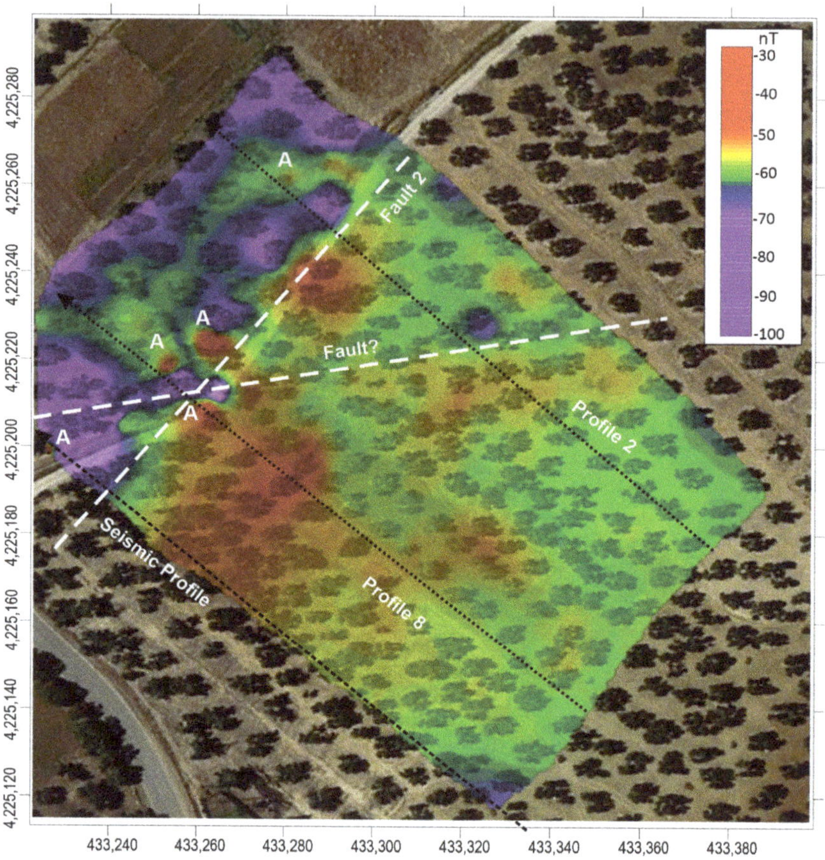

Figure 6. A map of the total field magnetic anomalies in the sector is shown in Figure 2. The positions of two of the magnetic profiles (2 and 8) are indicated, as well as the position of the seismic profile. A: small dipoles associated with ferromagnetic elements of an anthropic nature. Coordinates are in m.

Figure 7 shows two profiles of magnetic anomalies (Profile 2 and Profile 8), whose positions are indicated in Figure 6. In Profile 2, two anomalies associated with the two faults mentioned above are detected, where greater amplitudes are associated with fault 2. These two structures can be correlated with those indicated in the seismic profile of Figure 3. In Profile 8, as the two faults are very close, the magnetic anomalies appear to overlap. Anomalies of short wavelengths superimposed on the previous ones are also pointed out, suggesting very shallow bodies.

Like previous geophysical techniques, magnetic methods are capable of detecting faults that affect both the base and the sedimentary cover, even if they are not detectable in geological mapping. In addition, the orientation of the dipoles that form these magnetic anomalies allows us to find the strike of these structures, information that had not been obtained with seismic and TDEM techniques.

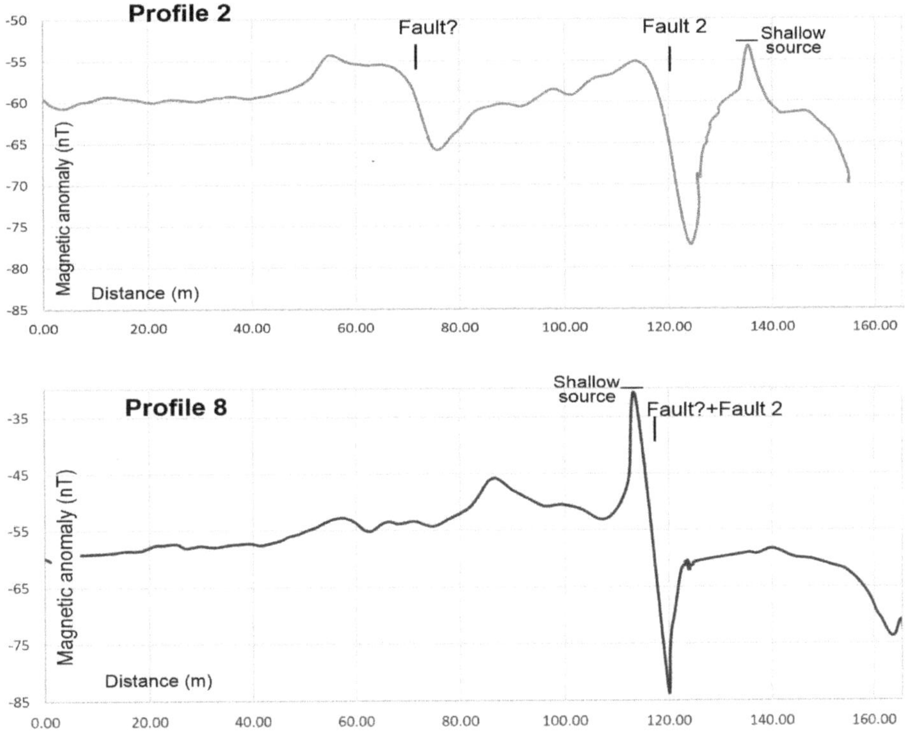

Figure 7. Total field magnetic anomaly profiles. Their positions are shown in Figure 6.

5. Conclusions

In this work, different direct and indirect methods were used in an integrated way to analyse the capacity of each method in the modelling of geological structures. For this purpose, the northern boundary of the Bailén basin in the vicinity of Baños de la Encina (southeast Spain) was chosen as study site. The following three different geophysical techniques were used: seismic reflection, time-domain electromagnetic (TDEM) and magnetic methods. These geophysical techniques were calibrated with available direct information, both from outcrop data and from boreholes.

The effectiveness of seismic reflection in characterizing sedimentary basins lies in its ability to provide high-resolution images of geological structures. In the seismic reflection profiles, abrupt changes are observed in the amplitude of the wave associated with the different lithologies of the sedimentary cover, as well as the acoustic basement associated with the Palaeozoic basement. By means of this method, two faults are detected that caused the subsidence of the eastern sector of the basin (the Bailén graben).

The TDEM technique is very versatile and can be used with different configurations and ranges of amplitude, voltage, time or device and loop size, depending on the depths at which the investigation is focused. Although its spatial resolution is lower compared to seismic reflection, it is a method that is easier to implement than the previous one. However, the interpretation of TDEM data requires complex mathematical models that do not provide a unique solution, so it requires the integration of the results obtained with other geophysical methods or direct geological data. In the case of TDEM, the best solution would be, a priori, the one that presents a better fit between measured and calculated data, characterized by low residual error values. The validation by means of direct geological data, from boreholes, and indirect data, obtained from the reflection seismic profile, confirms the adequacy of the geoelectric model obtained. In this study, a good fit between the seismic and electromagnetic data was generally found. This method made it possible to locate the position of the upper aquifer, associated with calcarenites, breccias or polygenic conglomerates: the top of this unit is characterized by a sudden increase in resistivity values associated with these lithologies in comparison to the upper marls. However, the characterization of the lower aquifer under the Triassic clays was not so evident.

The TDEM method also made it possible to detect the depth of the Palaeozoic basement under the sedimentary cover, as well as the presence of the two aforementioned faults. The effectiveness of the TDEM analysis in calculating this depth is attributed to the notable resistivity contrast between the infilling materials and the underlying phyllitic basement rocks. This contrast in resistivity serves as a key parameter, enabling the differentiation and characterization of distinct geological layers also identified in the seismic reflection profile, thereby facilitating accurate depth assessments.

The magnetic method was chosen because of its efficacy to discern among rock bodies with different magnetic susceptibilities, and that provides a way of delineating subsurface geologic features as faults over broad areas. Magnetic prospecting revealed two alignments of anomalies in the geological map, which are associated with two fault phases (N30° E and N60° E). Both fault directions have been mapped in the boundaries of the Bailén basin when they affect the Palaeozoic basement, but they are not mappable in the Tertiary marly facies. These new data provide information on the orientation of linear structural features and allow to infer previously unknown faults.

Overall, a strong relationship is evident among the fractures inferred from seismic, electromagnetic and magnetic methods, indicating a robust fit between the different data sets. Therefore, the combined study of seismic, electromagnetic and magnetic techniques is of great interest in fracture analysis. The map of the magnetic anomalies and associated geophysical profiles across the study area show changes in bedrock depth and allow to infer concealed faults that are not observed at the surface.

Furthermore, this study allows us to better understand the complex fracturing phase of the northern boundary of the detrital aquifer associated with the Bailén basin. The obtained profiles of the basin show lateral and vertical variations in the bedrock topography, affecting both the hydrogeological behaviour of the units described and the discharge obtained by groundwater withdrawal. This is also important for other types of studies, e.g., for hydrogeological and geotechnical approaches.

Author Contributions: Conceptualization, J.R.; investigation, J.R., R.M., M.C.H. and B.M.; software, R.M. and J.R.; formal analysis, J.R., R.M. and M.C.H.; writing—original draft preparation, J.R; writing—review and editing, M.C.H. All authors have read and agreed to the published version of the manuscript.

Funding: Grant PID2021-123506OB-I00 was funded by the MICIU/AEI/10.13039/501100011033 and ERDF/EU. This study was partly funded by the University of Jaen.

Institutional Review Board Statement: Not applicable.

Informed Consent Statement: Not applicable.

Data Availability Statement: The data that support the findings of this study are available upon request from the corresponding author.

Conflicts of Interest: The authors declare no conflicts of interest.

References

1. Telford, W.M.; Geldart, L.P.; Sheriff, R.E. *Applied Geophysics*; Cambridge University Press: Cambridge, UK, 1990; p. 770.
2. Martínez-Moreno, F.J.; Galindo-Zaldívar, J.; Pedrera, A.; Teixido, T.; Ruano, P.; Peña, J.A.; González-Castillo, L.; Ruiz-Constán, A.; López-Chicano, M.; Martín-Rosales, W. Integrated geophysical methods for studying the karst system of Gruta de las Maravillas (Aracena, Southwest Spain). *J. Appl. Geophys.* **2014**, *107*, 149–162. [CrossRef]

3. Rey, J.; Martínez, J.; Mendoza, R.; Sandoval, S.; Tarasov, V.; Kaminsky, K.; Hidalgo, M.C.; Morales, K. Geophysical characterization of aquifers in southeast Spain using ERT, TDEM, and vertical seismic reflection. *Appl. Sci.* **2020**, *10*, 7365. [CrossRef]
4. Mendoza, M.; Rey, J.; Martínez, J.; Hidalgo, M.C.; Sandoval, S. Geophysical characterisation of geologic features with mining implications from ERT, TDEM and seismic reflection (Mining District of Linares-La Carolina, Spain). *Ore Geol. Rev.* **2021**, *139*, 104581. [CrossRef]
5. Carrasco, J.; Carrasco, P.; Porras, D.; Martín, I. Drone Magnetic and Time Domain Electromagnetic Exploration in Metamorphic Formations: Tool for the Identification of Strategic Sites for Aquifer Exploitation. *Appl. Sci.* **2023**, *13*, 10949. [CrossRef]
6. Reynolds, J.M. *An Introduction to Applied and Environmental Geophysics*; John Wiley and Sons Ltd.: London, UK, 2011; p. 796.
7. Radwan, A.A.; Nabawy, B.S.; Shihata, M.; Leila, M. Seismic interpretation, reservoir characterization, gas origin and entrapment of the Miocene-Pliocene Mangaa C sandstone, Karewa Gas Field, North Taranaki Basin New Zealand. *Mar. Pet. Geol.* **2021**, *135*, 105420. [CrossRef]
8. Olutoki, J.O.; Siddiqui, N.A.; Haque, A.-E.; Akinyemi, O.D.; Mohammed, H.S.; Bashir, Y.; El-Ghali, M.A.K. Integrated analysis of wireline logs analysis, seismic interpretation, and machine learning for reservoir characterisation: Insights from the late Eocene McKee Formation, onshore Taranaki Basin, New Zealand. *J. King Saud Univ. -Sci.* **2024**, *36*, 103221. [CrossRef]
9. Hu, Y.; Liu, W.; Wang, J.; Zhang, G.; Zhou, Z.; Han, R. Basin-scale structure control of Carlin-style gold deposits in central Southwestern Guizhou, China: Insights from seismic reflection profiles and gravity data. *Ore Geol. Rev.* **2017**, *91*, 444–462. [CrossRef]
10. Jenkins, A.P.; Torvela, T. Basin analysis using seismic interpretation as a tool to examine the extent of a basin ore 'play'. *Ore Geol. Rev.* **2020**, *125*, 103698. [CrossRef]
11. Nabighian, N.M. *Electromagnetic Method in Applied Geophysics*; Society of Exploration Geophysics: Tulsa, OK, USA, 1988; p. 971.
12. McNeill, J.D. *Principles and Application of Time Domain Electromagnetic Techniques for Resistivity Sounding*; Technical Note TN-27; Geonics Ltd.: Toronto, ON, Canada, 1994. Available online: http://www.geonics.com/pdfs/technicalnotes/tn27.pdf (accessed on 10 April 2023).
13. Danielsen, J.E.; Auken, E.; Jorgensen, F.; Sondergaard, V.; Sorensen, K. The application of the transient electromagnetic method in hydrogeophysical survey. *J. Appl. Geophys.* **2003**, *53*, 181–198. [CrossRef]
14. Kafri, U.; Goldman, M. Are the lower subaquifers of the Mediterranean coastal aquifer of Israel blocked to seawater intrusion? Results of a TDEM (time domain electromagnetic) study. *Isr. J. Earth Sci.* **2006**, *55*, 55–68. [CrossRef]
15. Farag, K.S.I.; Howari, F.M.; Abdelmalik, K.W. Imaging of Hydrothermal Altered Zones in Wadi Al-Bana, in Southern Yemen, Using Very Low Frequency–Electromagnetic and Remote Sensing Data. *Arab. J. Geosci.* **2019**, *12*, 554. [CrossRef]
16. Zhu, Z.; Shan, Z.; Pang, Y.; Wang, W.; Chen, M.; Li, G.; Sun, H.; Revil, A. The transient electromagnetic (TEM) method reveals the role of tectonic faults in seawater intrusion at Zhoushan islands (Hangzhou Bay, China). *Eng. Geol.* **2024**, *330*, 107425. [CrossRef]
17. Hallbauer-Zadorozhnaya, V.Y.; Stettler, E. Time Domain Electromagnetic Sounding to delineate hydrocarbon Contamination of Ground Water. In *Symposium on the Application of Geophysics to Engineering and Environmental Problems*; Environmental & Engineering Geophysical Society: Denver, TX, USA, 2009; pp. 241–251.
18. Sridhar, M.; Markandeyulu, A.; Chaturvedi, A.K. Mapping subtrappean sediments and delineating structure with the aid of heliborne time domain electromagnetics: Case study from Kaladgi Basin, Karnataka. *J. Appl. Geophys.* **2017**, *136*, 9–18. [CrossRef]
19. Bense, V.F.; Person, M.A. Faults as conduit-barrier systems to fluid flow in siliciclastic sedimentary aquifers. *Water Resour. Res.* **2006**, *42*, 1–18. [CrossRef]
20. Blakely, R. *Potencial Theory in Gravity and Magnetic Applications*; Cambridge University Press: Cambridge, UK, 1995; p. 441.
21. Abdelazeem, M.; Gobashy, M.M. A solution to unexploded ordnance detection problem from its magnetic anomaly using Kaczmarz regularization. *Interpretation* **2016**, *4*, 61–69. [CrossRef]
22. Abedi, M.; Fournier, D.; Devriese, S.; Oldenburg, D.W. Integrated inversion of airborne geophysics over a structural geological unit: A case study for delineation of a porphyry copper zone in Iran. *J. Appl. Geophys.* **2018**, *152*, 188–202. [CrossRef]
23. Benítez, M.E.; Prezzi, C.; Benallivián Justiniano, C.A.; Verdecchia, S.O.; De Martino, F.J.; Carlini, M.; Lanfranchini, M.E. Ground magnetic survey and 3D geophysical model of ultrabasic rocks from the Martín García Complex (Buenos Aires, Argentina). *J. S. Am. Earth Sci.* **2023**, *121*, 104117. [CrossRef]
24. Eldougdoug, A.; Abdelazeem, M.; Gobashy, M.; Abdelwahed, M.; Abd El-Rahman, Y.; Abdelhalim, A.; Said, S. Exploring gold mineralization in altered ultramafic rocks in south Abu Marawat, Eastern Desert, Egypt. *Abstr. Sci. Rep.* **2023**, *13*, 7293. [CrossRef]
25. Aminu, M.B.; Adiat, K.A.N.; Akinlalu, A.A.; Olomo, K.O.; Owolabi, T.O.; Aliyu, E.O. A review on the applications of airborne geophysical and remote sensing datasets in epithermal gold mineralisation mapping. *Geosyst. Geoenviron.* **2024**, *3*, 100284. [CrossRef]
26. Okiwelu, A.A.; Obianwu, V.I.; Eze Ohara, E.; Ude, I.A. Magnetic anomaly patterns, fault-block tectonism and hydrocarbon related structural features in the Niger Delta basin. *IOSR J. Appl. Geol. Geophys. (IOSR-JAGG)* **2014**, *2*, 31–46.
27. Nabighian, M.N.; Grauch, V.J.S.; Hansen, R.O.; LaFehr, T.R.; Li, Y.; Peirce, J.W.; Phillips, J.D.; Ruder, M.E. The historical development of the magnetic method in exploration. *Geophysics* **2005**, *70*, 33ND–61ND. [CrossRef]
28. Alva-Valdivia, L.M.; Guerrero-Díaz, P.; Urrutia-Fucugauchia, J.; Agarwal, A.; Caballero-Miranda, C.I. Rock-magnetism and magnetic anomaly modelling of Las Truchas, case study. *J. S. Am. Earth Sci.* **2020**, *97*, 102409. [CrossRef]
29. Akinlalu, A.A.; Adelusi, A.O.; Olayanju, G.M.; Adaiat, K.N.; Omosuyi, G.O. Aeromagnetic mapping of basement structures and mineralization characterization of Ilesa Schist Belt, Southwestern Nigeria. *J. Afr. Earth Sci.* **2018**, *138*, 383–391. [CrossRef]

30. Bouzekraoui, M.; Es-Sabbar, B.; Karaoui, B.; Essalhi, M.; Saadi, M. Structural analysis, tectonic fracturing modeling, and kinematic evolution along the South Atlas Fault at the northern border of the Tinghir-Errachidia-Boudenib basin (Pre-African Trough, Morocco). *J. Afr. Earth Sci.* **2024**, *212*, 105193. [CrossRef]
31. Castelló, R.; Orviz, F. *Mapa Geológico de España 1:50.000, Hoja nº 884 (La Carolina)*; Instituto Geológico y Minero de España: Madrid, Spain, 1976.
32. Larrea, F.J.; Carracedo, M.; Ortega Cuesta, L.; Gil Ibarguchi, J.I. El Plutón de Linares (Jaén): Cartografía, petrología y geoquímica. *Cuad. Lab. Xeológico Laxe Coruña* **1994**, *19*, 335–346.
33. Martín Parra, L.; Matas, J.; Roldán, F.J.; Martín-Serrano, A. *Mapa Geológico y Memoria Explicativa de la hoja 70 (Linares), Escala 1:200.000*; Instituto Geológico y Minero de España: Madrid, Spain, 2015.
34. I.T.G.E. *Atlas Hidrogeológico de la Provincial de Jaén*; Diputación Provincial de Jaén: Jaén, Spain, 1997; p. 260.
35. I.G.M.E. *Exploración Geológico-Minera de la Fosa Tectónica de Bailén (Jaén)*; Sismica de reflexión; Ministerio de Industria y Energía: Madrid, Spain, 1983; p. 64.
36. AIE-2 Instruments. Available online: http://zond-geo.com/english/services/equipment/aie-2-instruments/ (accessed on 1 August 2024).
37. GEM Systems. Available online: https://www.gemsys.ca/ultra-high-sensitivity-potassium/ (accessed on 1 August 2024).

Disclaimer/Publisher's Note: The statements, opinions and data contained in all publications are solely those of the individual author(s) and contributor(s) and not of MDPI and/or the editor(s). MDPI and/or the editor(s) disclaim responsibility for any injury to people or property resulting from any ideas, methods, instructions or products referred to in the content.

MDPI AG
Grosspeteranlage 5
4052 Basel
Switzerland
Tel.: +41 61 683 77 34

Applied Sciences Editorial Office
E-mail: applsci@mdpi.com
www.mdpi.com/journal/applsci

Disclaimer/Publisher's Note: The title and front matter of this reprint are at the discretion of the Guest Editors. The publisher is not responsible for their content or any associated concerns. The statements, opinions and data contained in all individual articles are solely those of the individual Editors and contributors and not of MDPI. MDPI disclaims responsibility for any injury to people or property resulting from any ideas, methods, instructions or products referred to in the content.

www.ingramcontent.com/pod-product-compliance
Lightning Source LLC
LaVergne TN
LVHW070458061126
851858LV00011B/257